£45

Biogeomorphology

Biogeomorphology

Edited by
Heather A. Viles

Basil Blackwell

Copyright©Basil Blackwell Ltd 1988

First published 1988

Basil Blackwell Ltd
108 Cowley Road, Oxford, OX4 1JF, UK

Basil Blackwell Inc.
432 Park Avenue South, Suite 1503
New York, NY 10016, USA

British Library Cataloguing in Publication Data
Biogeomorphology.
  1. Geomorphology. Biological aspects
  I. Viles, Heather A.
  551.4

  ISBN 0-631-15405-1

Library of Congress Cataloging in Publication Data
Biogeomorphology.
  Includes index.
  1. Biogeomorphology.   I. Viles, Heather A.
QH542.5.B56   1988   574.5′ 22   88-10560

ISBN 0-631-15405-1

Phototypeset in 10 on 11½ Sabon
by Dobbie Typesetting Service, Plymouth
Printed in Great Britain by T. J. Press Ltd, Padstow, Cornwall

Contents

Part III   Coastal and Karst Environments

Preface

This volume aims to provide a survey of current knowledge on biological influences on geomorphological processes. The organic component of landscape development has often been ignored by geomorphologists, although much exciting work has been done on pertinent topics in recent years. As an introductory survey, it is hoped that this book will be of interest to undergraduates from many earth and environmental science disciplines, as well as to more advanced workers. The contributors to this book each present a different balance of review and new research findings, which will provide a valuable source of reference. There are many rich and rewarding topics to be pursued in the field of 'biogeomorphology' and I hope this volume will inspire new ventures.

The following are gratefully acknowledged for permission to reproduce material: John Wiley and Son for figure 3.1 from Kirkby, M. J. and Morgan, R. P. C. (eds) *Soil Erosion* (1980); Edward Arnold for figure 6.9 from Dorn, R. I. and Oberlander, T. M., Rock Varnish, *Progress in Physical Geography* (1982); Gebruder Borntraeger for figure 6.3 from Ash, J. E. and Wasson, R. J., Vegetation and sand mobility, *Zeitschrift für Geomorphologie* Supplementband 45 (1983); Allen and Unwin for figure 8.19 from Rhoads, D. C., Organism–sediment relations on the muddy sea floor, *Oceanography and Marine Biology* 12 (1974); Basil Blackwell for figure 9.1 from Jennings, J. N., *Karst Geomorphology* (1985).

Heather Viles

Introduction

I shall offer a few remarks on the superficial modifications caused directly by the agency of organic beings, as when the growth of certain plants covers the slope of a mountain with peat, or converts a swamp into dry land; or when vegetation prevents the soil in certain localities from being washed away by running water.

Lyell, 1835, p. 175

Both plants and animals aid to some extent in the work of rock disintegration. Plants are also not infrequently an important factor in promoting sedimentation, while burrowing insects and animals may exert an important influence upon the texture of soils and in bringing about a more general admixture by transferring to the surface that which is below.

Merrill, 1897, p. 201

Close to this place I wounded a water buck . . . About here the river runs through a succession of rocky gorges, dashing over huge boulders of granite (?) rock. Through these ravines hippopotami must have wandered for countless ages, for in one place where a ledge of rock ran along the bank of the river, they had worn a path for about twenty yards across it, at least four inches deep into the hard stone.

Selous, 1881, p. 426

The passages quoted above indicate that by the beginning of the twentieth century there was some awareness of the role played by plants and animals in geomorphology, although some of their effects were undoubtedly seen to be bizarre, rather than of fundamental significance. This book endeavours to provide some more recent examples of the geomorphological links between the organic and inorganic worlds, and to show that there may be some general and important relationships.

'Biogeomorphology' is a new term designed to encapsulate concisely the concept of an approach to geomorphology which explicitly considers the role of organisms. For many years geomorphologists in general have given only passing attention to biological factors. Several recent geomorphology texts give the impression that landform development occurs within a largely abiotic environment, although the importance of soils to geomorphology has been stressed in some compilations (Gerrard, 1981; Richards *et al.*, 1985), and the links between atmosphere, vegetation, soils and geomorphology investigated in the context of solute processes (Trudgill, 1986). Ecologists and biogeographers have also given little consideration to the interactions between their subject and geomorphology. A recent compilation on current themes in biogeography, for example, only hints at possible links with geomorphology, whereas links with climatology are explicitly discussed (Taylor, 1984).

Despite the general separation of studies of geomorphology and ecology, there has been an increasing number of studies in recent years on specific biological interactions with geomorphology. Papers have appeared on such diverse and diverting topics as the role of kangaroo rats in slope processes (Best, 1972), the role of pigeon droppings in rock weathering (Bassi and

Table 1 Some recent examples of work on biological influences on earth surface processes

Process	Organisms	Influence	Reference
weathering	lichens	direct weathering of Antarctic desert rocks	Friedmann and Weed, 1987
	bacteria	nitrification, leading to acidolysis weathering of forest soils	Berthelin et al., 1985
	organic solutes	weathering of volcanic ash	Antweiler and Drever, 1983
soil development	trees	tree fall leading to soil profile inversion	Schaetzel, 1986
	subterranean termites	alteration of desert grassland soil characteristics	Nutting et al., 1987
coastal erosion	parrot fish	erosion of Barbados reefs	Frydl and Stearn, 1978
	tile fish	erosion of sea floor	Twichell et al., 1985
	whales and walruses	erosion of sea floor	Nelson and Johnson, 1987
fluvial erosion	logs and debris	increased scour near debris jams	Duijsings, 1987
transport	vegetation	transport of dune sands	Buckley, 1987
sedimentation	tube-dwelling polychaetes	formation of intertidal mounds	Carey, 1987
	marine bacteria	precipitation of calcium carbonate	Novitsky, 1981
	blue-green algae	calcification in desert stromatolites	Krumbein and Giele, 1979
	green, blue-green algae	stabilization of coastal dunes	Van den Ancker and Jungerius, 1985

Chiatante, 1976), and the role of Brussels sprout plants in reducing soil erosion (Noble and Morgan, 1983). Some further examples are presented in table 1. An attempt has also been made to develop computer models relating vegetation and soil erosion in a dynamic way (Thornes, 1985). So far this work has remained disparate and no general synthesis has emerged. This book aims to provide such a synthesis in order to assess what is known about the biological component of geomorphology, and what the future holds for such studies. The rest of this introductory chapter provides some historical background into the relations between geomorphology and ecology/biogeography, elucidates some recent developments, and then outlines the approach and contents of the substantive part of this book.

Ecology, biogeography and geomorphology (as well as the other earth science disciplines) have developed from the natural history of the eighteenth and nineteenth centuries, which itself had important precursors (Tinkler, 1985). Many early natural historians pursued very wide-ranging investigations into the environment, and especially environmental history, at a time when there were no disciplinary boundaries to circumscribe their interests. Humboldt, Lyell and Darwin are all famous examples of such broad thinkers who explored many facets of the complexity of the environment. As the earth sciences became more institutionalized and organized into specific disciplines, each with a rapidly developing corpus of facts and theory, linkages between different subjects became more difficult. Specialization and professionalization became important goals. Some disciplines remained more interlinked than others, however, due to the importance of specific research questions. So, for example, geology and biology maintained links through a concern for elucidating stratigraphy and the fossil record.

Geomorphology developed as part of geology and only attained a separate identity at the end of the nineteenth century when geologists had largely become interested in other topics. In the British academic structure geomorphology has become institutionalized within geography departments, and has shared in geography's perceived isolation from other scientific disciplines. The exact subject matter of geomorphology has changed during its development as different research orientations have come and gone. A consistently important focus, however, has been the explanation of the evolution of landscapes, involving both the explanation of landforms and also of earth surface processes. Chorley et al. (1984) recognize two major components to geomorphological explanation, the historical and the functional. Varying stress has been placed on these two components over the history of geomorphology, but a consideration of both is necessary for any complete explanation.

Whilst the Davisian model of explanation dominated at least some parts of geomorphological inquiry, historical modes of explanation were dominant and the focus of research tended to be on large scale investigations. Biological influences on the landscape were not an important component of explanation because their effects are often more relevant at the micro- or meso-scale, although an organic orientation to geomorphological theory was apparent. Davis's model of the cycle of erosion, for example, was deeply rooted in the organismic tradition of scientific thought prevalent at the time, which also influenced plant ecology and pedology (Stoddart, 1986). In the years after World War II geomorphologists (at least in the English-speaking academic community) have become increasingly interested in using the methods and ideas of the physical sciences (and especially those of mathematics and physics) to provide a disciplinary framework. Quantification has become a key goal in the search for rigorous models which can be used to explain, predict and control landform development. A more functional, problem-solving orientation has become apparent in geomorphology. This focus has led to a concentration on those scales and types of landforms and processes which can be measured easily. As the geomorphological inquiry has tended towards the smaller scale, so the investigation of biological factors has been facilitated. However, biological factors have often proved difficult to quantify and fit into models based primarily on principles derived from physics.

Biogeography and ecology are closely allied subjects which occupy positions on the fringes of geography and biology. Both are concerned with the interaction and distribution of organisms on the earth's surface. Like geomorphology, these subjects have seen great changes in orientation over their history. Humboldt, Darwin and other natural historians have been seen as pioneers of biogeography; just as they have been regarded as making important contributions to the founding of other earth science disciplines. Biogeography and ecology received an important stimulus in the work of Clements in the early twentieth century. His work developed two interrelated themes, i.e. the dynamics of ecological succession in the plant community, and the organismic character of the plant formation (Worster, 1985). Comparisons have been made between the ideas of Clements in ecology and those of Davis in geomorphology (Stoddart, 1986) and certainly they developed in a similar scientific climate.

Reformulations and reassessments of Clementsian ideas resulted in a more mechanistic view of ecology, which became most clearly expressed in the ecosystem concept. Ecosystems became a major focus of ecological and biogeographical studies by the mid-1950s (Evans, 1956); paralleling to a certain extent the acceptance of systems as explanatory tools in geomorphological studies during the 1960s (Chorley, 1962). In both ecology and geomorphology there have been more recent doubts about the utility of a systems approach when dealing with the complexity of the natural environment. A recent development in biogeography, involving especially those biogeographers working in geography departments, has been an increasing concentration on historical biogeography, involving the study of environmental change over a range of time scales (Flenley, 1984). This subject of environmental change has also been a continuing focus of geomorphological investigations.

The histories of biogeography, ecology and geomorphology, therefore, illustrate some common concerns and shared methodologies, which have developed out of the early unity of natural history. Similar ideas can be seen to become fashionable within the ecological and geomorphological literature, suggesting some cross-fertilization of thought. So, for example, the concept of punctuated equilibria in biology (Gould, 1984) may be compared with neo-catastrophist ideas in geomorphology (Dury, 1980). Ecology and geomorphology are both characterized by having functional and historical approaches, and both share a concern with a wide range of temporal and spatial scales. At some points in their history they have been more closely linked than at others, and similarly some areas of research have provided more obvious links, such as the coastal environment.

The question now arises as to why geomorphological studies have begun to consider ecological factors in more depth. The earth sciences are continually developing and changing, spurred on by technological and theoretical innovations, as well as the emergence of new problems to study. Of importance to the development of biogeomorphological studies are the growth of biogeochemistry and geomicrobiology (Krumbein, 1978, 1983; Ehrlich, 1981). These new subjects, firmly interrelated with other established disciplines, investigate functional links between organic and inorganic components of the earth's surface at the micro-scale. A special concern of these subjects is the role of microorganisms and lower plants (which have been much neglected by biogeographers and ecologists) in geological and other earth surface processes.

Other advances which have encouraged the study of biological factors in geomorphology in recent years include technological ones such as the development of improved data collection, storage and manipulation facilities. For example, multivariate analysis techniques have become more sophisticated and at the same time more accessible to earth scientists, and now permit the manipulation of large quantities of diverse data. Many large-scale data collection exercises have taken place, producing many important ecological data, especially under the auspices of the International Biological Programme. The applied, management, focus of much contemporary geomorphology, allied with a desire to improve the power of geomorphological theory, has also led to an increased consideration of biological factors.

Biogeomorphology is not intended to be a long-lived term describing a separate field of research, but rather is used here simply to focus attention. There are two linked foci of interest that can be identified in this biogeomorphology:

1 The influence of landforms/geomorphology on the distributions and development of plants, animals and microorganisms;
2 The influence of plants, animals and microorganisms on earth surface processes and the development of landforms.

These two foci can be seen as end members of a spectrum developed according to the scale involved, following the ideas of the scale dependence of causality in the environment put forward by Schumm and Lichty (1965). In most situations, however, both these influences will be apparent, leading to an interdependence between the organic and inorganic components of the environment. The rest of this book will be largely concerned with the second part of biogeomorphology, as defined above. Another approach to plant–landform relationships has been presented in a recent volume (Howard and Mitchell, 1985), which deals primarily with larger scale problems.

There are numerous possible ways to present a collection of material on biological influences in geomorphology in order to provide a comprehensive, wide-ranging survey. Potential approaches would be to present the material arranged according to organism, or according to earth surface processes, or on a regional basis. In all cases there is bound to be some overlap. The approach followed in this volume is to focus upon particular geographical areas, or environments, and to review the range of organic influences on earth surface processes found within each environment. These environments are informally defined primarily on climatic criteria, i.e. temperate, tropical, arid and semi-arid, and arctic and alpine environments. Two other areas are also considered here, coastal and karst environments, which are included because of their biogeomorphological importance. The material has been presented in this rather simplistic 'environmental' arrangement in order to make it easily accessible, and not because there is any fundamental, significant difference between environments. Part I concentrates upon temperate fluvial environments, where a large amount of work on biological effects upon geomorphology has been carried out. Part II presents evidence from the other major climatic zones, and Part III deals with coastal and karst environments.

This arrangement of material into 'environments' obviously leads to some overlap between chapters; and there are also some gaps in coverage which only partly reflect gaps in knowledge. More importantly, perhaps, this environmental arrangement means that some associations of organisms and geomorphology are relatively poorly covered, e.g. biological influences upon lake processes, or are documented from a limited range of environments only. Biological influences upon rock weathering, for example, are of general importance, but are here discussed primarily in the context of karst areas. The final chapter makes some general points about biological influences upon different earth surface processes and their wider implications.

References

Antweiler, R. C., and Drever, J. I. 1983: The weathering of a late Tertiary volcanic ash: importance of organic solutes. *Geochimica et Cosmochimica Acta* 47, 623–9.

Bassi, M. and Chiatante, D. 1976: The role of pigeon excrement in stone biodeterioration. *International Biodeterioration Bulletin* 12, 73–9.

Berthelin, J., Bunne, M., Belgy, G. and Wedraogo, F. X. 1985: A major role for nitrification in the weathering of minerals of brown acid forest soils. *Geomicrobiological Journal* 4, 175–90.

Best, T. L. 1972: Mound development by a pioneer population of the banner-tailed kangaroo rat, *Dipodomys spectabilis baileyi* Goldman, in eastern New Mexico. *American Midland Naturalist* 87, 201–6.

Buckley, R. 1987: The effect of sparse vegetation on the transport of dune sand by wind. *Nature* 325, 426–8.

Carey, D. A. 1987: Sedimentological effects and paleoecological implications of the tube-dwelling polychaete *Lanice conchilega* Pallas. *Sedimentology* 34, 49–66.

Chorley, R. J. 1962: Geomorphology and general systems theory. *United States Geological Survey, Professional Paper* 500-B, 1–10.

——, Schumm, S. A., and Sugden, D. E. 1984: *Geomorphology*. London: Methuen.

Duijsings, J. J. H. M. 1987: A sediment budget for a forested catchment in Luxembourg and its implications for channel development. *Earth Surface Processes and Landforms* 12, 173–84.

Dury, G. H. 1980: Neocatastrophism, a further look. *Progress in Physical Geography* 4, 391–413.

Ehrlich, H. L. 1981: *Geomicrobiology*. New York: Marcel Dekker.

Evans, F. C. 1956: Ecosystem as the basic unit in ecology. *Science* 123, 1127–8.

Flenley, J. R. 1984: Time scales in biogeography. In Taylor, J. A. (ed.) *Themes in Biogeography*, pp. 63–105. London: Croom Helm.

Friedmann, E. I. and Weed, R. 1987: Microbial trace-fossil formation, biogenous and abiotic weathering in the Antarctic cold desert. *Science* 236, 703–4.

Frydl, P. and Stearn, W. C. 1978: Rate of bioerosion by parrotfish in Barbados reef environments. *Journal of Sedimentary Petrology* 48, 1149–58.

Gerrard, A. J. 1981: *Soils and Landforms*. London: George Allen and Unwin.

Gould, S. J. 1984: Toward the vindication of punctuational change. In Berggren, W. A. and Van Couvering, J. A. *Catastrophes and Earth History*, pp. 9–34. Princeton, New Jersey: Princeton University Press.

Howard, J. A. and Mitchell, C. W. 1985: *Phytogeomorphology*. New York: John Wiley.

Krumbein, W. E. (ed.) 1978: *Environmental Biogeochemistry and Geomicrobiology*, 3 vols. Ann Arbor: Ann Arbor Science Publisher.

——, (ed.) 1983: *Microbial Geochemistry*. Oxford: Blackwell Scientific Publications.

——, and Giele, G. 1979: Calcification in a coccid cyanobacterium associated with the formation of desert stromatollites. *Sedimentology* 26, 593–604.

Lyell, C. 1835: *Principles of Geology*, 4th edn. London: John Murray.

Merrill, G. P. 1897: *A Treatise on Rock-weathering and Soils*. New York: The Macmillan Co.

Nelson, C. H. and Johnson, R. K. 1987: Whales and walruses as tillers of the sea floor. *Scientific American* 256, 74–82.

Noble, C. A. and Morgan, R. P. C. 1983: Rainfall interception and splash detachment with a Brussels sprouts plant: a laboratory simulation. *Earth Surface Processes and Landforms* 8, 569–78.

Novitsky, J. A. 1981: Calcium carbonate precipitation by marine bacteria. *Geomicrobiological Journal* 2, 375–88.

Nutting, W. L., Haverty, M. I. and LaFage, J. P. 1987: Physical and chemical alteration of soil by two subterranean termite species in Sonoran Desert grassland. *Journal of Arid Environments* 3, 233–40.

Richards, K. S., Arnett, R. R. and Ellis, S. (eds) 1985: *Geomorphology and Soils*. London: George Allen and Unwin.

Schaetzel, R. J. 1986: Complete soil profile inversion by tree uprooting. *Physical Geography* 7, 181–8.

Schumm, S. A. and Lichty, R. W. 1965: Time, space, and causality in geomorphology. *American Journal of Science* 263, 110–19.

Selous, F. C. 1881: *A Hunter's Wanderings in Africa*. R. Bentley & Son.

Stoddart, D. R. 1986: *On Geography*. Oxford: Basil Blackwell.

Taylor, J. A. (ed.) 1984: *Themes in Biogeography*. London: Croom-Helm.

Thornes, J. B. 1985: The ecology of erosion. *Geography* 70, 222–35.

Tinkler, K. J. 1985: *A Short History of Geomorphology*. London: Croom-Helm.

Trudgill, S. T. (ed.) 1986: *Solute Processes*. Chichester: John Wiley.

Twichell, D. C., Grimes, C. B., Jones, R. S. and Able, K. W. 1985: The role of erosion by fish in shaping topography around Hudson Submarine canyon. *Journal of Sedimentary Petrology* 55, 712–19.

Van den Ancker, J. A. M. and Jungerius, P. D. 1985: The role of algae in the stabilisation of coastal dune blowouts. *Earth Surface Processes and Landforms* 10, 189–92.

Worster, D. 1985: Nature's Economy: *A History of Ecological Ideas*. Cambridge: Cambridge University Press.

Part I
Temperate Fluvial Environments

1 Vegetation and river channel form and process

K. J. Gregory and A. M. Gurnell

The influence of vegetation upon river channel form and process has to be visualized in the context of the drainage basin. Although it is not absolutely clear when the drainage basin was first acknowledged as the fundamental unit appropriate for analysis in hydrology and fluvial geomorphology, it has been employed implicitly since the mid-nineteenth century (Gregory, 1976a). Formal explicit acknowledgement of the drainage basin as the fundamental geomorphic unit came with a paper by Chorley (1969). More recently it has been appreciated that the drainage basin is significant not only for analysis of contemporary processes but also for the management of river systems.

Since 1967 there has been a considerable research interest directed towards hydrological, hydrogeomorphological and fluvial geomorpho-logical aspects of the drainage basin, but the significance of vegetation has not been considered in an integrated way. Instead vegetation has been considered as a factor pertinent to several aspects of drainage basin morphology and process. This may have been an inevitable consequence of the way in which reductionist approaches have separated hydrological processes from branches of hydrology such as forest hydrology, and from the way in which both have been somewhat divorced from river channel morphology and fluvial process studies in hydrogeomorphology. In this chapter it is argued that it is now timely for more coordinated consideration of the significance of vegetation upon river channel form and process, and that this is a necessary prerequisite for the further development of river management procedures.

1.1 Vegetation in the drainage basin

The drainage basin is the locus for a cascade of water and sediment transfer through a series of stores or subsystems. Figure 1.1 is a simplified representation of the hydrological cycle. The stores are shaded and it is evident that the nature of vegetation influences the amount of moisture supplied to each store, it determines the amount retained in some of the stores, and it also affects the sequence of flows that can take place.

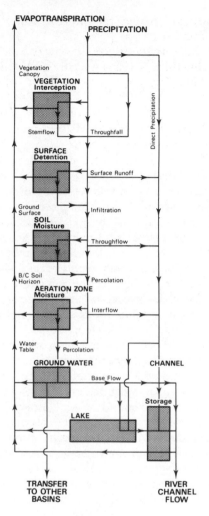

Figure 1.1   Sequence of the hydrological cycle

The first store (figure 1.1) is represented by interception on vegetative surfaces including leaves, branches and tree trunks. A certain amount of precipitation can immediately be stored on these surfaces and until precipitation amounts exceed critical values there will not be transfer of water to stores lower in the system. By throughfall and stemflow, precipitation moves through the vegetation to the ground surface. On the surface the accumulation of moisture as surface detention depends upon the amount of vegetation cover, which acts to bind the surface, and on the amount of the litter layer. The roughness of the vegetation cover will also influence the extent to which surface runoff or infiltration takes place.

The magnitude of interception storage varies very considerably with type of vegetation cover, and there is a strong contrast between the interception storage of woodland in comparison with moorland, grassland and arable areas as well as variations within these categories. However, Roberts (1983) reports that total annual losses from European forests are remarkably similar regardless of variations in both tree species and climate. This may be a result of the buffering of tree canopy interception losses by losses from understorey and leaf litter interception. It may also result from negative feedback effects between climatic variations and the vegetation surface resistance to transpiration loss. The enhanced losses of water from a Sitka spruce plantation in comparison with adjacent grassland, reported in the classic papers by Law (1956, 1957), have been confirmed by many catchment experiments as a typical impact of afforestation (Courtney, 1981). Intercepted water that is not evaporated passes through the forest canopy as stemflow and throughfall. These processes produce complex spatial patterns of water delivery to the forest floor which may be reflected in the fine root structures of the trees (Ford and Deans, 1978), so ensuring an efficient supply of soil moisture for transpiration. Thus, the presence of the tree and understorey canopies and the litter layer provide substantial interception storage for precipitation so that a large proportion may evaporate. High surface roughness can mean that the portion of precipitation which reaches the ground is mainly held in surface detention storage and infiltrates instead of forming overland flow.

In the last three decades research has demonstrated in detail the way in which vegetation influences the hydrological cycle (Penman, 1963). Many research investigations in instrumented catchments (e.g. Ward, 1971) have highlighted differences in the water balance and in the processes of runoff production according to contrasts in vegetation cover and management (e.g. Sopper and Lull, 1967). One example is the catchment experiment undertaken in central Wales by the Institute of Hydrology which compares hydrological processes in two small catchment areas at the head of the Rivers Severn and Wye. The Severn catchment was largely coniferous woodland (to be clear felled later) and the Wye mainly grassland, and the comparison of the two basins has produced a number of conclusions summarized by Newson (1985) as:

1 Evaporation from forest in the Upper Severn is double that in the Wye which is mainly grassland.
2 Interception losses are double those by the accepted evaporation route via transpiration.
3 Interception produces a drier soil in the forested catchment. However, the degree of canopy closure and the extent of drainage

ditches need to be considered in relation to the generation of flood runoff.

4 In times of drought interception is not active, and there are signs that conifers are slightly better in conserving moisture by controlling transpiration than is grass.

Such experiments have clearly demonstrated that changes in vegetation cover can give changes in the frequency and magnitude of peak discharges and of base flows. Studies in experimental basins also show how land-use change can have substantial influences on sediment yield. Thus Hadley *et al.* (1985) summarize how clear felling can give increases in sediment yield of 8, 39 or 310 times according to area studied.

Because vegetation determines the amount and the routes taken by water and sediment through the fluvial system, it is a major control upon the fluid that is available to affect river channels. W. L. Graf (1979) has argued that the balance between force and resistance lies at the heart of explanation in geomorphology. Vegetation contributes much of the resistance, and furthermore the removal or the reduction of vegetation cover can not only decrease this resistance but by increased runoff can increase the force of water on the land surface. Vegetation is a significant influence upon the water and sediment inputs to the river channel system, but in addition it is important to remember that the drainage basin is itself a dynamic system. Since 1965 runoff generation from drainage basins has been conceived as dependent upon the dynamic contributing areas, as indicated in figure 1.2. During precipitation over a drainage basin the area which contributes quickflow to the stream

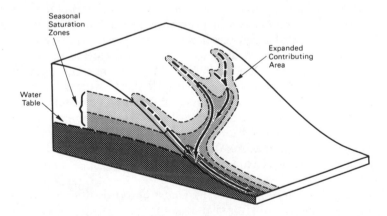

Figure 1.2   Dynamic contributing area model of runoff generation
The expanded contributing area is partly related to seasonal saturation zones and the influence of vegetation upon expansion and contraction of the zones is explained in the text

channel expands, and after precipitation ceases that area will slowly contract.

Therefore the prime determinant upon the generation of discharge hydrographs from drainage basins is the rate and degree to which such dynamic contributing areas expand and contract. This dynamic contributing area model, which has come to have a great influence upon the interpretation of discharge generation, involves the notion that in many temperate areas at least there will be only a comparatively small portion of the drainage basin, perhaps 10 or 15 per cent after dry antecedent conditions, which contributes to the generation of discharge hydrographs.

The drainage network in any drainage basin is made up of a number of components which may be classified according to the frequency of flow. A classification usually employed (Gregory and Walling, 1973) includes perennial streams which flow all the time, intermittent streams which flow when the water table is seasonally high, and ephemeral streams which flow only during or immediately after periods of significant precipitation. In some areas near-surface pipe networks have similar properties (Jones and Crane, 1984). It is therefore possible to envisage a drainage network which expands and contracts associated with expanded contributing areas as shown in figure 1.2. Particular types of vegetation and specific vegetation communities are associated with these expanded contributing areas (Gurnell, 1978) and very often in a drainage basin the pattern of vegetation communities relates specifically to several saturation zones. In the northern Shenandoah Valley, Virginia, the type of vegetation relates to the pattern of land forms in the drainage basin, and indicator species were shown to associate with particular land forms (Hupp, 1986). Studies in the UK have shown how different types of channel compose a drainage network (e.g. Gurnell and Gregory, 1984), how these types reflect vegetation character, and how they affect flow routing (Gardiner and Gregory, 1982).

Heath communities are particularly sensitive to the soil moisture regime and so a research investigation in the Highland Water basin (11.4 km^2), New Forest, Hampshire was undertaken to derive quantitative means of linking contributing area to heathland vegetation composition. The Highland Water basin is in an area of near-horizontally bedded Eocene Barton sand and clay, with a capping of Pleistocene gravel on the interfluves (figure 1.3). The northern part of the basin is covered mainly by heathland, whereas the southern part is under woodland. A detailed ground vegetation survey of the heathland areas was analysed by detrended correspondence analysis (Hill and Gauch, 1980). This ordination technique revealed a soil-moisture-related axis in the vegetation data which was used to classify the heath into six moisture-related classes. The boundaries of the classes were identified and mapped using air photos (Gurnell et al., 1985). An analysis of a year's field measurements at over 160 well sites in the basin has shown that distinctive water table level exceedance probabilities are associated with the categories of vegetation composition (Gurnell and Gregory, 1986), and can be used to identify the area contributing runoff after specific antecedent conditions (Gurnell and Gregory, 1987).

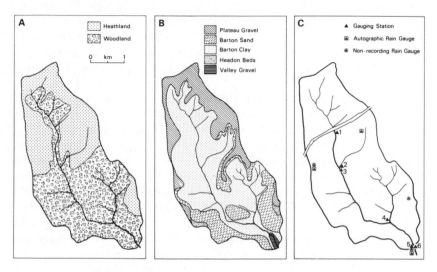

Figure 1.3 Highland water catchment
The distribution of vegetation (A), of geology (B) and of research design (C)

Thus a simple classification of vegetation composition can yield a hydrologically useful indication of zones of different soil-moisture regime, and Gurnell and Gregory (1986) concluded that parameters of vegetation composition in heath catchments could afford useful additional catchment characteristics to complement the more usual indices of catchment geometry – size, rock type and soils. Such an approach would require rapid and accurate means of mapping vegetational categories. Analysis of remotely-sensed high-resolution digital data appears to present a means of producing such maps, and even in the complex patchwork of the New Forest heaths analysis of simulated 20 m resolution SPOT data provided maps of acceptable accuracy (Gurnell and Gregory, 1987). Such analyses permit results of the Highland Water study to be extrapolated to the whole of the New Forest.

Links between vegetation and river channel form and process can be visualized through the sequence of the hydrological cycle and in relation to the dynamic contributing area model. Although these two aspects determine the amount of water and sediment available to the river channel system, the remainder of this chapter is devoted to river channel processes approached by looking at river channel morphology (section 1.2), at river channel processes (section 1.3), at the way in which vegetation relates to river channel changes and to management (section 1.4), and finally to the way in which a more integrated view of vegetation in relation to river channel dynamics in a drainage basin context can be the basis for enhanced river management (section 1.5). Although this separation is adopted for convenience, it is important to remember that morphology, process, and temporal change are all interrelated. Furthermore, it is important to recall that not only is vegetation itself an influence upon river channels and river channel processes, but in addition the processes, and particularly water and sediment availability, exercise an influence upon vegetation distribution.

In studies of river channel cross-sections channel capacity needs to be identified consistently, and the methods available for channel capacity determination (Gregory, 1976b; Williams, 1978a) include the use of vegetation to indicate the limits of channel capacity or the bankfull stage. Some channels have compound channel cross-sections, and each element in the channel cross-section can be associated with different types of vegetation. Thus, in the case of three perennial streams of northern Virginia, Osterkamp and Hupp (1984) identified the surfaces of depositional bars, active channel shelf, floodplain and terraces, and established that each of these surfaces supports characteristic species, some of which are nearly unique to the surface. They inferred that plant distributions are largely controlled by flow frequency and intensity and that plants may help to identify geomorphic levels and potential for flood damage. In some areas it has been possible to use lichen limits as indicators of consistent levels for the definition of river channel dimensions. In south-eastern Australia it was demonstrated that not only is there a hydrological significance of the lower lichen limit below which crustose lichens do not grow, but also that a series of higher lichen limits can be identified and may be associated with specific large floods (Gregory 1976c). Vegetation adjacent to river channels can also give important clues to river discharge history. This has been demonstrated by Yanosky (1982), and also in Wisconsin by Wendland and Watson-Stegner (1983). They obtained cores from several tree species located both on a river floodplain and on a nearby terrace, and demonstrated that the ratio of annual tree growth on flood plains to terrace could be related to the annual river discharge. Such growth ratios from the time prior to continuous hydrological records can be used to reconstruct river discharge and to infer flood frequency.

1.2   River channel morphology

In a comprehensive review of vegetation and river channel dynamics Hickin (1984) argued that fluvial geomorphology has not coped well with processes that are not easily quantifiable and physically or statistically manipulable. In his review he suggested that the influence of vegetation is a problem of this kind and has often been dealt with by excluding it from research undertaken.

Vegetation affects the overall channel morphology and the detailed features of the river channel, and it also collectively contributes to the overall channel roughness.

Overall channel dimensions are affected because the influence of vegetation combines with other controls to determine the size and shape of river channels. An important paper (Zimmerman *et al.*, 1967) indicated how vegetation influenced the form of small stream channels in the Sleepers river basin of Northern Vermont. They concluded that vegetation influences channel form by altering the roughness and the shear strength of the bed and banks. They further suggested that there are two thresholds along small streams of the Sleepers river basin. The first occurs where the drainage area is $0.5-2 \, km^2$ and channel form, size and location are greatly influenced by non-fluvial processes such as tree blow-down, damming by debris, and extension of roots. However, at drainage areas exceeding $0.5-2 \, km^2$ channel width increases, but both mean width and ranges of width vary greatly according to the type of vegetation lining the stream. With drainage areas exceeding $10-15 \, km^2$ the influence of vegetation on channel form becomes marginal.

In forested areas it therefore appears that vegetation continues as an influence on overall channel form until the width of the channel equals or exceeds the height of the trees growing in the forest. Comparing shapes of small stream channels Zimmerman *et al.* (1967) suggested that the width–depth ratios of streams cut through grassland are generally lower than those of channels through forest, which they took to indicate that grass sod behaves more like cohesive sediment than does bank material consolidated by tree roots. However, they recognized that differences in channel form under different types of vegetation are caused mainly by relative disturbance of the vegetation above ground, rather than by differences in shear strength resulting from different root systems.

In other areas it has usually been argued that tree-lined channels are deeper and narrower and therefore possess a lower width–depth ratio than do channels which have grassland on their banks (Maddock, 1972). A study of gravel bed rivers in Britain (Charlton *et al.*, 1978) showed that channels with grassy banks are 30 per cent wider than the average and that tree-lined channels are 30 per cent narrower than the average. Such contrasts are usually attributed to the fact that in forest, tree roots penetrate to lower levels in the channel banks and therefore increase the shear strength of the banks compared with grass covered banks where the shear strength is correspondingly less. Grass-lined channels thus tend to be undercut rather more easily than do tree-lined channels. There are exceptions to this, and in coniferous forest areas with characteristically shallow rooted species it was found (Murgatroyd and Ternan, 1983) that tree-lined channels can be wider than grass-lined ones. It may be possible to reconcile the apparently conflicting views of Zimmerman *et al.* (1967) in respect of small channels and of these later workers in relation to larger stream channels. In small stream channels rooting depths of grass sod are a significant proportion of the channel depth so that grass-lined channels may be the narrowest. However, further downstream grass rooting depths are very small in relation to the channel depth and hence grassed channels have the highest width–depth ratios.

Such overall channel dimensions are complemented by the way in which vegetatation influences the detail of river channel geometry. This has been considered generally by Rachocki (1978) in his study of four rivers in the Cassubian Lakeland area of Poland. He attributes the significance of the influence of vegetation on the channel to three groups of plants, namely aqueous plants, riverside plants, and overchannel plants. The influence of these three categories extends from the influence upon deposition of bed material, to accumulation of bars in the channel and to the rate of bank erosion. Along the Kicking Horse River in British Columbia, Smith (1976) found that the erodibility of bank alluvium varied inversely and exponentially with root density. This gave an excellent example of one of the few quantitative relationships established between vegetation control and channel morphology. Because of the existence of such relationships it would be expected that channel migration should be greater where the influence of vegetation is least. Studies of western Canadian rivers by Hickin and Nanson (1984) indicated that river banks which are well bound by roots offer much greater resistance to lateral erosion than unvegetated banks of alluvium exposed to the same forces. They found that rates of channel migration through unvegetated alluvium could be double those through naturally forested flood plain.

Two other ways in which the detail of river channel morphology relates to vegetation concern the accretion of concave-bank benches and the accumulation of channel bars (Hickin, 1984). Concave-bank benches develop on the outer bank of a river bend and have been identified in a number of areas. Because the bench forms on the outer bank that has overtightened by impinging on a resistant obstacle such as a valley wall, there is an area to which bed material has no access and where a high proportion of organic matter made up of floating trees, logs and leaf litter occurs within the depositional material that accumulates to build up the concave-bank bench (e.g. Page and Nanson, 1982). Vegetation remnants that have been transported along the channel can also be significant in building up various types of channel bar. For example, along rivers in British Columbia and Alberta it was suggested that scroll bars have developed around trees and logs (Nanson, 1981). Although elsewhere such

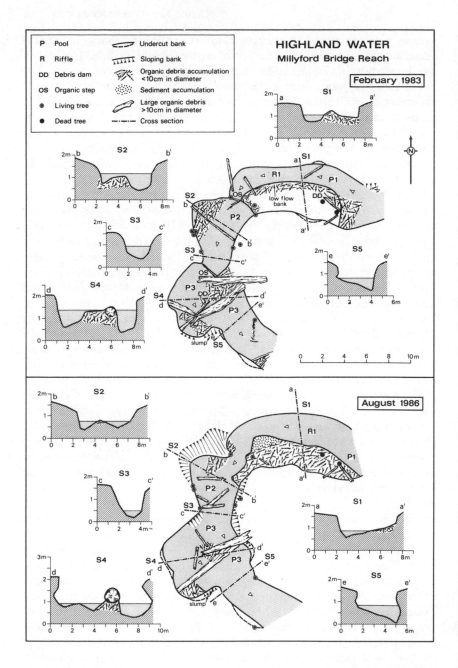

Figure 1.4 Vegetation influence upon channel morphology of a reach of the Highland Water
The study reach is located near the outlet of the basin (figure 1.3) and the 1983 survey was undertaken with Professor E. A. Keller. The changes by August 1986

trees and logs may not initiate bars, nevertheless Hickin (1984) has concluded that such vegetative debris can be very important to the growth and accretion of channel bars. In the small channel of the Konczak stream in western Poland, Witt (1985) found that aquatic vegetation has a significant influence on the deposition of bars and on the structure which is developed. He found that the vegetation not only stabilizes the bars that exist but also by acting as traps for coarse sediment tends to increase accretion on the bar.

Perhaps the most significant way in which channel morphology is influenced by vegetation is through the effect that vegetation debris has in the development of log and debris dams and jams. Although the significance of such features was noted in general by Zimmerman *et al.* (1967), research in forested stream channels before the late 1970s tended to focus on parts of the stream channel between the dams, rather than upon the effects of the dams themselves. Subsequently there has been considerable research interest in the significance that debris dams have for channel morphology and channel processes, and such studies have been undertaken in North America (Keller and Tally, 1979; Heede, 1981; Marston, 1982; Keller and Swanson, 1979; Likens and Bilby, 1982; Macdonald *et al.*, 1982), in New Zealand (Mosley, 1983), and in the UK (Gregory, Gurnell and Hill, 1985). Debris dams can be very significant, and in the western USA Heede (1981) indicated that in headwater channels a dam occurs every 1.5–8 m of channel. In a study of the Highland Water drainage basin (Gregory, Gurnell and Hill, 1985), it was found that there were 270 dams in the 11.4 km^2 drainage basin indicating an average of 1 dam for every 27 m of main channel. Such spacing is much less than the 395 m of channel per dam noted in the central Oregon coast range (Marston, 1982) but is similar to the range of log step spacing between 3.1 and 26 m in Arizona and Colorado (Heede, 1981) and the 2.5–5 m average spacing along New England streams (Likens and Bilby, 1982).

Stream channel morphology in the vicinity of debris dams is greatly affected by the presence of the dam, and detailed analyses have been undertaken by Keller and Tally (1979), who have recognized the morphological significance that dams have. In the Redwood Creek drainage basin of California, they showed how debris dams affect channel depth and width and channel slope. An example of the way in which a compound dam influences the detail of channel morphology is shown in figure 1.4, which is constructed using the techniques employed by

---

were instigated by large discharges in 1986 and these led to removal of the dam between S1 and S2, removal of organic debris at S2, accumulation of logs upstream of S3, partial breakup of dam at S4, and associated morphological changes with bank erosion at S1, S2, S3, S4 and bed lowering at S3, S4 and S5

Figure 1.5 Active debris dam on Highland Water, New Forest, UK

Professor E. A. Keller and also indicates how organic debris influences the detail of morphological changes. The slope of a river channel can be significantly affected by the presence of debris dams, and along Little Lost Man Creek in the coastal Redwood environment of California 60 per cent of the total fall in height occurs at the site of such debris dams. This significance arises because the debris dams are often responsible for impoundment of water upstream of the dam, so that there is then a clear fall over the dam. In Western Oregon it was found that the precise location of between 30 and 80 per cent of the fall in height along small streams was determined by the presence of these debris dams (Keller and Swanson, 1979).

Such debris dams have a significant influence upon inchannel processes (section 1.3) and are important in relation to management of river channels (section 1.4). However, debris dams are also very significant in forested areas because of their persistence. The residence time of Redwood debris in stream channels in north-western California can exceed 200 years (Keller and Tally, 1979), and accumulations of Douglas Fir are known to remain in channels for more than 100 years (Swanson *et al.*, 1976). In the survey of debris dams throughout the Highland Water drainage basin (figure 1.3), three types of dam were recognized. Active dams formed a complete barrier to water and sediment movement and also created a distinct step or fall in the channel profile (e.g. figure 1.5); complete dams provided a complete barrier across the channel but did not interrupt the channel profile; and partial dams represented a partial barrier across the channel because they were either incomplete or had been partly destroyed. Throughout the Highland Water basin there were 270 dams in 1982–83, compared with 294 in 1984 (Gregory, Gurnell and Hill, 1985).

Whereas 64 per cent of the original dams were in the same position, the remaining 36 per cent had changed in character and overall there had been some form of change either of dam type, of removal or of creation of a debris dam at 166 sites. This indicates that on average there had been some form of change within one year for every 48 m of channel in the Highland Water drainage basin: detailed changes are shown in figure 1.4. It is therefore important to appreciate that although there are variations from one forested area to another depending upon forest type and particularly tree height, organic debris in river channels has a residence time of up to 100 or 200 years although in detail the exact location of debris dams changes during fairly short periods. Accumulations of organic debris are not confined to small headwater river channels. Hickin (1984) has shown how log jams affecting part of the river channel along Canadian rivers can exercise an influence upon major changes of channel pattern. In a reach of the Squamish River, between 1960 and 1980, channel abandonment and flow diversion have been associated with the formation and destruction of such log jams.

The way in which channel morphology is affected by vegetation is included in the roughness of the stream channel. Hydraulic approaches to the investigation of river channels require an expression of the resistance to flow which is afforded by the overall character of the river channel. The roughness of any specific river channel reflects the cross-sectional channel geometry, the detailed channel cross-section, and the skin resistance associated with specific particles in the channel cross-section. These are just three aspects of the way in which roughness is compounded, in itself a very complex subject for analysis (e.g. Hey et al., 1982). In hydraulic calculations it is necessary to quantify the roughness of river channels and, for example, the Manning roughness coefficient n is used to express the relative roughness of a channel in relation to velocity (V), hydraulic radius (R) and slope (S) in the Manning equation

$$V = \frac{R^{2/3} S^{1/2}}{n}$$

Although roughness coefficients can be estimated with experience (e.g. Barnes, 1967), it has also been recognized that a roughness coefficient such as Manning n can be based upon the Cowan method of estimation (Cowan, 1956) which includes consideration of materials involved (n_0), degree of channel cross-section irregularity (n_1), variation in channel cross-section shape and area (n_2), relative effects of obstructions (n_3), and vegetation (n_4), together with the degree of meandering (m). Although the values associated with each of the components in the equation

$$n = (n_0 + n_1 + n_2 + n_3 + n_4)m$$

need to be considered carefully from one area to another, this approach does allow the several components of channel roughness to be considered, and it also explicitly recognizes the significance that vegetation can have. There have been developments of this approach as applied to particular areas and Jarrett (1985), for example, has determined roughness coefficients for streams in Colorado in a way that allows the development of the Cowan approach in relation to higher gradient streams with slopes greater than 0.002.

1.3 River channel processes

Roughness of a river channel in turn influences the processes that operate in that channel. Effects on river channel process can be thought of as due to the overchannel vegetation, the vegetation that occurs alongside the channel including riparian tree growth, and the within-channel vegetation (Rachocki, 1978). The last category includes plants which Sculthorpe (1967) has classified as emergents (rooted in the soil close to or below water level for much of the year); floating leaved (rooted in the bed of the channel although the leaves predominantly float on the water surface); free floating (not attached to the bed although roots may be entwined in other plants); and submerged (attached to submerged surfaces or materials and the substrate by roots, rhizoids, or the whole thallus). Study of freshwater plants is embraced within the fields of hydrobiology and freshwater ecology (e.g. Haslam, 1978). However, it is necessary for a geomorphologist to understand how such aquatic vegetation contributes to overall channel roughness. In their theory of vegetative channels, Petryk and Bosmajian (1975) predicted Manning n in terms of vegetation density, hydraulic radius, and the Manning n value that would obtain for the channel if the vegetation was not present. They calculated vegetation density from the drag coefficient of the vegetation (Cd), the area of vegetation exposed to flow in the channel cross-sectional area, and the channel reach length being studied.

In addition to the way in which aquatic plants contribute to river channel roughness, it is important to recognize the way in which the plants relate to roughness as discharge varies and also the way in which seasonal variations occur according to plant growth. The significance of aquatic plants for channel roughness decreases as river flow increases. Thus

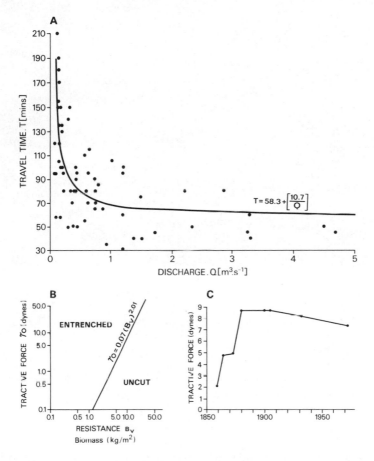

Figure 1.6 Changes of process and vegetation influences
(A) shows the travel time T (minutes) between 2 gauging stations 4,028 m apart in the Highland Water Basin (figure 1.3) plotted against discharge for 67 hydrographs at the downstream gauging station. At higher discharges the effect of 93 debris dams is much less than at lower flows
(after Gregory, Gurnell and Hill, 1985).
(B) shows a critical function relating tractive force and biomass on the valley floor of the Central City District Colorado, and (C) shows how calculated levels of tractive force for discharge events with a recurrence interval of 10 years changed after 1850 with changes in vegetation cover affected by mining activity
(after Graf, 1979)

Figure 1.7 Bank erosion influenced by vegetation, Highland Water, New Forest, UK
Upstream of the debris dam trees and tree roots have limited bank erosion, whereas upstream of the two trees channel bank erosion has occurred where there are grass banks .

Dawson and Robinson (1985) showed how plant roughness can be drowned out at higher flows, and it is possible to establish washout velocities for different types of river plant. There are also variations during the annual cycle of plant growth. It was shown along the River Skerne, County Durham, that the profuse summer growth of *Sparganium erectum* led to a significant decrease in the travel times of flood waves coming down the river in summer compared with the situation in winter (Hamill, 1983). Detailed investigations concerned with interrelations between vegetation and river flow have shown how this interaction can be the basis for an understanding of river management at different scales, and a biological classification of river plants in British water courses was proposed for this purpose by Haslam and Wolseley (1981). In headwater areas, particularly in the United States, considerable research has been devoted to the relationship between water flow and the character of grassed channels. This has been undertaken for example by the US Soil Conservation Service to establish the water flow velocities that can take place without erosion breaking the grass cover. Design of channels with vegetal linings can be based on the physical parameters found to control flow resistance in open channels (Kouwen and Li, 1980).

The presence of vegetation along river channels or within river channels has a significant influence on the processes of water flow and the transport of sediment. However, in this chapter attention is limited to the way in which vegetation influences specific processes and also influences the overall expenditure of energy in the fluvial system.

Vegetation in and along river channels determines the rate and amount of water and sediment that is transported through the river channel. As indicated above, vegetation debris as well as aquatic plants can influence the accumulation of sediment in the form of various types of channel bar and they can also reduce water velocities (e.g. figure 1.6A). Vegetation can also affect the distribution of bank erosion (e.g. figure 1.7). In West Tennessee Hupp (1986) demonstrated that establishment of vegetation, usually willow, river birch, box elder or sycamore, is an important indicator of bank stability. Many of the streams in West Tennessee have been dredged for flood control and can experience widening rates of up to 2.4 m per year along highly unstable reaches. Such unstable reaches are largely devoid of establishing woody vegetation, but where vegetation is restabilizing channel banks then accretion rates of 5 to 7 cm a year have been reported.

The debris dams that are a feature of many stream channels in forested areas also have a significant effect upon channel processes by reducing stream velocity and discharge (Heede, 1981), by affecting flow routing, and by increasing sediment storage. For example, up to 123 per cent of the mean annual sediment discharge is stored behind log steps in the Oregon Coast range (Marston, 1982), and 15 times the average annual sediment yield is held behind dams on small streams on the Idaho Batholith (Megahan, 1982). Just as sediment storage behind debris dams provides a buffer in the sediment routing system, so the stream energy can be dissipated by the presence of such dams. During the flow of water along a stream channel there is a constant use of energy. In the research on the Highland Water, it was shown that the travel times of 67 well defined hydrographs relate to the incidence of the debris dams. At high discharges the influence of the debris dams is minimized, but at lower flows debris dams have a significant influence on travel times (figure 1.6A).

The geomorphic significance of log steps was demonstrated for streams in the central Oregon Coast range by Marston (1982), when he quantified the dissipation of potential energy in various portions of the stream network. Because log steps cause a reduction in the potential energy which would otherwise be converted to kinetic energy used for sediment transport, the log steps affect energy consumption along stream channels. Whereas at low flow dissipation of potential energy amounts to 12 per cent of the total stream relief, boulder and bedrock falls account for about half of this dissipation and the remainder is contributed by energy dissipation at log steps. These values were based solely upon the energy expenditure at log steps, but if in addition the dissipation of energy due to instream woody debris is included then there can be dissipation of up to 60 per cent of stream energy in some areas, and between 50 and 100 per cent dissipation of stream energy can be accounted for by log steps and gravel bars (Heede, 1981).

A more general way of expressing the power involved in the interrelationship between vegetation and channel processes is exemplified in the study undertaken by Graf (1979) in the Central City district of Colorado. By analysing the relationship between tractive force representing fluvial processes and resistance reflected largely by vegetation biomass (figure 1.6B), Graf was able to demonstrate the way in which this critical relationship varied due to mining activities in the mid-nineteenth century compared with the present (figure 1.6C).

1.4   Changes of channel form and process in relation to vegetation

The example from the Central City area of Colorado exemplifies how vegetation influence can affect river channel change. Miners occasioned the removal of trees, and this in turn increased runoff amounts and decreased the resistance to surface runoff (figure 1.6C), so that channel changes occurred downstream exemplified by arroyo development (Graf, 1979). In this and other cases extensive modification of vegetation in catchment areas changes the production of water and sediment, and these in turn have effects on the downstream channel morphology. However, changes induced by vegetation are clearly interrelated to other changes in drainage basins, so that a measure of uncertainty still characterizes our ability to explain river channel changes (e.g. Burkham, 1981). In areas with a long settlement history, changes of vegetation which occurred long ago may have had effects on rivers and river channels which are now not easy to disentangle. Thus deforestation in China was responsible for triggering the incidence of extensive gully erosion in areas such as the middle Huang He basin, and in less sensitive areas such as Britain the removal of the originally more extensive forest cover induced short periods of greater runoff and sediment accumulation. Recent developments in an enormous field of research on channel changes are summarized by discussion of deliberate changes of the vegetation of river channels; of incidental changes which have arisen from the effects of human activity; and of long-term changes in fluvial processes related to vegetation.

Management of the vegetation adjacent to river channels is central to an understanding of recent river channel changes. In the Appalachians of Tennessee at times of peak discharge Guntersville reservoir backed up along Cow Creek and inundated farmland along the Creek for a distance of 38 km. Local watershed districts prevailed upon the local conservation service to straighten, widen and deepen the 72 km of the creek. Work started in 1970 and although some trees should have been left along the Creek, in 1971 the result was described as an ecological disaster (Leopold, 1977). The soft clay banks had crumbled into a muddy stream, there were now no rooted plants in the streambed and there were no fish or animals. It was estimated that the benefits accruing to agriculture by the control of occasional flooding were not greatly in excess of the costs to agriculture as a result of the ecological disaster that had been created (Leopold, 1977). This is just one dramatic example of the effects of river channelization. Channelization of rivers has been undertaken to protect land from flooding, to improve navigation, to assist drainage, and to reduce bank erosion. In the United States channelization procedures have been undertaken since the time of the early settlers and were particularly extensive in the nineteenth and twentieth centuries, so that Leopold (1977) suggested that 26,500 km of river channel in the USA were already improved and a further 16,000 km of channel were destined for improvement.

Channelization procedures (Brookes *et al.*, 1983; Brookes, 1985) by modification of existing channels or creation of new ones has often involved the removal of riparian vegetation. Consequences of river channelization have attracted considerable public outcry as exemplified by extensive reports by US Government Committees (Committee on Government operations, 1973), and also books by concerned individuals: Heuvelmens (1974) published a book dramatically entitled *The River Killers*.

Removal of some riparian vegetation is not always an unwise measure. During the late 1970s there was major flooding in the American south west at the end of a drought. In some areas such as Maricopa County in Arizona, estimated damages exceeded $77 million (Graf, 1980), and occurred because river flows could not be impounded by the extensive control structures that had been designed for irrigation storage rather than for flood control. In many reaches dense growth of phreatophytes in the river channels exacerbated the problem. The phreatophytes impeded flood waters, accelerated sedimentation, and trapped flood debris. Graf (1980) drew attention to the dilemma which managers faced – either to clear the vegetation despite its valuable habitat for wildlife, or alternatively to clear some of the riparian phreatophytes. Tamarisk or saltcedar is the main phreatophyte involved and in management of river channels in the South West, it is suggested that some sort of channel clearing and maintenance is inevitable; that in the short term mechanical removal and control of phreatophytes is possible with regrowth control by chemicals or by mowing and grazing; and that in the long term restoration of riparian habitats without dense phreatophyte growth by groundwater control requires further investigation (Graf 1980). Elsewhere subtle changes have taken place as a result of changes in vegetation, and thus Haslam (1981) indicated that throughout Britain many stretches of rivers have become shallower and swifter and that such changes have had an influence upon the distribution and character of aquatic vegetation.

Experience of the consequences of channelization (e.g. Gregory, Hockin, Brooker and Brookes, 1985) has led to the development of new techniques for working with the river rather than against it. Such innovations have been developed in a number of areas and all involve careful management of channel vegetation to some degree at least. At the simplest level this may necessitate leaving all riparian vegetation to minimize the extent of channel change. However, in other areas, it is necessary to develop techniques which emulate the natural river channel, and this requires the management and engineering of river channels to imitate natural ones in such a way that they accommodate the problems of flooding and erosion that have been increasing in recent years. In North America stream renovation has been offered as an alternative to channelization (Nunnally, 1978) and involves leaving as many trees as possible, minimizing channel reshaping, the early reseeding of grass in disturbed areas, judicious placement of riprap, and emulation of natural river channel form. Stream restoration (Keller and Hoffmann, 1976) is the process of altering urban stream channels so that behaviour is analogous to natural streams, while providing some measure of flood prevention or control and producing a positive aesthetic experience. Thus on Brier Creek it has been found that this method of working with nature adequately meets the requirements of channelization without some of the major disadvantages (Nunnally and Keller, 1979). Similar sympathetic methods of river management have been advocated in the UK (Nature Conservancy Council, 1983), and also in West Germany where biotechnical engineering involves the emulation of natural channel morphology, preservation or creation of natural habitats for flora and fauna, and bank stabilization with living vegetation (Binder, 1979). In Denmark, recommendations for the management of river channels utilize techniques of natural and induced restoration and in such channel improvement trees and vegetation play a major part (Brookes 1984).

A number of incidental changes to river channels have affected the relation between ecology and river channel form and process. The effects of such changes have yet to be fully established but drainage networks in temperate areas have been metamorphosed in recent decades. Drainage network changes since the nineteenth century are significant in extent and have occurred as a result of the replacement of flushes by clearly defined stream channels (Ovenden and Gregory, 1980). Flushes do not contain a definite stream channel and are depressions in headwater areas along which there is water-loving vegetation, often including *Juncus*, and along which water flows as either saturated overland flow or saturated throughflow. There are clear indications, based upon comparison of Ordnance Survey maps of different dates, that such flushes were more extensive in the nineteenth than in the twentieth century. In parts of upland Britain stream channels have developed along depressions that originally contained flushes and this in effect has extended the density of the network of stream channels. Because water flow is more rapid in open channels than it is through flushes, such extension of the network of drainage channels can be responsible for more rapid hydrograph rise and for greater flood peaks. Whereas many changes in drainage network extent in such temperate areas can be deliberate, resulting from extensive drainage and ditching (Burt and Gardner, 1982), there is evidence of additional metamorphosis of components of the stream network.

Metamorphosis of stream channels has also occurred in south-east Australia (Eyles, 1977), and in seasonal tropical areas the long seasonally-inundated depressions characterized by a distinctive vegetation and known as dambos (Whitlow, 1985) have changed due to degradation. In Zimbabwe dambos are valley grasslands which are seasonally waterlogged and easily distinguished by a characteristic grass and sedge flora and a general absence of woody species (Whitlow, 1985). Desiccation of the area draining to the dambos by land-use pressure induces channel incision, so that there is a significant change from a depression which was seasonally inundated to one with a stream channel.

River channel changes can often be influenced by changes in vegetation distribution. Thus along the Turanganni River in New Zealand, the planting of willow shrubs at channel bends instigated a change in river channel pattern from a braided to a single thread channel (Nevins, 1969). A number of studies have demonstrated how river channel vegetation is associated with phases of river channel change. The Gila River in Arizona was a narrow stable stream channel which meandered through a flood plain covered with willow, cottonwood and mesquite from 1846 to 1904 (Burkham, 1972). Subsequently, however, there was major destruction of the flood plain from 1905 to 1917 with an average width increase to 600 m, from 45 m in 1875 and 90 m in 1903. Reconstruction occurred between 1918 and 1970, the stream channel narrowed to a width of less than 60 m in 1964 and the floodplain was densely covered with saltcedar between 1918 and 1970. An earlier study of the Cimarron River (figure 1.8) in south-western Kansas (Schumm and Lichty, 1963) showed that the channel width averaged 15 m in 1874, and that beginning in 1914 and continuing until 1942 there was extensive channel widening until almost all the floodplain was destroyed, culminating in an average channel widening of 350 m. After 1942 the channel became narrower and averaged 165 m. Whereas it had originally been proposed that accelerated erosion had

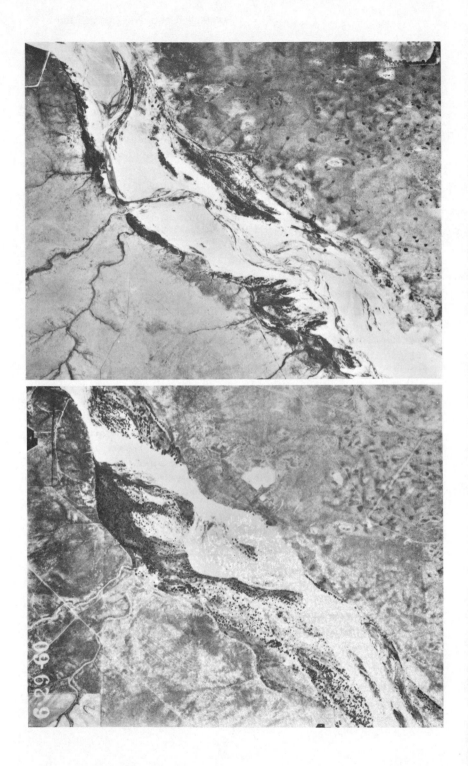

been prompted by over-cultivation and reflected agricultural history, Schumm and Lichty concluded that the period of channel widening was accompanied by a period of below-average precipitation with floods of a high peak discharge, whereas the period of flood plain construction was a period of above-average precipitation and floods of low peak discharge. These two climatic phases affected vegetation growth, and Schumm and Lichty concluded that this was the key to river behaviour. Wet years and low water allowed vigorous growth of perennial vegetation which stabilized existing deposits and encouraged further sediment deposition.

It is not always easy to be certain that channel changes are influenced by vegetation, because some channel changes may be inspired by hydrological changes which give rise to new habitats which in turn encourage the spread of vegetation. Tamarisk was introduced into the American south-west in the late 1800s, and spread throughout the Colorado plateau region occupying islands, sandbars and beaches along rivers and streams (Graf, 1978). Graf concluded that the tamarisk spread about 20 km per year, and suggested that the plant trapped and stabilized sediment causing an average reduction in channel width of 27 per cent. However, the extent to which such spread of vegetation is the cause or the effect of river channel change has been debated (Everitt, 1979).

Changes of river channels associated with changes of vegetation and channel ecology are particularly evident downstream of major dams and reservoirs. In a review of ecological effects Brooker (1981) concluded that changes in the aquatic environment arise from three principal modifications, namely the barrier to migrations imposed by the dam and the reservoir, particularly related to fish; changes in the downstream water quality caused by storage of the water in the reservoir; and changes in the river flow regime downstream of the reservoir. Because impoundments on the rivers of the world are so extensive (Petts, 1984), ecological effects on regulated rivers have also been very substantial. Thus downstream of Dartmoor reservoirs, Petts and Greenwood (1981) identified the disruptions that occurred in downstream ecosystems and noted major differences between the benthic micro-invertebrate populations of the natural and regulated sites, although they appreciated that a diverse range of habitats is still available downstream of the dams.

Figure 1.8   Cimarron River and flood plain, Kansas (US Geological Survey) Photograph (A) was taken in 1936 towards the end of the phase of channel widening (see pp 33). Photograph (B) shows the same area in 1960 indicating how vegetation extension on the flood plain influenced flood plain construction and channel stabilization

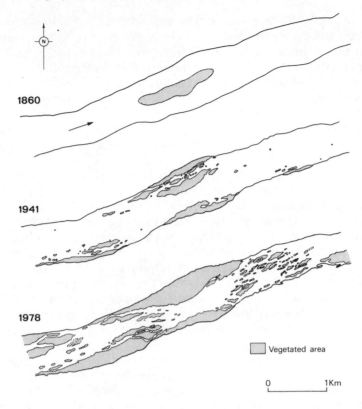

Figure 1.9 Channel changes of the Platte River, 1860–1978
The changes occurred after extensive water-use projects which reduced the flood
discharges. The channel dimensions were substantially reduced and the extent of
vegetated area increased, as shown by the shaded areas
(After Eschner *et al.*, 1983)

Major changes in river character involving changes of vegetation and
habitat are now well reported. The North Platte and Platte rivers in
Nebraska have been described as a case of shrinking channels (Williams,
1978b) because the channel width in 1969 was 10 to 20 per cent of the
channel width in 1865. Such reduction has occurred as a result of the
regulating effects of upstream dams and the greater use of water by man
for irrigation and for powerplants and reservoirs. It has been necessary
to identify the way in which vegetation has changed because originally,
in the nineteenth century, the Platte was a wide 2 km shallow river with
bankfull spring flows and low summer flows, and the channels had
hundreds of timbered islands (Eschner *et al.*, 1983). However, in the
twentieth century as the channel has reduced in width, there has been
progressive vegetation encroachment (figure 1.9). This has been possible

because the present hydrological regime allows vegetation germination and present flood discharges, which are lower than in the nineteenth century, do not scour away the seedlings and the habitat in which they grow. It has also been suggested, however, that overbank flows are now more common probably because as the channel has narrowed and vegetation encroachment has increased, the roughness of the channels has also increased so that the flood flows can less easily be accommodated within the river channels (Eschner *et al.*, 1983). It has been necessary to undertake a number of interrelated studies on the Platte river basin because of the way in which changes of the river channels have affected the habitats downstream, particularly for migratory water fowl (US Geological Survey, 1983).

Although the interrelationship between vegetation and stream channel form and process is subtle and complex, increased knowledge of the relationship helps to understand longer term changes. It has been accepted that changes in the vegetation cover have a significant influence upon palaeohydrology of the geologic past (Schumm, 1968). However, in temperate latitudes in the late Quaternary, it is evident that vegetation has been a significant regulator upon river channel behaviour. In Britain, much of the landscape prior to the last 5,000 years was extensively forested, and forest removal gave rise to significant changes along river channels. These included changes due to aggradation, as phases of Holocene alluviation have been recognized along valleys such as those of the upper Thames (Robinson and Lambrick, 1984). When forest was more extensive, debris dams and organic debris in river channels would have had a greater influence upon river channel morphology and river channel processes. It is therefore likely that the incidence of such organic debris has considerable significance in explaining not only the pattern of river channel behaviour, but also flood plain sedimentation during the Holocene (Gregory, 1983).

1.5   Conclusions

Vegetation influences the supply of water and sediment to the river channel network; the morphology of river channels; and the routing of water and sediment through the network of river channels. An understanding of these influences can assist our interpretation of past river behaviour. Sufficient examples have been provided to indicate how recent research is clarifying the intricate way in which vegetation is encompassed within the pattern of variables in the drainage basin system. Although it is not yet possible to achieve the global assessment of the influence of vegetation on river channels requested by Hickin (1984), the strands necessary for such an assessment are beginning to be attained. Understanding the role of vegetation and ecology in contemporary river channels facilitates our understanding of the present and improves interpretation of the past. However, it is upon the past and the present that we consider the future, and two aspects of future river channels concern habitats and aesthetics.

In viewing the ecological component of the fluvial system it is important to think of the instream uses. Thus methods have been developed for analysis of aquatic habitats (Stall and Herricks, 1982). In New Zealand Mosley (1983) has demonstrated that increasing pressure on rivers increases need for consideration of uses which include recreation, provision of fish and wildlife habitat as well as enhancement of scenic beauty. Factors controlling such instream uses have been reviewed, and relationships with discharge discussed. Although assessment of instream flow needs is multi-disciplinary in scope, input from those with a geographical or geomorphological background is very desirable to ensure that adequate consideration is given to the river as a whole, and to the interrelationship between changes in flow regime and changes in channel morphology (Mosley, 1985). In some areas particular procedures have been developed to design in-channel habitats: for example, in the Pacific North-West Lisle (1985) showed how the addition of more roughness elements by boulders and wooded debris can provide a diversity of habitats which are a key resource for fish habitat. Thus in future management of rivers it is not only important to think of the vegetation and the effect that it has directly and indirectly upon river behaviour, but also to think of the ways in which river management can provide habitats for fauna and particularly for fish in the river environment. Although not mentioned in detail above there are some indications that in the past beavers, by construction of beaver dams, may have had an important influence upon the processes operating in river channels.

In addition to thinking of rivers of the future as providing instream uses, it is also imperative that we consider the overall aesthetics of the river and its channel. Some progress has been made in devising schemes of assessment of the aesthetic quality of rivers and river channels (e.g. Dunne and Leopold, 1978), and any further progress in this direction must take account of the very significant impact that organic influences have upon the appearance of the river and its channel. In 1653 Izaak Walton wrote in *The Compleat Angler* of 'Hampshire, which I think exceeds all England for swift, shallow, clear pleasant brooks' and appreciated rivers for the fishing that they provided. In the last three centuries, during which more than 300 reprints of *The Compleat Angler* have appeared, rivers in all countries have been changed dramatically. But if we can understand the precise way in which vegetation affects river channel form and process, we will be able to manage the rivers of the future more effectively and so conserve the river environment appreciated and valued by Izaak Walton.

References

Barnes, H. H. 1967: Roughness characteristics of natural channels. *US Geological Survey Water Supply Paper* 1849.

Binder, W. 1979: *Grundzuge der Gewasserpflege*. Munich: Schriftenreiche Bayer. Laudesant fur Wasserwirtschaft.

Brooker, M. P. 1981: The impact of impoundments on the downstream fisheries and general ecology of rivers. In T. H. Coaker (ed.), *Applied Biology 6*, pp. 91–152. London: Academic Press.

Brookes, A. 1984: Recommendations bearing on the sinuosity of Danish stream channels. *Technical Report no. 6: Natural Agency of Environmental Protection, Freshwater Laboratory*. Silkeborg Denmark, 130 pp.

—— 1985: River channelization. *Progress in Physical Geography* 9, 44–73.

——, Gregory, K. J. and Dawson, F. M. 1983: An assessment of river channelization in England and Wales. *The Science of the Total Environment* 27, 97–112.

Burkham, D. E. 1972: Channel changes of the Gila River in Safford Valley, California 1846–1970. *US Geological Survey Professional Paper*, 655 G.

—— 1981: Uncertainties resulting from changes in river form. *Journal of Hydraulics Division, Proceedings American Society of Civil Engineers* 107, 593–610.

Burt, T. P. and Gardner, A. T. 1982: The permanence of stream networks in Britain: some further comments. *Earth Surface Processes and Landforms* 7, 327–32,

Charlton, F. G., Brown, P. M. and Benson, R. W. 1978: *The hydraulic geometry of some gravel rivers in Britain*. Report INT-180, Wallingford; Hydraulics Research Station.

Chorley, R. J. 1969: The drainage basin as the fundamental geomorphic unit. In R. J. Chorley (ed.) *Water, Earth and Man*, pp. 77–100. London: Methuen.

Courtney, F. M. 1981: Developments in forest hydrology. *Progress in Physical Geography* 5, 217–41.

Cowan, W. L. 1956: Estimating hydraulic roughness coefficients. *Agricultural Engineering* 37, 473–5.

Dawson, F. M. and Robinson, W. N. 1985: Submerged macrophytes and the hydraulic roughness of a lowland chalk stream. *Verhandlungen der Internationalem Vereiningung für theoretische und angewandte Limnologie* 22, 1944–1948.

Committee on Government operations 1973 Fifth Report. *Stream Channelization: What Federally financed draglines and bulldozers do to our nations streams*. House Report, 93–530, 139 pp. US Government Printing Office.

Dunne, T. and Leopold, L. B. 1978: *Water in Environmental Planning*. San Francisco: W. H. Freeman & Co.

Eschner, T. R., Hadley, R. F. and Crowley, K. D. 1983: Hydrologic and morphologic changes in channels of the Platte River Basin in Colorado, Wyoming, and Nebraska: A historical perspective. *US Geological Survey Professional Paper* 1277A.

Everitt, B. L. 1979: Fluvial adjustments to the spread of tamarisk in the Colorado plateau region: Discussion and reply. *Geological Society of America Bulletin* 90, 1183–4.

Eyles, R. J. 1977: Changes in drainage networks since 1820, Southern Tablelands, NSW. *Australian Geographer* 13, 377–86.

Ford, E. D. and Deans, J. D. 1978: The effects of canopy structure on stemflow, throughfall and interception loss in a young Sitka spruce plantation. *Journal of Applied Ecology* 156, 905–17.

Gardiner, V. and Gregory, K. J. 1982: Drainage density in rainfall–runoff modelling. In V. P. Singh (ed.), *Rainfall–Runoff Relationship*, pp. 449–76. Mississippi State University: Water Resources Publications.

Graf, W. L. 1978: Fluvial adjustments to the spread of Tamarisk in the Colorado Plateau Region. *Geological Society of America Bulletin* 89, 1491–1501.

—— 1979: Mining and channel response. *Annals Association of American Geographers* 69, 262–75.

—— 1980: Riparian management: A flood control perspective. *Journal of Soil and Water Conservation* 35, 158–61.

Gregory, K. J. 1976a: Changing drainage basins. *Geographical Journal*, 142, 237–47.

—— 1976b: *The Determination of Channel Capacity*. New England Research Series in Applied Geography, Armidale: University of New England, 60 pp.

—— 1976c: Channel capacity and lichen limits. *Earth Surface Processes* 1, 273–85.

—— 1983: Human activity and palaeohydrology: a review. In S. Kozarski (ed.), *Palaeohydrology of the Temperate Zone*. Quaternary Studies in Poland, 4, 73–80.

—— and Walling, D. E. 1973: *Drainage Basin Form and Process*. London: Arnold.

——, Hockin, D. L., Brooker, M. P. and Brookes, A. 1985: The impact of river channelization. *Geographical Journal* 151, 53–74.

——, Gurnell, A. M. and Hill, C. T. 1985: The permanence of debris dams related to river channel processes. *Hydrological Sciences Journal* 30, 371–81.

Gurnell, A. M. 1978: The dynamics of a drainage network. *Nordic Hydrology*, 6, 207–21.

—— and Gregory, K. J. 1984: The influence of vegetation on stream channel processes. In T. P. Burt and D. E. Walling (eds), *Catchment Experiments in Geomorphology*, pp. 515–35. Norwich: GeoBooks.

—— and Gregory, K. J. 1986: Water table level and contributing area: The generation of runoff in a heathland catchment. In *Conjunctive Water Use Proceedings of the Budapest Symposium*, International Assocation of Hydrological Sciences, 156, 81–95.

—— and Gregory, K. J. 1987: Vegetation characteristics and the prediction of runoff: analysis of an experiment in the New Forest. *Hydrological Processes* 1, 125–142.

——, Gregory, K. J., Hollis, S. and Hill, C. T. 1985: Detrended correspondence analysis of heathland vegetation: the identification of runoff contributing areas. *Earth Surface Processes and Landforms* 10, 343–51.

Hadley, R. F., Lal, R., Onstad, C. A., Walling, D. E. and Yair, A. 1985: *Recent Developments in Erosion and Sediment Yield Studies*. Paris: Unesco, 127 pp.

Hamill, L. 1983: Some observations of the time of travel of waves in the River Skerne, England, and the effect of aquatic vegetation. *Journal of Hydrology*, 66, 291–304.

Haslam, S. M. 1978: *River plants*. Cambridge: Cambridge University Press.

—— 1981: Changing rivers and changing vegetation in the past half century. *Proceedings Aquatic Weeds and their control*, 49–57.

—— and Wolseley, P. A. 1981: *River Vegetation: Its identification, assessment and management*. Cambridge: Cambridge University Press.

Heede, B. M. 1981: Dynamics of selected mountain streams in the western United States of America. *Zeitschrift für Geomorphologie 25*, 17–32.

Heuvelmans, M. 1974: *The River Killers*. Harrisburg, Pa: Stackpole Books.

Hey, R. D., Bathurst, J. M. and Thorne, C. M. 1982: *Gravel Bed Rivers*. Chichester: Wiley and Sons.

Hickin, E. J. 1984: Vegetation and river channel dynamics. *Canadian Geographer* 28, 111–26.

—— and Nanson, E. G. 1984: Lateral migration rates of river bends. *Journal Hydraulic Engineering* 110, 1557–67.

Hill, M. O. and Gauch, H. G. 1980: Detrended correspondence analysis: an improved ordination technique. *Vegetatio* 42, 47–58.

Hupp, C. R. 1986: Vegetation and bank-slope development. *Proceedings of the Fourth Federal Interagency Sedimentation Conference*. Volume 2, Sub-Committee on Sedimentation of the Interagency Advisory Committee on Water Data. Las Vegas, Nevada, 5–83 to 5–92.

Jarrett, R. D. 1985: Determination of roughness coefficients for streams in Colorado. *US Geological Survey Water Resource Investigations Report* 85–4004, 54 ppT.

Jones, J. A. A. and Crane, F. G. 1984: Pipeflow and pipe erosion in the Maesnaut experimental catchment. In D. E. Walling and T. P. Burt (eds), *Catchment Experiments in Geomorphology*, pp. 55–72. Norwich: GeoBooks.

Keller, E. D. and Hoffman, E. K. 1976: A sensible alternative to stream channelization. *Public Works* October, 70–2.

—— and Swanson, F. G. 1979: Effects of large organic debris on channel form and fluvial processes. *Earth Surface Processes* 4, 361–80.

—— and Tally, T. 1979: Effects of large organic debris on channel form and fluvial processes in the costal redwood environment. In D. D. Rhodes and G. P. Williams (eds), *Adjustments of the Fluvial System*, pp. 169–97. Dubuque, Iowa: Kendall/Hunt.

Kouwen, N. and Ruh-Ming, Li 1980: Biomechanics of vegetative channel linings. *Journal of the Hydraulics Division Proceedings American Society of Civil Engineers*, 106, 1085–1103.

Law, F. M. 1956: The effect of afforestation upon the yield of water catchment areas. *Journal British Waterworks Association* 38, 480–94.

—— 1957: Measurements of rainfall, interception and evaporation losses in a plantation of Sitka spruce trees. *International Association of Scientific Hydrology*, Toronto Symposium 2, 397–411.

Leopold, L. B. 1977: A reverence for rivers, *Geology*, 429–30.

Likens, G. E. and Bilby, R. E. 1982: Development, maintenance and role of organic debris dams in New England streams. In F. J. Swanson *et al.* (eds), *Workshop* on sediment Budgets and Routing in Forested Drainage Basins. USDA Forest Service, Technical Report PNW-141, 122-8.

Lisle, T. M. 1985: Roughness elements: a key resource to improve anadromous fish habitat. US Forest Service, Redwood Sciences Laboratory, 93-98.

Macdonald, A., Keller, E. A. and Tally, T. 1982: The role of large organic debris in stream channels draining redwood forests in north-western California. In D. K. Hardey, D. C. Marm and A. Macdonald (eds), Friends of the Pleistocene 1982. Pacific Cell field trip guidebook, Late Cenozoic history and Forest Geomorphology of Humbold Co California, pp. 226-45.

Maddock, T. 1972: Hydraulic behaviour of stream channels. Transactions 37, North American Wildlife Conference and Natural Resources Conference, 366-74.

Marston, R. A. 1982: The geomorphic significance of log steps in forest streams. Annals Association of American Geographers 72, 99-108.

Megahan, W. F. 1982: Channel sediment storage between obstructions in forested watersheds draining the granitic bedrock of the Idaho Batholith. In F. J. Swanson et al. (eds), Workshop on Sediment Budgets and Routing in Forested Drainage Basins, USDA Forest Service, Pacific Northwest Forest and Range Experiment Station, Gen. Tech. Reprint. PNW-141.

Mosley, P. M. 1981: The influence of organic debris on channel morphology and bedload transport in a New Zealand forest stream. Earth Surface Processes 6, 571-9.

—— 1983: Flow requirements for recreation and wildlife in New Zealand Rivers – a review. Journal of Hydrology (N2) 22, 152-74.

—— 1985: River channel inventory, habitat and instream flow assessment. Progress in Physical Geography 9, 494-523.

Murgatroyd, A. L. and Ternan, J. L. 1983: The impact of afforestation on stream bank erosion and channel form. Earth Surface Processes and Landforms 8, 357-69.

Nanson, G. C. 1981: New evidence of scroll bar formation on the Beatham River. Sedimentology 28, 889-91.

Nature Conservancy Council 1983: Nature Conservation and River Engineering. London: NCC, 39 pp.

Nevins, T. H. F. 1969: River training – the single thread channel. New Zealand Engineering, December, 367-73.

Newson, M. D. 1985: Forestry and water in the uplands of Britain – the background of hydrological research and options for harmonious land use. Quarterly Journal of Forestry 79, 113-20.

Nunnally, N. R. 1978: Stream renovation: an alternative to channelization. Environmental Management 2, 403-11.

—— and Keller, E. A. 1979: Use of fluvial processes to minimise adverse effects of stream channelization. Water Resources Research Institute of the University of North Carolina, Report 144, 115 pp.

Osterkamp, W. R. and Hupp, C. R. 1984: Geomorphic and vegetative characteristics along three northern Virginia streams. Geological Society of America Bulletin 95, 1093-1101.

Ovenden, J. C. and Gregory, K. J. 1980. The permanence of stream networks in Britain. Earth Surface Processes 5, 47-60.

Page, K. and Nanson, G. C. 1982: Concave bank benches and associated flood plain formation. *Earth Surface Processes* 7, 529–43.
Penman, H. L. 1963: *Vegetation and Hydrology.* Technical Communication 53. Commonwealth Bureau of Soils Harpenden, Commonwealth Agricultural Bureau.
Petryk, R. S. and Bosmajian, G. 1975: Analysis of flow through vegetation. *Journal Hydraulics Division proceedings. American Society Civil Engineers* 101, 871–84.
Petts, G. E. 1984: *Impounded Rivers* Chichester: Wiley.
—— and Greenwood, M. 1981: Habitat changes below Dartmoor reservoirs. *Reports Transactions Devon Association for the Advancement of Science* 113, 13–27.
Rachocki, A. 1978: Wplyw räslinnosci na ksztaltawanie koryt i brzegow raek (The impact of plants on the formation of river banks and channels) *Przeglad Geograficzny* 4, 469–81.
Roberts, J. 1983: Forest transpiration: a conservative hydrological process? *Journal of Hydrology*: 66, 133–41.
Robinson, M. A. and Lambrick, G. H. 1984: Holocene alluviation and hydrology in the upper Thames basin. *Nature* 306, 809–14.
Schumm, S. A. 1968: Speculations concerning palaeohydrologic controls of terrestrial sedimentation. *Geological Society of America Bulletin* 79, 1573–88.
—— and Lichty, R. W. 1963: Channel widening and flood plain construction along Cimarron River in south-western Kansas. *US Geological Survey Professional Paper* 352D, 71–88.
Sculthorpe, C. D. 1967: *The Biology of Aquatic Vascular Plants.* London: Arnold.
Smith, D. G. 1976: Effect of vegetation on lateral migration of glacial meltwater river. *Geological Society of America Bulletin* 87, 857–60.
Sopper, W. E. and Lull, H. W. 1967: *Forest Hydrology.* Oxford: Pergamon.
Stall, J. B. and Herricks, E. E. 1982: Evaluating aquatic habitat using stream network structure and streamflow predictions. In R. G. Craig and J. L. Craft, *Applied Geomorphology,* pp. 240–53. London: George Allen and Unwin.
Swanson, F. J., Lienkaemper, G. W. and Sedell, J. R. 1976: History, physical effects and management implications of large organic debris in western Oregon streams. *USDA Forest Service, Pacific Northwest Forest and Range Experiment Station,* Gen Tech Rep PNW-69, 15 pp.
US Geological Survey 1983: Hydrologic and geomorphic studies of the Platte River Basin *US Geological Survey Professional Paper* 1277.
Walton, I. 1653: *The Compleat Angler* or *The Contemplative Man's Recreation.* London: T. Maxey.
Ward, R. C. 1971: Small Watershed Experiments: An appraisal of concepts and research developments. *University of Hull Occasional Papers in Geography* 18.
Wendland, W. M. and Watson-Stegner, D. 1983: A technique to reconstruct river discharge history from tree-rings. *Water Resources Bulletin* 19, 175–81.
Whitlow, J. R. 1985: Dambos in Zimbabwe: a review. *Zeitschrift für Geomorphologie* Supplementband 52, 115–46.
Williams, G. P. 1978a: Bank-full discharge of rivers. *Water Resources Research* 14, 1141–53.
—— 1978b: The case of the shrinking channels – the North Platte and Platte Rivers in Nebraska. *US Geological Survey Circular* 781.

Witt, A. 1985: Vegetational influences on intrachannel deposition: evidence from the Konczak stream, Greater Poland Lowlands, Western Poland. *Questiones Geographicae* 9, 145–60.

Yanosky, T. M. 1982: Hydrologic influences from ring widths of flood damaged trees, Potomac River, Maryland. *Environmental Geology* 4, 43–52.

Zimmerman, R. C., Goodlett, J. C. and Comer, G. H. 1967: The influence of vegetation on channel form of small streams. *International Association of Scientific Hydrology: Symposium River Morphology* Publication 75, 255–75.

2   The influences of vegetation, animals and micro-organisms on soil processes

Peter Mitchell

Introduction

Virtually all the models of soil genesis proposed in the past 100 years have acknowledged the importance of five factors of soil formation: lithospheric material, topography, climate, biosphere and time, the general acceptance of which might lead one to assume that their relative importance has been established and that a substantive body of knowledge exists about them. An examination of soils textbooks will reveal that this is not true and will show that the factor most commonly neglected is the biosphere.

Four biases are common in texts discussing this factor:

1 The biosphere is often only considered as vegetation;
2 Texts which recognize the role of animals often limit their discussion to the effects of soil dwelling invertebrates;
3 Most texts ignore any physical soil movement by biospheric agents and deal only with element cycling and the accumulation of organic matter;
4 The few texts which do recognize soil movement almost always treat the process as one of mixing and homogenization, rather than as a process leading to horizonation.

Paton (1978), Hole (1981) and Bal (1982) have all partly reviewed the literature concerning the role of the biosphere. Humphreys (1985) and Mitchell (1985) looked specifically at the physical movement of soil by organisms in field studies in eastern Australia. This chapter is based on these works and will only deal with the literature pertaining to fluvial (temperate) environments.

The literature

Much of the literature concerning the relationships between organisms and soils has been produced by biologists or ecologists and the most studied topics concern soil microbiology with emphasis on the litter decay and the nutrient cycles. Summaries of this material may be found in Rodin and Bazilevich, 1965; Russell, 1973; Dickinson and Pugh, 1974; N. Walker, 1975; Lohm and Persson, 1977; and Smith, 1982.

The rhizosphere, or the soil/root interface has received some attention (Harley and Scott Russell, 1979; Böhm, 1979). However, even this work has largely ignored the physical movement of soil material caused by root growth and decay, although engineers and foresters (O'Loughlin, 1974; Burroughs and Thomas, 1977; Gray and Leiser, 1982) have shown interest in tree roots and slope stability. Numerous studies also exist on the influence of stemflow and leaf drip on podzolization, for example Ryan and McGarity, 1983.

The ecology of soil organisms was reviewed by Wallwork (1970, 1976) and Leadley Brown (1978). Bal (1982) made a comprehensive study of Dutch polders, and Zlotin and Khodashova (1980) of the forest-steppe of central Russia. This latter work illustrates the value of a total assessment of biological processes in an ecosystem and achieved a remarkable degree of quantification.

The most complete view of the influence of animals on soil was presented by Hole (1981), who established twelve somewhat arbitrary sets of effects. The same sets apply to plants (Crocker, 1959), yet although they are comprehensive, they are also non-exclusive because of the complexity of the real biosphere. This complexity may be reduced by combining Hole's

Table 2.1   The effects of bioturbation on soils

Group 1 Processes
    1   Regulation of the soil biota
    2   Regulation of the nutrient cycle
    3   Regulation of the soil environment; pH, temperature and composition of the soil air
    4   Regulation of animal and plant litter
    5   Regulation of air and water movement through the soil

Group 2 Processes
    6   Producing special constituents or soil features
    7   Regulation of soil erosion
    8   Forming and destroying soil fabric
    9   Mixing soil horizons or creating soil layers
   10   Creating micro-relief by mounding and excavation
   11   Forming and filling soil voids

Source: Modified from Hole, 1981 and Crocker, 1959.

Table 2.2 Measured or calculated rates of soil bioturbation by invertebrates in temperature environments reported in the literature

Species	Location	Mean rate (t ha^{-1}yr^{-1})	Reference
Earthworms			
Not identified	England	34.97	Darwin, 1881
Allolobophora noctum	England	19.21	Evans, 1948
A. noctum and Lumbricus spp.	England	25.24	Evans, 1948
Not identified	France	36.6	Darwin, 1881
Not identified	Germany	5.94	Evans and Guild, 1947
Not identified	Poland	41.76	Evans and Guild, 1947
Not identified	Switzerland	43.94	Evans and Guild, 1947
Lumbricus terrestris	Luxembourg	15.0	Hazelhoff et al., 1981
Not identified	Indiana	21.72	Thorp, 1949
Pheretima hupiensis	Japan	38.3	Watanabe, 1975
Microchaetus spp.	S. Africa	114.7	Ljungström and Reinecke, 1969
Cryptodrilus and/or Oreoscolex spp.	Australia	0.69	Humphreys and Mitchell, 1983
Ants			
Lasius flavus	England	8.24	Waloff and Blackith, 1962
Formica cinerea	Wisconsin	1.33	Baxter and Hole, 1967
F. exsectoides	Wisconsin	11.36	Salem and Hole, 1968
F. fusca	British Columbia	19.6	Wiken et al., 1976
Iridomyrmex purpureus	Australia	0.006	Greenslade, 1974
Aphaenogaster spp.	Australia	8.41	Humphreys, 1981
Camponotus intrepidus	Australia	<0.05	Cowan et al., 1985
I. purpureus	Australia	<0.05	Cowan et al., 1985
Termites			
Nasutitermes exitiosus	Australia	0.14	Lee and Wood, 1968
Cicadas			
Psaltoda moerens	Australia	0.2	Humphreys and Mitchell, 1983
Thopa saccata	Australia	0.03	Humphreys and Mitchell, 1983
Crustacea			
Crawfish	Indiana	8.0	Thorp, 1949
Eustacus kierensis	Australia	7.30	Young, 1983
Scarabs			
Peltotrupes	Florida	2.61	Kalisz and Stone, 1984

sets into two larger groups (table 2.1). Group 1 processes modify the soil as a growth medium by changing its status with respect to air, water, and nutrients. Group 2 processes involve the physical movement of mineral or soil particles (bioturbation) on a scale large enough to contribute to profile homogenization or horizonation. Organisms largely responsible for Group 1 effects fall into Wallwork's (1970) micro- and meso-fauna and mainly inhabit pre-existing soil voids. Group 2 effects involve the mega-fauna which are too large to inhabit existing soil voids and affect the soil by directly rearranging soil components. This chapter will concentrate on Group 2 effects since this provides a convenient means of limiting the discussion. The literature, however, deals with individual groups of organisms rather than processes and the following brief review will adopt that frame.

Invertebrates

There is a considerable literature concerning terrestrial bioturbation by earthworms, ants and termites. Soil turnover by invertebrates in fluvial (temperate) environments is summarized in table 2.2.

Earthworms

Worms have the distinction of being the species on which the pioneering work on terrestrial bioturbation was conducted by Charles Darwin (1881). This oft quoted and justly famous text has provided a model for many subsequent studies, but also shares several weaknesses seen in those studies. Darwin did not identify the species he was dealing with. In most of his studies it was apparently the common European worm *Lumbricus terrestris*, but in some cases other species were probably involved. Another basic weakness of this and many subsequent studies is that too few experiment replications were used, and many rate calculations are based on limited observations often made on unusual circumstances.

Darwin used three approaches to determine the role of worms in soil turnover: short term plot studies, medium term observations of the burial of a recognizable layer in the soil and an examination of buried Roman ruins in order to assess the possible role of worms. Each of these involved different assumptions and different types of experimental error, yet the results were remarkably consistent and agree in order of magnitude with the more reliable calculations of Evans and Guild (1947), Evans (1948) and Guild (1948). Darwin's techniques have all been used by subsequent workers who have reported rates of soil turnover, and the most popular approach has been to collect all the material cast on the surface of a marked plot over a period of time. This simple technique has a number of disadvantages that have not been satisfactorily addressed by any worker. The task of cast collection is actually quite difficult especially on a grass or litter covered surface. Losses due to erosion or cast breakdown by dessication, rain or frost between collection periods may be high, and no assessment has been made as to whether the collection of casts has any effect on worm activity. With all of these problems the potential for error is high and all reported rates need to be accepted with some degree of caution.

Earthworm biology is reviewed by Edwards and Lofty (1972), Satchell (1983) and Lee (1985). Several important points may be drawn from these works. In any one soil several worm species may occur and they will normally be stratified with depth. Litter-inhabiting species have little impact on the soil, topsoil species are more likely to encourage profile horizonation and subsoil species encourage homogenization. Worm casting and burrowing are seasonal activities dependent on soil temperature, moisture regimes and the seasonal availability of litter. Soils with high worm populations are generally better structured and have higher infiltration rates. In places worms may be so abundant in the soil surface that they create a distinctive micro-relief or patterned ground (Ljungström and Reinecke, 1969). Nielsen and Hole (1964) drew attention to worm induced soil layering and many researchers, including Darwin (1881), Ljungström and Reinecke (1969), Hazelhoff et al. (1981), Jungerius and van Zon (1982), noted the destruction of surface casts by erosion processes.

In Australia, Lee et al. (1981) estimated that there are several hundred species of earthworm in regions with more than 250 mm annual rainfall. Few of these are well known but there are close relationships between Australian and New Zealand worms and the altitudinal zonation in worm populations noted in the Southern Alps (Lee, 1959) is paralleled in the Snowy Mountains (Wood, 1974), where Mitchell (1985) found worm mixing and casting to be very important in soil genesis.

Ants

Ants are social insects with colonies that vary in size from a few hundred to 500,000 or more individuals. They have a wide range of life styles and almost all construct nests of some type, often in the soil. Texts dealing with the general ecology of ants from which useful soils data may be drawn include the works of Wheeler (1910), Sudd (1967), Brian (1978) and Dumpert (1981).

Ground nests vary in shape from simple excavations on the surface or under stones, to crater forms and masonry domes which may reach considerable dimensions, even as large as the better-known termite mounds of the tropics (see chapter 5). There have been few attempts to measure soil turnover rates, most workers having simply reported mound sizes and estimated mound longevity. Since longevity varies from a few days or weeks for the easily eroded small crater forms, to more than 100 years reported by Greenslade (1974) for the meat ant (*Iridomyrmex purpureus*) it would seem that estimates of turnover rate could have a large error margin. Ants have a worldwide distribution, and Greenslade and Greenslade (1983) estimated that there are 3–5,000 species in Australia. Local diversity can be very high: for example, Greenslade and Thompson (1981) recorded more than 280 species in a single collection period from a limited area of southern Queensland. Australian ants play an unusually important role in plant pollination and seed dispersal. Berg (1975) observed that more than 1,500 species of Australian plants were myrmecochorous. Ants collect seed from these species, remove the elaiosomes and dispose of the viable seed in refuse heaps, under litter, adjacent to stones and other places where germination might be enhanced. On the other hand Ashton (1979) recorded that ant harvesting and consumption of *Eucalyptus regnans* seed amounted to 60 per cent of the annual seedfall, and this effectively prevented tree regeneration except after wildfire. The importance of these ant–plant relationships is a debated issue (Buckley, 1982) and the implications with respect to longer term vegetation and soil changes are unknown.

Particular ecological issues such as the above have attracted the attention of research workers and are reasonably investigated. However, the great majority of Australian ants are very poorly known and even the taxonomy is incomplete and confused. One consequence of this patchy interest is that there are few works dealing with ant–soil relationships, although studies on the meat ant are an exception. Even though this is one of the better-studied genera it was thought until recently to consist of three varieties, but has now been shown to consist of ten sibling species (Greenslade and Halliday, 1982). Meat ant nests include conspicuous soil mounds which are commonly covered with fine gravel and/or plant fragments. Typical mounds have diameters of about 2.0 m and are 20–30 cm high. Greaves (1939) recorded one nest with plan dimensions of 9.1 × 11.3 m, which Cowan *et al.* (1985) calculated could have had a volume of about 800 l whereas the typical nest is about 20 l. Despite the nest size and wide distribution of this ant, Greenslade (1974) showed that the species was not important as an agent of soil bioturbation. This conclusion was confirmed in another environment by Cowan *et al.* (1985) and also applied to a similar mound building species *Camponotus intrepidus*. These works provide a salutary lesson on the risk of exaggerating the importance of conspicuous phenomena.

Humphreys (1981, 1985) found that soil mounding by the funnel ant (*Aphaenogaster longiceps*), was important in the Sydney Basin. Three species are known in Australia, of which *A. longiceps* is the most widely distributed from Victoria to dry parts of Queensland and Papua New Guinea. This ant was important at the Killonbutta site described below. It is a nocturnal insectivore which constructs numerous interconnected funnel-shaped mounds on the surface which may act as pit-fall traps. A second species *A. pythia* is common in wetter parts of NSW and Queensland, where it is a pest in pasture and sugar cane plantations because of its habit of removing soil from around plant roots (Saunders, 1966).

Termites

Termites are another group of social insects which are important agents of bioturbation, particularly in tropical environments (see chapter 5). The biology of the group is reasonably known and Lee and Wood (1971) have dealt extensively with their relationship with the soil. About 200 species are found in Australia, 40 of which build soil mounds, but they are relatively unimportant in the temperate environments.

Other invertebrates

Of the numerous other invertebrates which spend some or all of their life in the soil and which are known to physically move soil about, only two organisms have caught the attention of pedologists. Infilled cicada burrows are common in soils under prairie grassland in the western United States, and Hugie and Passey (1963) described extensive intercrossing burrow systems forming 20 mm cylindrical peds and krotovina (pedotubules). However the only attempt to estimate soil turnover rates by these insects is that reported from the Sydney Basin by Humphreys and Mitchell (1983).

Thorp (1949) mentioned soil turnover by crustacea, and Young (1983) presented some data for a small freshwater crayfish which occupied patterned ground near Sydney. She acknowledged that these animals performed an important maintenance role on the micro-relief, but discounted their possible role in constructing the pattern. This difficulty of disentangling cause and effect is very common in bioturbation studies.

Vertebrates

The pedological and geomorphic literature dealing with vertebrates is every bit as patchy as that for invertebrates. Representatives of every terrestrial class have been recognized as playing a role in this area, but there is no comprehensive review of all species or relationships. The burrowing activities of the common mole (*Talpa europaea*) have been examined by Godfrey and Crowcroft (1960) and there is an extensive literature on burrowing rodents, much of which relates to desert or arctic areas. The problem of 'mima mounds' has been widely debated (Ross *et al.*, 1968; Mielke, 1977) and some of these enigmatic features are attributed to the activities of gophers or ground squirrels, groups which have also received attention from biologists (Hickman and Brown, 1973; Anderson and McMahon, 1981). Animal–vegetation relationships have often been examined, but despite the importance of ground cover to soil erosion the pedological importance of the relationships are not often explored. Notable exceptions to this statement are the works of Ellison (1946), Reig (1970), Zlotin and Khodashova (1980).

Summary data on rates of soil turnover by a range of vertebrates are presented in table 2.3.

Table 2.3 Measured or calculated rates of soil bioturbation by vertebrates in temperature environments reported in the literature

Species	Location	Mean rate (t ha^{-1} yr^{-1})	Reference
Spalax micropthalmus	USSR	2.25	Zlotin and Khodashova, 1980
Citellus pygmaeus	USSR	1.5	Abaturov, 1972
Talpa europaea	USSR	10.83	Abaturov, 1972
Talpa europaea	Czechoslovakia	55.0	Abaturov, 1972
Talpa europaea	Poland	1.0–3.0	Jonca, 1972
Talpa europaea and voles	Luxembourg	20.0	Imeson, 1976
Wild pigs	Luxembourg	0.34	Imeson, 1976
Manitoba toad	Minnesota	3.45–6.1	Yair and Rutin, 1981
Not identified	Washington	0.48–25.0	Yair and Rutin, 1981
Geomys pinetus	Florida	0.29–8.16	Kalisz and Stone, 1984

Birds

The role of sea birds that nest in large colonies on islands and inaccessible headlands has been noted with respect to the nutrient cycle and accumulation of guano and rock phosphate (for example Blakemore and Gibbs, 1968), but studies which involve the physical soil cycle are few. Some data is available from the study of commercial muttonbirds (shearwaters and petrels) of southern Australia and New Zealand. Breeding burrow densities of up to 4 m^{-2} have been recorded (Crook, 1975) and rookeries may extend over more than 200 ha. Fineran (1973) and G. R. Evans (1973) found that two New Zealand species so destroyed their environment by undermining and trampling the vegetation that the colonies constantly shifted as the site became untenable. Kirkpatrick *et al.* (1974) commented that muttonbird burrows were the most important soil genesis factor on many Bass Strait islands, but despite a considerable biological literature on this species there are few details on burrow dimensions or patterns of use. Burrow densities on Bass Strait islands average 0.72 m^{-2} and in the 150 known Tasmanian rookeries there are an estimated 16 million birds (Naarding, 1981), using about eight million burrows in any year. Birds return to the same burrow to breed every year, but some burrows are destroyed by erosion and even burrow cleaning will move a considerable volume of soil. The total bioturbation must be significant but it is important to realize that it is geographically restricted.

Australia has more than 700 bird species and of these at least ten species extensively probe the soil during feeding, 11 rake forest litter, 26 nest in soil burrows, 44 nest in shallow scrapes on the soil, three build large incubating mounds of soil and litter, and ten construct mud nests in trees or on cliffs. Not all are likely to be important as soil bioturbators but some occur in such large numbers or disturb such large areas that they may be significant. The only published rates of soil turnover for any bird are reports on the feeding activity of the lyrebird *Menura novaehollandiae* by Ashton (1975). This bird was important at the Blackheath site described below.

Lyrebirds are moderately large ground feeding birds found in a range of vegetation types through eastern Victoria to southern Queensland up to 1,500 m altitude. They are well known for their vocal mimicry and courtship dancing displays. The most detailed account of their natural history is by Robinson and Frith (1981). In the months before breeding the male bird constructs up to 83 display mounds by clearing vegetation and litter from a circle about 1 m in diameter and raking up soil to form a mound 10–15 cm high. The number of mounds in use varies with the season and the size of the birds' territory. Robinson and Frith (1981) recorded a mean of 42, and a mean territory size of 2.4 ha, at Tindinbilla in the Australian Capital Territory.

Mammals

From a total of 777 genera of modern terrestrial mammals (excluding bats) listed by E. P. Walker (1968), 329 or 42 per cent can be regarded as potentially significant soil bioturbators. Of the mammals in temperate Australia only the spiny ant-eater or echidna (*Tachyglossus aculeatus*) was important in the studies described below, although Mitchell (1985) also found that soil disturbance by the common wombat (*Vombatus ursinus*) was significant in the Snowy Mountains of southern New South Wales.

The spiny ant-eater is found in all Australian environments from sub-alpine regions to the tropics and the arid zone. The natural history of the species has been described by Griffiths (1968, 1978). The animal is known to excavate a considerable volume of soil during feeding but no previous work has attempted to quantify this.

Plants

The roles of plants in rock and mineral weathering, accelerating soil leaching, the soil nutrient cycle, reducing erosion by providing ground cover, stabilizing bare ground with surface algae, stabilizing slopes with tree roots and mixing soil through treefall and root wedging have been widely recognized. Of all these relationships, treefall and root wedging have the greatest potential for moving soil and although foresters have a considerable interest in this, few studies have considered the importance of treefall in soil and slope genesis.

Tree roots comprise 25 to 35 per cent of the plant's biomass and occupy a considerable soil volume which is displaced as they grow. Blevins *et al.* (1970) showed that this root wedging operates even at the root hairs where platy minerals and clays are jostled into closer contact with the root. On a larger scale the common occurrence of butt hillocks (Butuzova, 1962) at the base of tree trunks is a clear indication of soil displacement. In environments where decay rates are low and fire is uncommon, these hillocks include a large quantity of organic matter that may be mixed into the mineral soil when the tree falls. In other environments the hillock may be bare mineral soil with a steeper micro-slope that is subject to erosion by stemflow. The displacement of stones by the roots and trunk may create a stone ring around the tree base which on the death of the tree can look like a periglacial stone circle (Denny and Goodlett, 1968).

Single large storms can cause treefall over thousands of hectares disturbing enormous volumes of soil and commonly creating a micro-relief of pits and mounds (Lutz and Griswold, 1939; Lutz, 1940; Neuwenhuis and van den Berg, 1971). The pits slowly infill with eroded soil and litter and the changed soil conditions may mean that the mounds are favoured as the point for establishment of the next generation of trees (Hutnick, 1952; Denny and Goodlett, 1956; Veneman et al., 1984). If this is a general pattern of forest replacement it implies that soil overturn by treefall will be a systematic process eventually covering the entire forest area. In the Appalachians, E. P. Stephens (1956) observed that 14 per cent of the forest floor had been disturbed in 500 years through six age classes of events. Lorimer (1980) recognized eight events in 250 years in a similar forest each of which involved about 10 per cent of the area, and Falinski (1978) found the average rate of fall in a Polish forest was 2–4.5 trees $ha^{-1} yr^{-1}$. Despite this interest there have been no rates of bioturbation by treefall published, although estimates of $1.4 t ha^{-1} yr^{-1}$ can be made from Stephens' (1956) data and $2.0 t ha^{-1}$ for a single event described by Kotarba (1970).

A very important indirect influence of vegetation on soil is the frequency and intensity of wildfire. Erosion usually increases substantially (Blong et al., 1982), and there are also chemical, structural and other hydrological effects (F. R. Humphreys and Craig, 1981; Clinnick, 1984). Fire disrupts the entire ecosystem and may also cause long term change to animal populations and the nature of their activities (Newsome et al., 1975).

Summary points from the literature

The scattered literature mentioned here clearly shows the reality of bioturbation in temperate terrestrial environments and may be used to make a number of generalizations that could provide a guide to further work.

1 Bioturbation is important in three ways:

   a   Through mineral turnover in the nutrient cycle, where the flow of elements is most often accelerated but can also be retarded when long term stores of organic matter are involved as in the case of termite mounds;

b   In the physical movement of soil material by mixing and mounding;
c   In the creation of micro-relief, ranging from ephemeral worm casts
     through ant and termite mounds with a longevity in excess of
     100 years, to treefall pits and mounds that may still be recognizable
     after more than 500 years.

2 The common effect of soil turnover is profile homogenization, but
   contrary to the views expressed in much of the literature this is not
   the inevitable effect. Many organisms are size selective in the material
   they move, and several studies have shown that soil layering could
   be created by fauna or combinations of fauna and erosion.

3 Many invertebrates do not cast soil on the surface, form distinctive
   micro-relief or have obvious burrow openings. These species have
   been largely overlooked as bioturbators.

4 The spatial patterns of biotic activity are not usually random. Ant
   and termite nests tend to be uniformly distributed as a result of
   competition. Vertebrates usually have established territories and
   preferred micro-environments for burrow location and feeding, and
   even treefall events may have a more systematic pattern of distribution
   than might be expected. Depth stratification and/or seasonal or
   periodic migration, particularly of invertebrates, is probably universal
   yet has only been noted for a few species. This has been observed
   with earthworms, but the only determined rates of bioturbation are
   for surface casting species. Over time all of these processes will move
   across the entire landscape and in this larger temporal frame the effects
   of Pleistocene climatic changes need to be considered, since
   bioturbation agents no longer present in an area may formerly have
   been important and have left their legacy in the present landscape.

5 Body size, population, and feeding habits are all likely to be important
   in determining the maximum volumes of soil moved in a given environment. Species which gain their sustenance from within the soil and
   occupy it for most or all of their life are volumetrically more important
   bioturbators, thus worms appear to be generally more important than
   ants or termites. Fossorial vertebrates are more important than most
   obligate burrowers and it could be suggested that tree root wedging
   is more important than treefall, although there are no data to support
   the point.

6 At any locality, more than one species will be involved in bioturbation
   and the soil dwelling/mixing niche is likely to be shared or stratified.
   These relationships are very complex and will usually involve
   feedback whereby one species or mechanism affects another. No
   studies of bioturbation have attempted an overall assessment of these
   patterns in any one environment.

7 Most calculations of soil mixing rates are of limited reliability. They
   often involve obvious but perhaps atypical species, usually assume
   a longevity of structures such as mounds that is debatable, rarely deal

with sub-surface mixing and often involve imprecise measurement of too few examples.

8   All existing studies can be viewed as fitting a developmental series depending on the level of observation involved in them:

a   Stage 1 where the obvious processes are identified and described;

b   Stage 2 where simple attempts have been made to quantify the observations of stage 1;

c   Stage 3 in which some attempt has been made to describe the dynamics of the situation;

d   Stage 4 in which an attempt has been made to put all the processes into their proper pedologic, geomorphic and hydrologic context. Rates have been measured with some precision and with reasonable control over variance.

Few studies have reached stage 4, and in fluvial environments only the team research conducted in Luxembourg (Imeson 1976, 1977; Imeson and Kwaad 1976) has attempted that level. Most other studies have not exceeded stage 1.

Field studies in eastern New South Wales

The environment and background to the studies

Fluvial environments may be approximately equated with the temperate zone, where there is a moderate temperate range, permanent streams, high soil moisture status and a continuous cover of forest or woodland vegetation. The micro-climate in which soil organisms live will generally be very uniform and not subject to the extremes of tropical, polar or desert areas. However, in eastern Australia the temperate zone differs from its northern hemisphere counterparts and since these differences are important in soil genesis they need to be recognized:

1   Australian forests and woodlands are dominated by eucalyptus which have less seasonality in litterfall than deciduous forests;

2   Climatically there is greater variation in rainfall and temperature and the soils and biota are often exposed to drought conditions;

3   Periodic wildfire is a major factor influencing soils and interrupting the decay cycle;

4   Contrary to the view of Dunne (1978), overland flow or specifically 'rainwash' (drop-agitated surface flow) is the main agent of erosion on hillslopes (Walker and Butler, 1983; Bishop et al., 1980);

5   Soils with a strong texture contrast between A and B horizons are very common and these were recognized as a separate group, the

duplex soils, by Northcote (1971). These soils provided an additional focus for the field studies described below.

There has been extended debate about the origin of the texture contrast in duplex soils. It was recognized as a characteristic of 12 great soil groups and explained in terms of clay eluviation and illuviation by C. G. Stephens (1962). Stace *et al.* (1968) claimed that a texture contrast was present in 14 groups and listed five additional possible mechanisms for its genesis: clay destruction in the A horizon, differential weathering, lateral clay transfer, layering of parent material and, in one example only, the addition to the A_1 of coarse material from upslope. A seventh hypothesis was proposed by Stewart (1959) who suggested that selective turnover by termites might be involved, but this was largely ignored by subsequent authors.

During fieldwork in south eastern Queensland in 1963–65, Thompson and Paton (1980) developed a model explaining the texture contrast in soil profiles on hillslopes as due to surface creep processes exceeding the rates of pedogenesis. This model was applied to hillslopes in the Sydney Basin by Bishop *et al.* (1980) and duplex soils were shown to be the result of the superposition of two unlike materials through the action of contemporary lateral movement of sediment down the slope. Rainwash was the process moving the sediment and bioturbation was implicated in preparing the sediment for erosion. This chapter reports the extension of these studies to a larger range of hillslopes and materials and is based on the testing of a specific hypothesis framed by Humphreys and Mitchell (1983).

The hypothesis

Bioturbation and the physical disturbance of surface soil material by organisms will allow rainwash to affect a greater thickness of the soil over time than rainwash on an undisturbed surface, where its effects will be confined to the litter and the soil surface. Any preferential removal of fines from the surface will coarsen the bioturbated layer.

The effectiveness of this mechanism will depend on at least three variables:

1 The potential for the soil material to be sorted according to particle size;
2 The rate and depth of bioturbation exposing new material to rainwash;
3 The rate and effectiveness of rainwash.

There are clearly many feedback effects in these relationships and the hypothesis provides not only an explanation of duplex soils but an integration of geomorphic process and bioturbation that can effectively explain many profile types through the selective loss of fines. Depending on the variables at any place, several end effects are conceivable. First, where there is considerable soil mixing by the biota, little surface mounding and little rainwash, the biota will tend to homogenize the soil. Second, where there is considerable shallow mixing or surface mounding, and rainwash is effective, the combined processes will lead to profile horizonation. In this event if the original soil materials are of uniform grain size no significant texture differentiation is likely, but the profile may exhibit a strong contrast in fabric between the A and B horizons. Paton (1978) proposed the term 'fabric contrast soil' for such profiles and an example on dolerite has been described by Hart (1988). Alternatively, if the original soil material contains non-uniform grain sizes, texture contrast development is probable and the duplex soil is an extreme example.

The hypothesis outlined above involves both rainwash erosion and bioturbation. Both processes could be expected to vary in intensity with the climatic environment, and the hypothesis could be assessed by examining the soils along an environmental gradient. The gradient selected was a complex one, involving several different bedrocks on slopes at three locations along a decreasing altitude and rainfall sequence west from Sydney. The sites were: on Triassic quartz sandstone at Blackheath in the Blue Mountains, on Permian sandstones and Devonian argillite at Lidsdale on the central tablelands, and on granite at Killonbutta on the central western slopes (figure 2.1 and table 2.4). All the sites were in relatively

Table 2.4 Summary of the environmental parameters at each study site

	Blackheath	Lidsdale	Killonbutta
Location	85 km WNW of Sydney	117 km WNW of Sydney	245 km WNW of Sydney
Geology	Triassic quartz sandstone	Lower Permian sandstone over Devonian argillite	Devonian granite
Topography	Plateau and slopes to 26°, 900-80 m above sea level	Slopes 1-26°, 900-60 m above sea level	Slopes 2-6°, 640-60 m above sea level
Vegetation	Open forest: Eucalyptus oreades, E. piperita, E. sieberi	Open forest: E. rossii, E. mannifera, E. dives	Open forest to woodland: E. macrorhyncha, E. polyanthemos, Callitris endlicheri
Average annual rainfall (mm)	1027	886	704
Average number of raindays	129	128	88
Mean daily temperatures for the hottest and coldest months (°C)			
MINIMUM	1.9/12.4	– 0.8/12.0	0.0/13.7
MAXIMUM	8.9/22.0	9.0/23.5	13.2/30.6

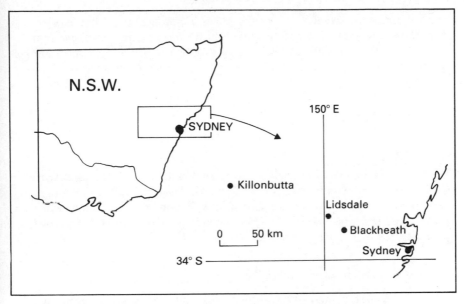

Figure 2.1   General location of the study sites in eastern New South Wales

undisturbed natural forest and were selected because bioturbation and geomorphic processes were obvious and judged to be amenable to simple attempts at quantification. This bias was deliberate and the results presented here need to be considered with this in mind. Various experiments were conducted to quantify the rates of the observed processes. Wherever possible, more than one technique was used to measure any one process: for example, the measurement of slope erosion rates employed splash trays, Gerlach overland flow troughs, and the assessment of sediment volumes held in natural traps of known or reliably estimated ages. In most cases the order of magnitude of the figures obtained was the same. Only summary results are reported here. Full details are presented in Mitchell, 1985, and will be published elsewhere. It should be recognized that there has been little uniformity in the techniques employed in other studies of this type and there are difficulties in comparing these results with other works. It should also be noted that all the studies included two periods of severe drought in 1979–80 and 1982 when it is probable that both bioturbation and erosion rates were lower than normal.

The approach used for the examination of the soils at each site was stratigraphic because Bishop *et al.* (1980) had shown that A horizons of duplex soils in the Sydney Basin were best interpreted as an uncomformable surface layer, and from this and other studies a methodology was adopted for the recognition of discrete soil layers (as opposed to horizons). In the field any one layer had to be seen to have an independent existence, that is, to be a genuine stratum, and this was most often done by tracing that layer over more than one substrate. Layers could show both lateral and vertical variations in their properties and as a matter of convenience were numbered from the surface down with subscripts for pedological or sedimentological variations within any one layer. The use of the same numbers at different sites does not imply any form of correlation.

The Blackheath study

The soil materials

Four distinct surface materials were recognized on the site, each with distinctive composition and fabric. Three of these could be traced over several substrates and therefore shown to be legitimate strata. The fourth material did not exist independently of underlying bedrock and was regarded as an altered portion of that stratum. A brief description of each layer follows:

Layer 0   Litter of eucalyptus debris.

Layer 1   A loamy sand or sandy loam topsoil with some comminuted organic matter and charcoal fragments. Fabric varied from incoherent single grains to weakly aggregated clumps. Mean composition 9 per cent granules and pebbles of iron cemented quartz sandstone, 87 per cent sub-angular quartz sand and 4 per cent silt/clay. Approximately 14 per cent organic matter. Moist colour 10YR5/3. The presence of organic materials and the field evidence for transport and some sorting indicated that layer 1 was not developed *in situ*. Three variants were recognized:
    Layer 1_i, the normal layer on the plateau surface.
    Layer 1_{ii}, coarser grained lyrebird display mounds.
    Layer 1_{iii}, coarser slope debris which was more intimately mixed with
        layer 0 material.

Layer 2   The sandstone bedrocks of the site and their *in situ* alteration products. Two slightly different sandstones were recognized and numbered layers 2_i and 2_{ii}. Both weathered to an essentially similar subsoil material, layer 2_{iii}, with the following properties: sandy loam to clayey sand with a porous, brittle, earthy fabric; mean composition 1 per cent granules and pebbles, 88 per cent sand and 11 per cent silt/clay. Moist colour 10YR6/6. No organic matter was present except as partial filling of pedotubules in the upper 10 cm.

Stratigraphically there were 15 possible vertical combinations of these materials (figure 2.2). The three most common combinations formed; yellow earths or Uc5.22 on the plateau, earthy sands or Uc5.11 and lithosols or Uc1.21 on the slopes (terminology of Stace *et al.*, 1968, and Northcote, 1971). These profiles were morphologically equivalent to

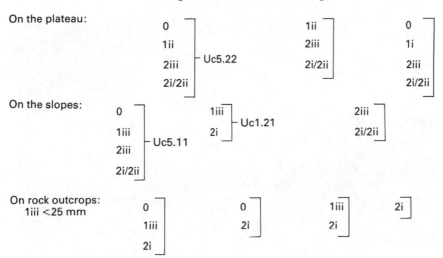

Figure 2.2   All possible vertical combinations ('profiles') of soil layers in relation to topography at Blackheath

profiles in the entisol order of *Soil Taxonomy* (Soil Survey Staff, 1975), particularly the psamments.

Bioturbation and erosion

Lyrebird display mounds

At Blackheath 22 mounds were located at the plateau edge on the 1.8 ha site. The mean mound size was: area 1.03 m², height 12.4 cm and thickness 4.7 cm, containing 24.0 kg of mineral soil and 1.5 kg of organic matter, mostly charcoal and comminuted leaf and bark fragments less than 4 mm in diameter. Mounds were circular in plan with an asymmetric pattern of upslope cut and downslope fill when located on slopes of more than about 4°. Soil excavation could produce a face up to 23 cm high and around the perimeter of the mound the soil was disturbed to an average depth of 9 cm. The total area of disturbance for each mound was 1.26 m² and this involved the full depth of layer 1_i on the plateau edge.

The intensity of mound use varied considerably. Two were clearly favoured by the resident male bird and expanded in area as he worked the perimeter and raked new soil to the centre, two others were abandoned and the remainder were only occasionally worked over. An annual rate of mound replacement of one in ten was estimated.

On the plateau edge mounds occupied 1 per cent of the area and a calculated soil turnover rate was 0.4 t ha^{-1} yr^{-1}. After simplistic extrapolation, with the assumption that there would be no repeated use

Figure 2.3 Excavation of soil layer 1$_{iii}$ by feeding lyrebirds, beneath a sandstone block on a 15° slope at Blackheath (scale is 10 cm long)

Figure 2.4 Extensive excavation of soil layer 1$_{iii}$ by feeding lyrebirds, from around the base of a large eucalypt exposing near-surface roots: view looking upslope (scale is 10 cm long)

of a site until the entire area had been worked over, this means that total soil turnover to a depth of 10 cm could occur in 1,000 years.

Soil disturbance by feeding activity

Feeding was conducted almost exclusively on the slopes and involved the regular shifting of soil and litter to a depth of 5–10 cm. On slopes steeper than about 8° the feeding birds always faced uphill and kicked the excavated debris about 1 m downslope. Birds were seen to shift rocks weighing up to 2 kg and they often worked up the slope on a limited front creating a low cliff of soil 12–15 cm high which slowly advanced uphill. Some feeding was concentrated at the edges of rock outcrops (figure 2.3), around the base of trees (figure 2.4) and along the downslope side of fallen logs. Rates of soil disturbance by feeding were quantified by measuring the volume of soil and litter shifted during several feeding sessions by single birds under observation, and by measuring the volume of freshly disturbed soil from five plots (each 49 m²) 100 hours after a heavy rainfall.

Measurements of ground cover on the feeding slopes showed 7–14 per cent living vegetation, 12 per cent loose rocks, 29–32 per cent bare soil recently disturbed by birds, and 49–57 per cent mixed litter and loamy sand previously disturbed by birds. Virtually all the vegetation less than 1 m high had been uprooted except in patches of dense fern and *Callicoma* spp. in which there was no lyrebird activity. Several micro-relief forms were created by feeding. Concentrated excavation around the base of rocks and logs undercut them to the extent that they often subsequently collapsed

Figure 2.5 Two cantilevered litter plumes built on an 18° slope by feeding lyrebirds. The large log fragments in the debris were moved by the birds (scale is 10 cm long)

downhill. The continuous downslope movement of soil and litter formed a series of discontinuous terraces and plumes the ultimate size of which varied with slope angle. Plumes of debris accumulated against shrubs, logs and rocks and were often cantilevered over the slope (figure 2.5).

Birds were found to disturb the soil by feeding at a rate of $63\,t\,ha^{-1}\,yr^{-1}$, 45 t of which was mineral soil. At this rate the soil on 1 ha of the slope could be overturned to a depth of 8 cm in 13 years. This is within the lifespan of a single bird and the process tended to be progressive because new areas were worked before old workings were reactivated.

Erosion

The processes of downslope movement observed on the site were lyrebird feeding, rainsplash, litter flow, and some limited rainwash. The total effect of all these erosion processes was calculated to produce a sediment yield from the slope of between 7.13 and $9.0\,t\,ha^{-1}\,yr^{-1}$.

Interpretation of the Blackheath study

The stratigraphic and topographic relationships of all the materials on the site (figure 2.2) illustrates a situation where there are two input sources for materials (sandstone bedrock and litter) and two interacting processes (normal weathering and lateral surface movement). Lyrebirds were central to the differentiation of the layer variants and essential to the effective downslope movement of layer 1. No evidence of any clay illuviation was found in thin sections of layer 2_{iii} and it was concluded that wherever layer 1 was mixed by birds the opportunity existed for the removal of fines from the full thickness of the layer by the erosion processes.

Profiles found on the slope were genetically (but not morphologically) similar to the duplex soils described by Bishop et al. (1980), and were examples of fabric contrast soils. The genesis of these profiles supported the hypothesis erected above.

The Lidsdale study

The soil materials

Several different bedrock lithologies complicated the interpretation of layers at this site. Seven layers were recognized and six of these were subdivided. A brief description of each follows:

Layer 0   A litter layer of eucalyptus debris. Together with living vegetation this layer provided an average ground cover of 65 per cent on Permian bedrock areas and 51 per cent on Devonian bedrock areas.

Layer 1 was the most extensively distributed material in the catchment and usually formed the A horizons of all soils. It varied in character depending on its proximity to different rock types and the degree of mixing which had occurred between the two recognized end members.

Layer 1_i was derived from the Permian rocks and variously overlay layers: 3_i, 4_i, and 5_i. This was a sandy loam with a mineralogy reflecting the Permian bedrocks. Moist colour 10YR4/3 to 5Y4/1 in poorly drained areas. A bleach of 10YR7/3 (dry), at the base of the layer.

Layer 1_{ii} was derived from the Devonian bedrock and overlay layers 6 and 6_i, but was not regarded as separate from layer 1_i because the two were similar and both comprised a single continuous unit of contemporary slope sediment. The mineralogy and texture increasingly reflected the Devonian bedrock downslope of the Permian/Devonian contact. Moist colour 10YR5/3 with a thicker bleach.

Layer 2 was of limited distribution and comprised crudely bedded lower slope colluvial rubble and alluvial sediments along the streams.

Layers 3, 4 and 5 were Permian conglomerate and sandstones which weathered to gravelly, red or yellow, sticky clays with strong pedality (layers 3_i, 4_i, and 5_i).

Layer 6 was Devonian argillite and shale. Layer 6_i was only sporadically developed on the slope.

The soil profiles formed by the combination of the various layers varied from yellow podzolics (Dy2.31, 2.41 and 3.41) where layers 3_i, 4_i, 5_i, or 6_i occurred beneath layer 1, to varied lithosols or earthy sands (Gn2.34, 3.04, and Um2) where layer 1 overlay layer 2, or where layer 1_{ii} occurred directly over layer 6. The texture contrast soils were morphologically equivalent to profiles in the ultisol order of *Soil Taxonomy*.

Bioturbation and erosion

Treefall was the most important form of bioturbation with additional mixing and movement caused by funnel ants, mound building ants (*Camponotus intrepidus*), termites (*Nasutitermes exitiosus*), the spiny ant-eater, a bandicoot (*Isodon* spp.) and unexpectedly the grey kangaroo (*Macropus major*). Also present in the soil were an unidentified earthworm, cicadas, beetle larvae and spiders. These invertebrates were not monitored, as they were judged to be unimportant relative to the other organisms. Estimated rates of soil disturbance by all fauna were $0.048\,t\,ha^{-1}\,yr^{-1}$ in the Permian bedrock areas and $0.025\,t\,ha^{-1}\,yr^{-1}$ in

Figure 2.6 A side view of a 3-month-old root slab failure in *Eucalyptus mannifera* on a 3° slope in the Permian bedrock area at Lidsdale. Note the debris from soil layer 1ᵢ resting on the trunk (trunk diameter approximately 50 cm)

the Devonian bedrock areas. These figures refer to the direct movement of soil, but some species were also noted to have a much larger effect on soil movement because they kept an area of ground clear of vegetation and litter. This was particularly true of the termite mounds and the kangaroo resting sites. Rainwash trays located adjacent to these features accumulated up to five times the amount of sediment in other trays on equivalent vegetated plots.

Treefall

The treefall study involved 145 fallen trees on the 22 ha site with ages up to six years. Year to year variation was substantial and the data included one major storm event.

Five common failure modes were recognized:

1 High breaks at a weak point on the trunk (windsnap). These usually did not involve any soil disturbance and were more common in trees growing on layer 1_{ii} where sinker roots penetrated the near vertical bedding planes of the Devonian bedrock. They also occurred in areas of layer 1_i when the ground was dry.

2 Dead falls which disturbed very little soil, since most of the roots had decayed before failure. The soil pit formed was approximately the same diameter as the tree trunk and no mound was created.

3 Root slab failures were very common on areas of duplex soil and occurred most often on the Permian bedrock areas when layer 1_i was saturated. The example in figure 2.6 is a *Eucalyptus mannifera* which fell 3 months prior to the date of the photograph. The roots of this tree were almost entirely developed in layer 1_i with only a few short sinkers into layer 3_i. The slab was turned up on the edge of an asymmetric pit with a maximum diameter of 4.0 m at 90° to the trunk and a minimum diameter of 2.60 m. Twisted lateral roots ruptured the soil surface for a further 1–1.5 m on either side of the pit. The pit was 10 cm deep at the edges and reached a maximum of 35 cm at a point immediately below the original position of the trunk. The total amount of soil disturbed was calculated to be 2.47 t, only 1.03 t of which was attached to the root slab at the time of measurement. Intact clods of soil 10–17 kg in weight were thrown as much as 2.8 m directly downslope of the slab.

4 Root ball failures involved the disturbance of a sphere- or cone-shaped body of soil. They were most common on duplex soils of the Permian bedrock areas and when seen on Devonian areas were often distorted by the presence of large rocks entangled in the roots. Figure 2.7 illustrates a typical example of the same species as in figure 2.6. This tree fell approximately 2 years prior to the photograph and about 30 per cent of the soil originally on the root ball had fallen back into the pit. A characteristic of root ball failures in this forest was that

Figure 2.7   An end view of a 2-year-old root ball failure in *Eucalyptus mannifera*. Note the rotation of the root ball in the pit and the limited volume of the pit in relation to the amount of soil remaining on the tree (card scale is 13 cm long)

they involved rotation of the tree base in the pit so that the ball occupied the pit (except on steeper slopes). As the ball eroded most of the soil was returned to the pit.

5 Leaning trees were trees that had fallen but were arrested by hanging in adjacent trees or were propped above the ground by large branches. When root damage was not severe the tree could survive and one example of *Eucalyptus rossii* was noted with epicormic branches growing on the topside of a near horizontal trunk. The branches had 25 annual growth rings. The uplifted soil at the base of these trees was slowly eroded into cavities beneath the roots.

The major factors controlling treefall were soil layering, soil moisture conditions, wind direction and strength, aspect, and the tree root pattern. Most trees were predisposed to fall downslope because they tended to grow out from the slope in response to maximum sunlight, and on the east side of the catchment 83 per cent of the trees fell downslope into the direction of the stronger winds.

Pit and mound micro-relief was not a common feature in this forest. These features were formed, but in the absence of a rotting trunk they were very difficult to recognize. This is attributed to rotation of the root ball in the pit (noted above), rapid filling of the pit by soil erosion from the root slab and by sediment and litter from upslope. Material filling young pits had essentially the same composition as layer 1 material in the vicinity.

Most mature trees in the forest had a butt hillock of soil at their base up to 15 cm high. This was considered to be the effect of root wedging and on steeper slopes the hillock was distinctly asymmetric with an accumulation of layer 0 on the upslope side and erosion of layer 1 where it was unprotected by layer 0, on the downslope side.

Across the whole of the study catchment the average depth of soil disturbance by treefall was 25 cm, and the mean rate of disturbance was $1.04\,t\,ha^{-1}\,yr^{-1}$ with a standard deviation of 0.76. There was a difference in the rate of total bioturbation for Permian ($1.40\,t\,ha^{-1}\,yr^{-1}$) and Devonian ($0.75\,t\,ha^{-1}\,yr^{-1}$) bedrock areas but this may not be significant.

Erosion

Rainwash and rainsplash erosion rates were measured over 4.26 years with sediment collection troughs at 54 points within the study catchment. The mean rates of rainwash on the two bedrock areas were significantly different (Permian bedrock areas $0.818\,t\,ha^{-1}\,yr^{-1}$ with a standard deviation of 1.164 and Devonian bedrock areas $2.069\,t\,ha^{-1}\,yr^{-1}$ with a standard deviation of 2.816) and the prime cause of this difference could be attributed to the combined effect of ground cover and slope angle.

Data collected from rainsplash erosion boards were less satisfactory and although the importance of rainsplash as an agent of sediment redistribution was confirmed, no net downslope movement of sediment was identified. Rainsplash rates changed as the study progressed and this was found to be due to the interruption of the normal downslope movement of layer 0 as it accumulated against the splash boards. This independent movement of layer 0 downslope was not anticipated at the start of the study and no provision had been made to quantify it. However an indication of its importance was obtained from analysis of the weight of organic matter trapped in the rainwash troughs during their first year of operation. The figures are a minimum estimate and showed that the weight of layer 0 moved downslope in one year was the equivalent of between 23 and 37 per cent of the annual litter fall recorded in the forest by Stafford (1976) and Lambert (1979). Imeson and van Zon (1979) and van Zon (1980) were apparently the first workers to record the magnitude of downslope litter movement, and they noted in their studies that 24 per cent of the total sediment delivery to the streams was splashed soil rafted on the litter. It would appear that this is a largely unexplored process in biogeomorphology.

Interpretation of the Lidsdale study

The relationship between the layers of materials on this site, although complicated by the different lithologies present, was fundamentally the same as at Blackheath. Two input sets were present (rock weathering and litter fall) and the slope mantle (layer 1) was subject to both extensive bioturbation and rainwash. Layer 1 materials were all depleted in fines relative to their substrates and the simplest explanation of this texture difference is that the fines were removed from the slope as a suspended load whilst the sand fractions moved more slowly.

Little direct evidence for the export of fines is available: turbid water was seen flowing from open treefall pits and the initial surface flow from most storms was slightly turbid. However, the grain size of the sediment collected in the rainwash troughs was consistent with a saltating bed load of sands and a probable contact bed load of coarser sand and gravels, but a suspended load fraction was absent. The troughs were not designed to collect this fraction and, in any case, in short periods of time the actual amount of clay shifted would be limited by the amount brought to the surface by bioturbation in the interval since the last rainfall and runoff event. Not all of the clay and silt lost from layer 1 has left the catchment. A substantial volume has accumulated in layer 2 over an unknown period of time.

Some indication of the importance of the two main sets of slope processes can be obtained by extrapolating the measured rates. The calculation makes the simplistic assumption that there will be no reworking of the surface by treefall until the entire catchment is disturbed, and takes no account of large changes in rate that could be expected with high magnitude, low frequency climatic events or fires. Soil bioturbation processes would be capable of turning over the entire 25 cm thickness of layer 1 in 2,500 years in Permian bedrock areas and 4,800 years in Devonian bedrock areas. Comparing this with the erosion rate it can be shown that in one mixing cycle on Permian bedrock areas 2,050 t of soil, or a layer 14 cm thick would be stripped from each hectare of slope. In fact the 25 cm layer 1 would be winnowed of fines and a nominal 11 cm thick coarse residue would remain as the slope mantle.

On Devonian bedrock areas in one mixing cycle, 9,700 t of soil or a layer 68 cm thick would be stripped from each hectare of slope. The rate of mixing is much less than the rate of erosion in this case and layer 1_{ii} would move down through *in situ* weathered rock. This stripping rate is likely to be faster than the rate of rock weathering, and very soon rocks would outcrop on the slope and layer 6_{ii} would be confined to zones of deeper regolith or areas with slower removal rates. Layer 1_{ii} would still form on the slope as a discrete stratum, but it would be thinner and coarser than layer 1_i since it would be composed of material with the slowest rate of movement. This is the slow process domination principle described by Moss *et al.* (1979).

Layer 1 materials are clearly 'young' and in active transport over the substrate layers. Duplex soils only formed in those areas where stripping rates were less than bioturbation rates and there was an underlying stratum with an appropriately higher clay content.

The Killonbutta study

The soil materials

Four layers of material were recognized on the study sections:

Layer 0   A litter layer, the composition of which varied with changes in the dominant vegetation. Adjacent to the granite outcrops it was mainly dead moss and lichen. Under *Callitris endlicheri* it was about 3 cm thick and comprised matted branchlets, twigs and fruit, and under *Eucalyptus* spp. it was thinner, mainly composed of leaves and bark and only provided 60–80 per cent effective ground cover. On areas where rainwash was active, that is below rock outcrops and on the lower parts of the slope where layer 1 was of finer grain size and tended to develop a surface crust, layer 0 was organized into contour-parallel, arcuate downslope, litter dams which ponded micro-terraces of layer 1 sediment.

Layer 1   Comprised the A horizon of all the soil profiles but was recognized as a separate stratum because it overlaid layers 2, 3 and 3_i. It varied systematically in composition downslope but was not subdivided into variant facies because all the changes were continuous and gradual (figure 2.8). Texturally it ranged from a coarse loamy sand with a high felspar content derived from rock outcrops on the upper slope, to a fine

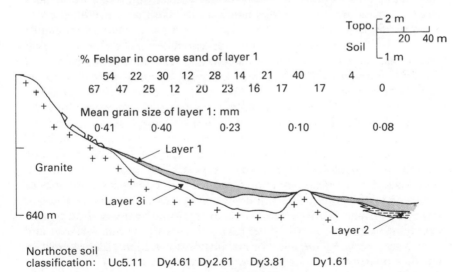

Figure 2.8   Slope section at Killonbutta showing the distribution of the soil layers in relation to topography and bedrock. Note the steady decrease in percentage felspar and decreasing grain size in layer 1 with increasing distance from granite outcrop areas

loamy sand with no felspar on the lower slope. Charcoal and invertebrate faecal pellets were present at all slope positions and a bleach developed at the base of the layer from the midslope to the lower slope.

Layer 2   was found only at the end of the section and was a bed of Jurassic(?) claystone, part of a sedimentary sequence which flanked the granite hills and which was not included in the study.

Layer 3   was the relatively coherent granite bedrock, and layer 3_i its *in situ* weathered equivalent which retained the fabric of the parent rock, had an abundance of felspar in the fine sand fraction and contained no charcoal except as a small part of burrow fillings. These properties were consistent down the slope although colour and thickness of the layer varied with drainage conditions. The texture varied from a sticky, sandy clay loam to a sandy light clay depending on the thickness of the layer.

Bioturbation and erosion

Invertebrates

Mound building ants had a very low population on the site and since Cowan *et al.* (1985) had shown that these species were unimportant elsewhere they were not specifically assessed.

The funnel ant was abundant, although patchily distributed over the site. The population was suspected to vary with seasonal conditions since larger numbers of active craters had been seen in the forest after wetter years in the early to mid-1970s. These ants constructed crater shaped mounds 5–25 cm in diameter and 0.5–11 cm in height. The mean size was 13 cm basal diameter, 8 cm rim diameter, 2.4 cm high, with a volume of 240 cm^3 containing 190 g of mineral soil. Mounds were commonly crowded together with densities of up to 25 m^{-2}, they often overlapped and there was a tendency for mounds to be more common along the margins of rocks, around tree bases (figure 2.9) and beneath fallen logs. The mounds were constructed on top of layer 0 and were ephemeral features being easily destroyed by rainsplash. Some larger mounds showed internal layering caused by three or four phases of partial destruction and reconstruction. Particle size analysis of the mounds showed an absence of grains larger than 2.5 mm and slightly less clay than in the adjacent layer 1. The mean density of mounds in the study area was 1,500 ha^{-1}. At any one time about 30–40 per cent of the mounds were inactive, and there was a marked preference for mound construction to occur more often on the upper parts of the slope where layer 1 was coarser and better drained.

A total of twelve 5 × 2 m plots were regularly sampled for worm casts and ant mounds and the mean mounding rates were 0.28 t ha^{-1} yr^{-1} for ants and 0.063 t ha^{-1} yr^{-1} for earthworms. These figures are much lower

Figure 2.9 Typical complex mounds of *Aphaenogaster longiceps* at the base of a tree. Note the deposition of excavated layer 1 material on top of layer 0 in a position where it is exposed to erosion (scale is 10 cm long)

than those reported for the same ant by Humphreys (1981) in a very different environment.

Spiny ant-eater

A total of 26 spiny ant-eater feeding scrapes were recorded on the study plots during the monitoring period. The typical scrape consisted of a curved vertical face cut into the surface soil or directly into an ant mound, which was backed by a splay of soil thrown back by the animal's forefeet. Single scrapes were 15–40 cm long, 12–25 cm wide and 2–20 cm deep. The mean size was 22 × 16 cm in area and 6 cm deep which involved the disturbance of 1,694 g of soil from layer 1. Several scrapes often occurred together and would be separated from nearby scrapes of similar age by several metres. All the meat ant mounds in the forest had been attacked at some time, but the majority of scrapes were located more or less at random and did not appear to concentrate on funnel ant mounds. Soil disturbance calculated from the 12 study plots was $1.65 \, \text{t ha}^{-1} \, \text{yr}^{-1}$ and this figure was checked by additional measurements over larger areas in which all recognizable scrapes were counted at the end of the study period. Scrapes of known age developed a lag granule surface due to rainsplash within four months of their excavation, were 50 per cent covered by litter in 12 months and had effectively vanished in 2 to 2.5 years. Using these observations and assuming all the scrapes on the larger plots were of average size the check figure for soil disturbance was $2.32 \, \text{t ha}^{-1} \, \text{yr}^{-1}$.

Other bioturbation processes

Soil turnover by bandicoots and treefall was assessed at $0.007 \, t \, ha^{-1} \, yr^{-1}$, and a number of other agents of soil bioturbation were recognized but not measured as these were considered to be insignificant in comparison to those discussed above. Virtually all fallen trees on the site fell as dead trees, after decay and termites had consumed the main roots. These disturbed very little soil. In some cases dead trees or stumps were burnt out in the occasional bushfire leaving a cavity in the soil surface which slowly filled with litter and layer 1 sediment.

Erosion

Rainwash was effective on these low angle slopes and the mean rate of sediment yield was $0.37 \, t \, ha^{-1} \, yr^{-1}$.

Interpretation of the Killonbutta study

The layers recognized on this site (figure 2.8) had the same stratigraphic relationships to one another as the layers recognized at Lidsdale and Blackheath. In all cases the same source sets were present and the same processes operating. In upper slope positions the combination of layers 1 and 3 formed a lithosol or Uc5.11 profile. This was equivalent to an orthent or psamment in terms of *Soil Taxonomy*. From the point on the slope where layer 3_i became important a solodic soil which was variously keyed as Dy4.61, 2.61, 3.81, or 1.61 occurred. Morphologically these are equivalent to ustults or ustalfs, but genetically there is no connection with the interpretations of *Soil Taxonomy* and these profiles have much more in common with the duplex soils described by Bishop *et al.* (1980). Layer 1 was transgressive over several substrates and was mixed by ants and spiny ant-eaters, moving downslope in such a way that size fractionation or sorting occurred. In none of the profiles was there any evidence of the vertical translocation of clay, and the progressive downslope development of a bleach at the base of layer 1 attested to the importance of lateral throughflow at the interface of layers 1 and 3_i.

Conclusions

The role of the biosphere on all the sites examined was to constantly return potential suspended load material to the surface. Here it was placed at risk to erosion and transport on the hillslope, and topsoil A horizons were developed in the actively mobile layer 1 materials which were moving as a bedload sediment mantle. The grain size of these topsoils was a function of available sizes, rates of mixing and rainwash, and the slow process domination principle of Moss *et al.* (1979). The thickness of the topsoils was primarily a function of the effective depth of mixing, but probably also varied with factors controlling the rainwash rate such as slope, percentage vegetation cover, grain size and rainfall characteristics. Major events such as fire, storms, drought and climatic change would all substantially alter the bioturbation and rainwash rates, and extrapolation of the results given here needs to be done with caution. However, in every case examined the present day rates, measured under conditions of relative site stability, are sufficient to lead to the development of duplex soil profiles in only a few thousand years.

In the situation where rainwash is known to be particularly effective, that is longer, low angle slopes (Moss *et al.*, 1979), it could be expected that a lateral fining sequence would occur in the surface mantle. This was observed by Milne (1947), predicted by Walker and Butler (1983) and was found on the Killonbutta section (figure 2.8).

The results obtained from each site supported the original hypothesis, and also the suggestion that horizonation or layering could be a common end effect of bioturbation not just homogenization. This horizonation took two forms. Firstly, where the bedrock of the site did not produce a wide range of grain sizes on weathering, uniform textured profiles were formed but the layering was evident as a distinct change in soil fabric as at Blackheath. Secondly, where the bedrock did produce a range of grain sizes on weathering, the combined processes of bioturbation and rainwash tended to remove the finer fractions from the surface layer of mineral soil, leading to the development of texture contrast soils such as those at Lidsdale and Killonbutta.

Two other scenarios are also recognized. The third situation is where the rate of erosion exceeds the rate of soil mixing, such as on the Devonian bedrock areas at Lidsdale. This may be a less common case since it depends partly on the presence of steeper slopes and results in the slope being covered in a coarse mobile mantle of material. The final scenario was not encountered on the sites described here, but has been described by Mitchell (1985) in the sub-alpine areas of the Snowy Mountains, where rainwash was ineffective because of a dense protective ground cover vegetation and soil mixing by earthworms was very effective and very deep. Uniform or gradational texture profiles developed there as a result of homogenization, despite the presence of a range of grain sizes in the weathered rock.

Implications

The work reported here has shown that the biosphere plays a very important role in soil genesis on hillslopes in the temperate areas of eastern New South Wales. The wider application of the hypothesis to other environments is probable and there is a need to further integrate bioturbation and geomorphic processes into landscape models.

There are also a number of specific questions that deserve attention. For example:

1 Why is rainwash such a universal process in this climatic environment where infiltration might have been expected? Two aspects of this need investigation, and both are biological:
   a Australian litter is physically coarser and perhaps less absorbent than litter from deciduous or coniferous forests of the northern hemisphere and this may create more litter flow.
   b Australian topsoils commonly have a strong water repellence (Bond, 1964; Roberts and Carbon, 1971), a property that was present in all the layer 1 materials on these study sites. Fungi and fire have both been implicated as causes, but little attention has been given to the importance of this property in hillslope hydrology.
2 On all of the study sites the litter was found to move downslope independently of the surface soil mantle. To date no litter breakdown studies have taken account of this movement, and since the estimated rate of movement at Lidsdale suggested that as much as one third of the annual litter fall could move down to the valley floor in one year it is clearly important in the nutrient cycle and probably important in geomorphic studies.
3 What part does the combined process of bioturbation and rainwash erosion play in soil creep? Since the studies reported here have shown that the effects of rainwash are distributed through the topsoil by bioturbation this is effectively a creep mechanism. If a velocity profile for the Lidsdale or Killonbutta sections could be drawn it would have two components: a vertical movement of particles from depth to the surface, and a rapid downslope movement of particles at the surface. No such profile or process combination was recognized by Carson (1976) in his review of soil creep mechanisms.
4 The vertical movement of fines in the soil, particularly by invertebrates with a limited capacity to move large particles, could allow the settling of stones down to the base of any bioturbation layer. This mechanism may account for the common occurrence of stone layers at the interface of A and B horizons in duplex soils. If it can be shown to be a real process it follows that stone lines *per se* should not be used as evidence of either a truncated surface or of a creeping mantle in the conventional sense.
5 The implications of this work with respect to soil classification are profound. Neither of the soil classification systems in use in Australia (Stace *et al.*, 1968; Soil Survey Staff, 1975) could comfortably deal with the duplex profiles created by the superposition of layers described here. The term 'podzolic' was unsatisfactory because it implied that the profiles had undergone podzolization for which there

was virtually no evidence, and the terms 'ultisol' and 'alfisol' were equally unsatisfactory because they implied that some process of clay accumulation had occurred. The basic problem underlying these difficulties in classification lies with the restrictive definition of a profile as a genetic entity in which A, B and C horizons are related by processes (transformations, transfers, additions and removals) which have only been considered to operate in a vertical plane. It is clear from this study that contemporary lateral transfers and removals must also be considered, and it is equally clear that the very presence of a well-differentiated profile is no indication of soil 'maturity', nor that a long period of landscape stability was involved.

Soil chrono-sequence studies on alluvial terraces by Walker and Coventry (1976) and Chittleborough *et al.* (1984), concluded that strong texture contrasts may take 30,000 or more years to develop. This is an order of magnitude longer than the rates reported here for hillslopes would suggest, and comparison of the soils and the soil forming processes in these different areas should be undertaken.

By way of a final statement on the importance of the fluvial hillslope soil genesis model presented here, it is of interest to consider the opinion of Moss (1979). When discussing the potentially high rates of sediment movement by rainwash and overland flow on slopes, he stated that the processes must be restrained (presumably by vegetation) in humid regions as evidenced by the common occurrence of mature soils on slopes. The studies in this chapter have shown that mature soils, that is duplex profiles, in eastern New South Wales, are in fact a consequence of the efficiency of sediment movement by rainwash when it is combined with sediment resupply by bioturbation.

Acknowledgements

The author is indebted to the National Parks and Wildlife Service and the Forestry Commission of NSW for permission to conduct research on their estates, and to Macquarie University for funding and facilities.

References

Abaturov, B. D. 1972: The role of burrowing animals in the transport of mineral substances in the soil. *Pedobiologia*, 12, 261–6.

Anderson, D. C. and MacMahon, J. A. 1981: Population dynamics and bioenergetics of a fossorial herbivore, *Thomomys talpoides* (Rodentia: Geomyidae) in a spruce-fir sere. *Ecological Monographs*, 51, 179–202.

Ashton, D. H. 1975: Studies of litter in *Eucalyptus regnans* forests. *Australian Journal of Botany*, 23, 413–33.

—— 1979: Seed harvesting by ants in forests of *E. regnans* F. Muell, in central Victoria. *Australian Journal of Ecology*, 4, 265–77.

Bal, L. 1982: *The zoological ripening of soils*. Wageningen: Centre for Agricultural Publishing and Documentation.

Baxter, F. P. and Hole, F. D. 1967: Ant (*Formica cinerea*) pedoturbation in a prairie soil. *Soil Science Society of America Proceedings*, 31, 425–8.

Berg, R. Y. 1975: Myrmecochorous plants and their dispersal by ants. *Australian Journal of Botany*, 23, 475–508.

Bishop, P. M., Mitchell, P. B. and Paton, T. R. 1980: The formation of duplex soils on hillslopes in the Sydney Basin, Australia. *Geoderma*, 23, 175–89.

Blakemore, L. C. and Gibbs, H. S. 1968: Effects of gannets on soil at Cape Kidnappers, Hawke's Bay. *New Zealand Journal of Science*, 11, 54–62.

Blevins, R. L., Holowaychank, N. and Wilding, L. P. 1970: Micromorphology of soil fabric at the tree root–soil interface. *Soil Science Society of America Proceedings*, 34, 460–4.

Blong, R. J., Riley, S. J. and Crozier, P. J. 1982: Sediment yield from runoff plots following bushfire near Narrabeen Lagoon, NSW. *Search*, 13, 36–8.

Böhm, W. 1979: *Methods of studying root systems*. Berlin: Springer-Verlag.

Bond, R. D. 1964: The influence of the microflora on the physical properties of soils. II. Field studies on water repellent sands. *Australian Journal of Soil Research*, 2, 123–31.

Brian, M. V. (ed.) 1978: *Production ecology of ants and termites*. Cambridge: Cambridge University Press.

Buckley, R. C. (ed.) 1982: *Ant–plant interactions in Australia*. The Hague: Dr. W. Junk.

Burroughs, E. R. and Thomas, B. R. 1977: Declining root strength in Douglas-fir after felling as a factor in slope stability. *USDA Forest Service Research Paper* INT-190.

Butuzova, O. V. 1962: Role of the root system of trees in the formation of micro-relief. *Soviet Soil Science*, 4, 364–72.

Carson, M. A. 1976: Mass-wasting, slope development and climate. In E. Derbyshire (ed.), *Geomorphology and climate*, pp. 101–36. London: Wiley.

Chittleborough, D. J., Walker, P. H. and Oades, J. M. 1984: Textural differentiation in chronosequences from eastern Australia. *Geoderma*, 32, 181–248.

Clinnick, P. F. 1984: A summary review of the effects of fire on the soil environment. *Soil Conservation Authority of Victoria. Technical report series*.

Cowan, J. A., Humphreys, G. S., Mitchell, P. B. and Murphy, C. L. 1985: An assessment of pedoturbation by two species of mound-building ants, *Camponotus intrepidus* (Kirby) and *Iridomyrmex purpureus* (Smith). *Australian Journal of Soil Research*, 22, 95–107.

Crocker, R. L. 1959: The plant factor in soil formation. *Presidential address, ANZAAS*.

Crook, I. G. 1975: The tuatara. In G. Kuschel (ed.), *Biogeography and ecology in New Zealand*, pp. 331–52. The Hague: Dr. W. Junk.

Darwin, C. R. 1881: *The formation of vegetable mould through the action of worms, with observations on their habits*. London: John Murray.

Denny, C. S. and Goodlett, J. C. 1956: Micro-relief from fallen trees. Surficial geology and geomorphology of Potter Co. Pa. *USGS Professional Paper*, 288, 59–68.

—— and 1968: Tree-thrown origin of patterned ground on beaches of the ancient Champlain sea near Plattsburgh, New York. *USGS Professional Paper*, 600B, B157–B164.

Dickinson, C. H. and Pugh, G. J. F. 1974: *Biology of plant litter decomposition*, 2 vols. London: Academic Press.

Dumpert, K. 1981: *The social biology of ants*. Translation by C. Johnson. Boston: Pitman.

Dunne, T. 1978: Field studies of hillslope flow processes. In M. J. Kirkby (ed.), *Hillslope hydrology*, pp. 227–93. Chichester: Wiley.

Edwards, C. A. and Lofty, J. R. 1972: *Biology of earthworms*. London: Chapman and Hall.

Ellison, L. 1946: The pocket gopher in relation to soil erosion on mountain range. *Ecology*, 27, 101–14.

Evans, A. C. 1948: Studies on the relationships between earthworms and soil fertility. II. Some effects of earthworms on soil structure. *Annals of Applied Biology*, 35, 1–13.

—— and Guild, W. J. McL. 1947: Studies on the relationship between earthworms and soil fertility. I. Biological studies in the field. *Annals of Applied Biology*, 34, 307–30.

Evans, G. R. 1973: Hutton's shearwaters initiating local soil erosion in the seaward Kaikoura Range. *New Zealand Journal of Science*, 16, 637–42.

Falinski, J. B. 1978: Uprooted trees, their distribution and influence in the primeval forest biotope. *Vegetatio*, 38, 175–83.

Fineran, B. A. 1973: A botanical survey of Seven Mutton-bird Islands, southwest Stewart Island. *Journal of the Royal Society of New Zealand*, 3, 475–526.

Godfrey, G. K. and Crowcroft, P. 1960: *The life of the mole*, Talpa europaea L. London: Museum Press.

Gray, D. H. and Leiser, A. T. 1982: *Biotechnical slope protection and erosion control*. New York: Van Nostrand Reinhold.

Greaves, T. 1939: The control of meat ants (*Iridomyrmex detectus* Sm.). *Journal of Commonwealth Scientific and Industrial Research*, 12, 109–14.

Greenslade, P. J. M. 1974: Some relations of the meat ant, *Iridomyrmex purpureus* (Hymenoptera: Formicidae) with soil in South Australia. *Soil Biology and Biochemistry*, 6, 7–14.

—— and Thomspon, C. H. 1981: Ant Distribution, vegetation and soil relationships in the Coloola–Noosa River area, Queensland. In A. N. Gillison and D. J. Anderson (eds.), *Vegetation classification in Australia*, pp. 192–207. Canberra; CSIRO and Australian National University Press.

—— and Halliday, R. B. 1982: Distribution and speciation in meat ants, *Iridomyrmex purpureus* and related species. (Hymenoptera: Formicidae). In W. R. Barker and P. J. M. Greenslade (eds), *Evolution of the flora and fauna of arid Australia*, pp. 249–55. Adelaide: Peacock Publications.

—— and Greenslade, P. 1983: Ecology of soil invertebrates. In CSIRO, *Soils: an Australian viewpoint*, pp. 645–69. Melbourne: CSIRO and Academic Press.

Griffiths, M. 1968: *Echidnas*. Oxford: Pergamon Press.

—— 1978: *The Biology of the Monotremes*. New York: Academic Press.

Guild, W. J. McL. 1948: Studies on the relationship between earthworms and soil fertility. III. The effect of soil type on the structure of earthworm populations. *Annals of Applied Biology*, 35, 181–94.

Harley, J. L. and Scott Russell, R. (eds.) 1979: *The Soil-root interface*. London: Academic Press.

Hart, D. M. 1988: A fabric contrast soil on dolerite in the Sydney Basin, Australia. *Catena*, 15, 27–35.

Hazelhoff, L., van Hoof, P., Imeson, A. C. and Kwaad, F. J. P. M. 1981: The exposure of forest soil to erosion by earthworms. *Earth Surface Processes and Landforms*, 6, 235–50.

Hickman, G. C. and Brown, L. N. 1973: Mound-building behaviour of the southeastern pocket gopher (*Geomys pinetis*). *Journal of Mammalogy*, 54, 786–9, 971–4.

Hole, F. D. 1981: Effects of animals on soil. *Geoderma*, 25, 75–112.

Hugie, V. K. and Passey, H. B. 1963: Cicadas and their effects upon soil genesis in certain soils in southern Idaho, northern Utah, and northeastern Nevada. *Soil Science Society of America Proceedings*, 27, 78–81.

Humphreys, F. R. and Craig, F. G. 1981: Effects of fire on soil chemical, structural and hydrological properties. In A. M. Gill, R. H. Groves, and I. R. Noble (eds), *Fire and the Australian biota*, pp. 177–200. Canberra: Australian Academy of Science.

Humphreys, G. S. 1981: The rate of ant mounding and earthworm casting near Sydney, New South Wales. *Search*, 12, 129–31.

—— 1985: Bioturbation, rainwash and texture contrast soils. Unpublished PhD thesis, Macquarie University.

—— and Mitchell, P. B. 1983: A preliminary assessment of the role of bioturbation and rainwash on sandstone hillslopes in the Sydney Basin. In R. W. Young and G. C. Nanson (eds.) *Aspects of Australian Sandstone Landscapes*. Australian and New Zealand Geomorphology Group Special Publication No. 1, 66–80.

Hutnik, R. J. 1952: Reproduction on windfalls in a northern hardwood stand. *Journal of Forestry*, 50, 693–4.

Imeson, A. C. 1976: Some effects of burrowing animals on slope processes in the Luxembourg Ardennes. 1. The excavation of animal mounds in experimental plots. *Geografiska Annaler*, 58A, 115–25.

—— 1977: Splash erosion, animal activity and sediment supply in a small forested Luxembourg catchment. *Earth Surface Processes*, 2, 153–60.

—— and Kwaad, F. J. P. M. 1976: Some effects of burrowing animals on slope processes in the Luxembourg Ardennes. 2. The erosion of animal mounds by splash under forest. *Geografiska Annaler*, 58A, 317–28.

—— and van Zon, H. 1979: Erosion processes in small forested catchments in Luxembourg. In A. F. Pitty (ed.), *Geographical approaches to fluvial processes*, pp. 93–107. Norwich: GeoBooks.

Jonca, E. 1972: Winter denudation of molehills in mountainous areas. *Acta Theriologica*, 17, 407–17.

Jungerius, P. D. and van Zon, H. J. M. 1982: The formation of the Lias cuesta (Luxembourg) in the light of present-day processes operating on forest soils. *Geografiska Annaler*, 64A, 127–40.

Kalisz, P. J. and Stone, E. L. 1984: Soil mixing by scarab beetles and pocket gophers in north-central Florida. *Soil Science Society of America Journal*, 48, 169–72.

Kirkpatrick, J. D., Massey, J. S. and Parsons, R. F. 1974: Natural history of Curtis Island, Bass Strait. 2. Soils and vegetation. *Proceedings of the Royal Society of Tasmania*, 107, 131–44.

Kotarba, A. 1970: The morphogenetic role of foehnwind in the Tatra Mountains. *Studia Geomorphologica Carpatho-Balcanica*, 4, 171–88.

Lambert, M. J. 1979: Sulphur relationships of native and exotic tree species. Unpublished M.Sc. thesis, Macquarie University.

Leadley Brown, A. 1978: *Ecology of soil organisms*. London: Heinemann.

Lee, K. E. 1959: *The earthworm fauna of New Zealand*. New Zealand Department of Scientific and Industrial Research, Bulletin 130.

—— 1985: *Earthworms. Their ecology and relationships with soils and land use*. North Ryde: Academic Press Australia.

—— and Wood, T. G. 1968: Preliminary studies of the role of *Nasutitermes exitiosus* (Hill) in the cycling of organic matter in a yellow podzolic soil under dry sclerophyll forest in South Australia. *Transactions 9th International Congress of Soil Science, Adelaide*, 2, 11–18.

—— and Wood, T. G. 1971: *Termites and soils*. London: Academic Press.

—— , Warcup, J. H. and Hutson, B. R. 1981: The soil biota. *Proceedings Australian Forest Nutrition Workshop*. Canberra: CSIRO Division of Forest Research.

Ljungström, P. O. and Reinecke, A. J. 1969: Ecology and natural history of the microchaetid earthworms of South Africa. IV. Studies on the influences of earthworms upon the soil and the parasitological questions. *Pedobiologia*, 9, 152–7.

Lohm, U. and Persson, T. (eds) 1977: *Soil organisms as components of ecosystems*. Stockholm: Swedish National Science Research Council, Bulletin no. 25.

Lorimer, G. G. 1980: Age structure and disturbance history of a southern Appalachians virgin forest. *Ecology*, 61, 1169–84.

Lutz, H. J. 1940: Disturbance of forest soil resulting from the uprooting of trees. *Yale University School of Forestry Bulletin*, no. 45.

—— and Griswold, F. S. 1939: The influence of tree roots on soil morphology. *American Journal of Science*, 237, 389–400.

Mielke, H. W. 1977: Mound building by pocket gophers (Geomyidae): their impact on soils and vegetation in North America. *Journal of Biogeography*, 4, 171–80.

Milne, G. 1947: A soil reconnaissance journey through parts of Tanganyika Territory, December 1935 to February 1936. *Journal of Ecology*, 35, 192–265.

Mitchell, P. B. 1985: Some aspects of the role of bioturbation in soil formation in south eastern Australia. Unpublished PhD thesis, Macquarie University.

Moss, A. J. 1979: Thin-flow transportation of solids in arid and non-arid areas: a comparison of processes. In *The hydrology of areas of low precipitation*. International Association of Hydrological Sciences. Publication no. 128, 435–45.

—— , Walker, P. H. and Hutka, J. 1979: Raindrop-stimulated transportation in shallow water flows: an experimental study. *Sedimentary Geology*, 22, 165–84.

Naarding, J. A. 1981: Study of the short-tailed shearwater (*Puffinus tenuirostris*) in Tasmania. *National Parks and Wildlife Service, Tasmania. Wildlife Division Technical Report*, 81/3.

Neuwenhuis, J. D. and van den Berg, J. A. 1971: Slope investigations in the Morvan (Haut Folin area). *Revue de Géomorphologie Dynamique*, 20, 161–76.

Newsome, A. E., McIlroy, J. and Catling, P. 1975: The effects of an extensive wildfire on populations of twenty ground dwelling vertebrates in south-east Australia. *Proceedings of the Ecological Society of Australia*, 9, 107–23.

Nielsen, G. A. and Hole, F. D. 1964: Earthworms and the development of coprogenous A_1 horizons in forest soils of Wisconsin. *Soil Science Society of American Proceedings*, 28, 426–9.

Northcote, K. H. 1971: *A factual key for the recognition of Australian soils*. Glenside, South Australia: Rellim Technical Publications.

O'Loughlin, C. L. 1974: A study of tree root strength deterioration following clearfelling. *Canadian Journal of Forest Research*, 4, 107–13.

Paton, T. R. 1978: *The formation of soil material*. London: George Allen and Unwin.

Reig, O. A. 1970: Ecological notes on the fossorial octodont rodent *Spalacopus cyanus* (Molina). *Journal of Mammalogy*, 51, 592–601.

Roberts, F. J. and Carbon, B. A. 1971: Water repellence in some sandy soils of SW Western Australia. 1. Some studies related to field occurrence. *Field Station Records. Division of Plant Industry*. CSIRO, 10, 13–20.

Robinson, F. N. and Frith, H. J. 1981: The superb lyrebird *Menura novaehollandiae* at Tindinbilla A.C.T. *Emu*, 81, 145–57.

Rodin, L. E. and Bazilevich, N. I. 1965: *Production and mineral cycling in terrestrial vegetation*. Translation by G. E. Fogg. Edinburgh: Oliver and Boyd.

Ross, B. A., Tester, J. R. and Breckenridge, W. J. 1968: Ecology of mima-type mounds in northwestern Minnesota. *Ecology*, 49, 172–7.

Russell, E. W. 1973: *Soil conditions and plant growth*, 10th edn. London: Longmans.

Ryan, P. J. and McGarity, J. W. 1983: The nature and spatial variability of soil properties adjacent to large forest eucalypts. *Soil Science Society of America Journal*, 47, 286–93.

Salem, M. Z. and Hole, F. D. 1968: Ant (*Formica exsectoides*) pedoturbation in a forest soil. *Soil Science Society of America Proceedings*, 32, 563–7.

Satchell, J. E. (ed.) 1983: *Earthworm ecology: From Darwin to vermiculture*. London: Chapman and Hall.

Saunders, G. W. 1966: Funnel ants (*Aphaenogaster* sp. Formicidae) as pasture pests in north Queensland. 1. Ecological background, status and distribution. *Bulletin of Entomological Research*, 57, 419–32.

Sheets, R. G., Linder, R. L. and Dahlgren, R. B. 1971: Burrow systems of prairie dogs in South Dakota. *Journal of Mammalogy*, 52, 451–2.

Smith, O. L. 1982: *Soil microbiology: a model of decomposition and nutrient cycling*. Boca Raton Fla.: CRC Press.

Soil Survey Staff, 1975: *Soil taxonomy: a basic system of soil classification for making and interpreting soil surveys*. Washington: USDA Handbook No. 436.

Stace, H. C. T., Hubble, G. D., Brewer, R., Northcote, K. H., Sleeman, J. R., Mulcahy, M. J. and Hallsworth, E. G. 1968: *A handbook of Australian soils*. Glenside, South Australia: Rellim Technical Publications.

Stafford, R. M. 1976: Litterfall in forests of *Pinus radiata* and mixed *Eucalyptus* species in Lidsdale State Forest, N.S.W. Unpublished M.Sc. thesis, University of New South Wales.

Stephens, C. G. 1962: *A manual of Australian soils*. Melbourne: CSIRO.

Stephens, E. P. 1956: The uprooting of trees: A forest process. *Soil Science Society of America Proceedings*, 20, 113–6.

Stewart, G. A. 1959: Some aspects of soil ecology. In A. Keast, R. L. Crocker and C. S. Christian (eds), *Biogeography and ecology in Australia*, pp. 301–14. The Hague: Dr. W. Junk.

Sudd, J. H. 1967: *An introduction to the behaviour of ants*. London: Edward Arnold.

Thompson, C. H. and Paton, T. R. 1980: Texture differentiation in soils on hillslopes, south-eastern Queensland. *CSIRO Division of Soils, Divisional Report*, 53.

Thorp, J. 1949: Effects of certain animals that live in soils. *Science Monthly*, 68, 180–91.

van Zon, H. J. M. 1980: The transport of leaves and sediment over a forest floor. *Catena*, 7, 97–110.

Veneman, P. L. M., Jacke, P. V. and Bodine, S. M. 1984: Soil formation as affected by pit and mound microrelief in Massachusetts USA. *Geoderma*, 33, 89–99.

Walker, E. P. (ed.) 1968: *Mammals of the world*, 2 vols. Baltimore: Johns Hopkins Press.

Walker, N. (ed.) 1975: *Soil microbiology*. Kent: Butterworths.

Walker, P. H. and Coventry, R. J. 1976: Soil profile development in some alluvial deposits of eastern New South Wales. *Australian Journal of Soil Research*, 14, 305–17.

—— and Butler, B. E. 1983: Fluvial processes. In CSIRO, *Soils: an Australian Viewpoint*, pp. 83–90. Melbourne: CSIRO and Australian National University Press.

Wallwork, J. A. 1970: *Ecology of soil animals*. London: McGraw Hill.

—— 1976: *The distribution and diversity of soil fauna*. London: Academic Press.

Waloff, N. and Blackith, R. E. 1962: The growth and distribution of the mounds of *Lasius flavus* (Fabricius) (Hym: Formicidae) in Silwood Park, Berkshire, *Journal of Animal Ecology*, 31, 421–37.

Watanabe, H. 1975: On the amount of cast produced by the Megascolecid earthworm *Pheretima hupeiensis*. *Pedobiologia*, 15, 20–8.

Wheeler, W. M. 1910: *Ants: Their structure, development and behaviour*. New York: The Columbia University Press.

Wiken, E. B., Broersma, K., Lavkulich, L. M. and Farstad, L. 1976: Biosynthetic alteration in a British Columbia soil by ants (*Formica fusca* Linne). *Soil Science Society of America Proceedings*, 40, 422–6.

Wood, T. G. 1974: The distribution of earthworms (Megascolecidae) in relation to soils, vegetation and altitude on the slopes of Mt. Kosciusko, Australia. *Journal of Animal Ecology*, 43, 87–106.

Yair, A. and Rutin, J. 1981: Some aspects of regional variation in the amount of available sediment produced by isopods and porcupines, northern Negev, Israel. *Earth Surface Processes and Landforms*, 6, 221–34.

Young, A. R. M. 1983: Patterned ground on the Woronora Plateau. In R. W. Young and G. C. Nanson (eds.), *Aspects of Australian Sandstone Landscapes*. Australian and New Zealand Geomorphology Group Special Publication No. 1, 48–53.

Zlotin, R. I. and Khodashova, K. S. 1980: *The role of animals in biological cycling of forest-steppe ecosystems*, N. R. French (ed.). Stroudsburg, Pa: Dowden, Hutchinson and Ross.

3 The impact of organisms on overall erosion rates within catchments in temperate regions

Stanley W. Trimble

The effects of organisms on erosion are considered here at three scales: (a) small-scale–large area (climatic–vegetation region), (b) meso- or stream-basin scale and (c) large-scale–small areas (hillslopes). The last gives the opportunity to examine processes and the interrelationships among many variables.

Small-scale–large area, or climatic–vegetation regions

It is difficult to relate natural or geologic erosion to controlling variables. This is not only because of the many natural variables involved, of which slope or relief tends to be the most important, but also because humans can increase erosion by orders of magnitude, thus confounding the data (Meade, 1969; Trimble, 1976). Moreover, such studies must depend on sediment yields (as compared to slope erosion) as the dependent variable. This introduces the sediment delivery problem (Walling, 1983) which itself involves many additional variables such as size of stream basin, relief, channel and valley morphology and land use, dams, texture of sediment, and streamflow, just to name a few. Additionally, stream valleys are sinks as well as sources of sediment so that channel and valley fluxes must be established in order to tell much about the value of sediment yields (Trimble, 1983). Nevertheless, there are some generalizations that can be made.

The subject area span of climatic geomorphology is beyond the scope of this essay and readers are referred to the standard works (Büdel, 1982; Derbyshire, 1976; Fournier, 1982). Works more specifically attuned to the relationship of erosion and sediment yield are Douglas, 1976; Jansen and Painter, 1974; Langbein and Schumm, 1958; and Wilson, 1973. All of these use sediment yields as the dependent variable, although some have been partially adjusted for area. For example, the well-known Langbein–Schumm

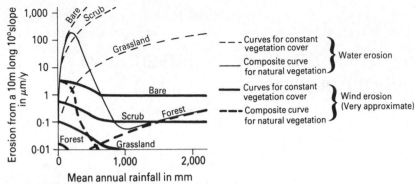

Figure 3.1   Estimated rates of soil erosion by wind and water as a function of rainfall and vegetation cover
(from Kirkby, 1980)

curve relates mean annual unit sediment yield to mean annual precipitation and temperature (see also Schumm, 1965, and Wilson, 1973).

Kirkby (1980) has combined much of the foregoing work with experimental plot data to create a composite which shows standardized slope erosion rates as a function of both mean annual precipitation and natural vegetation (figure 3.1). The Kirkby model is an improvement on earlier ones in that vegetation is an independent variable rather than being implicitly dependent upon average annual precipitation and temperature. His model thus shows the effects of artificially controlled vegetation under various amounts of precipitation. It also allows for wind erosion in humid climates which, as we shall see later in this essay, is a more important problem than many surmise.

Such conceptual models are thought-provoking and therefore useful, but there is in reality much variance from the representations shown on the graph. For example, the wet–dry climates such as Subtropical Dry Summer and Tropical Wet–Dry (Köppen: Csa and Aw or Am respectively) impose a seasonal stress on natural vegetation that allows more erosion than the annual average precipitation would indicate. Another climatic complication is the differences of kinetic energy among climates. Due to intensities, identical amounts of rainfall are delivered with much more kinetic energy in Humid-Subtropical (Caf) and Humid Continental (Daf) climates than in the Marine West Coast (Cbf). This translates directly into area differences of erosion for the same annual average precipitation, all other factors, including vegetation, held constant. For example, 75 mm of annual rainfall produces total kinetic energy of 4×10^9 J ha^{-1} in north-central Texas while 75 mm in the Puget South lowland delivers less than 5×10^8 J ha^{-1}. Since these values are factors in the Universal Soil Loss Equation (Wischmeier and Smith, 1978), the figures translate into 8 times as much annual erosion in Texas as in the Puget Sound area, all other variables including vegetation held constant. Many organism–physical relationships appear to be climate-specific and one must be careful about extrapolating a process outside the climate in which it is identified. Finally, figure 3.1 would be more realistic if the Forest and Grassland functions did not extend below about 300–400 mm precipitation.

Meso- or stream–basin scale

The meso scale permits climate to be held constant and thus allows closer examination of other variables, especially vegetation. Also, it is possible to recognize more interrelationships of vegetation and other variables, such as topography, in controlling erosion. Sediment yield is usually a surrogate for rate of erosion in basin studies so that the problem of storage fluxes always complicates matters (Meade, 1982; Trimble, 1975, 1977; Walling, 1983). The problem has led to the study of sediment budgets, an approach which separates erosion, sediment yield, and storage changes into their component parts (Swanson et al., 1982; Hadley, 1986), which may then be related to causative factors.

A review of all basin studies would be beyond the scope of this essay. Excellent reviews of the relation of vegetative cover and sediment yield at the basin scale are given by Leopold, (1956). An important and pioneering work was that of Brune (1948) in the central USA (Köppen: Daf). He found that not only was sediment yield increased by increasing the proportion of the basin in cultivated land, but that the rate of erosion increased exponentially with the percentage of cultivated land. A major reason for this exponential increase of erosion was topography: as grassland and forest was replaced by cropland, steeper slopes had to be cultivated and thus exposed to accelerated erosion. A similar exponential model was found to apply in subbasins of the Potomac River (Caf) (Sedimentation Task Force, 1967).

In the Coon Creek basin of Wisconsin (Daf), Trimble and Lund (1982) found that the Brune model was basically correct, but that the relationships were far more complex than an exponential curve (figure 3.2). They found a pronounced lag in the response of erosion and sedimentation to their independent variable, erosive land use (vegetative cover weighted with conservation practices). As agriculture spread in the basin (deteriorating phase), it took years for the soil to lose its structure, organic material, low bulk density, and high infiltration capacity. As fields were expanded, steeper slopes were cultivated so that increasing overland flow had longer flow routes and thus greater momentum, topsoils eroded, and there was a positive feedback so that erosion was accelerated. Conversely, as erosive land use decreased (improvement phase), it took time for soils to regain their good hydrologic condition and for upland waterways to heal so that, again, a lag was encountered. Thus, a given vegetative cover during the

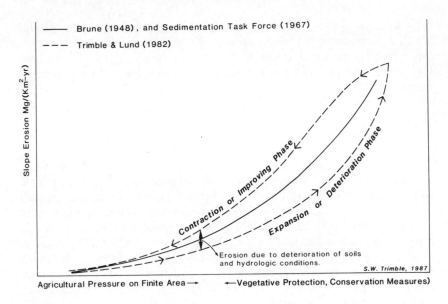

Figure 3.2 Relation of slope erosion to agricultural pressure on finite area (or vegetative protection, soil conservation measures)

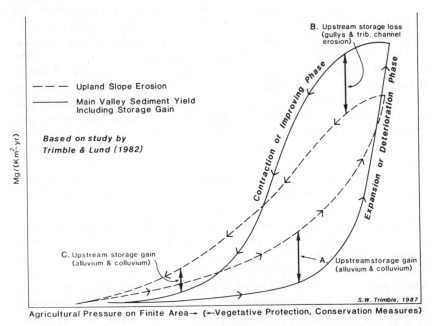

Figure 3.3 Differential relations of upland slope erosion and downstream sediment yield to agricultural pressure on finite areas (or vegetative protection, soil conservation measures). Severe scenario, Coon Creek, Wisconsin, 1853-1975

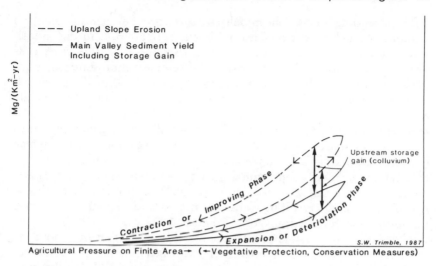

Figure 3.4 Differential relations of upland slope erosion and downstream sediment yield to agricultural pressure on finite areas (or vegetative protection, soil conservation measures). Mild scenario, hypothetical
Based on a study by Trimble and Lund (1982)

expansion phase generally caused less erosion than in the contracting phase. The overall result is a hysteresis loop.

Trimble and Lund also reconstructed valley sedimentation rates for the 125 years of agricultural occupance, finding that they were quite different and usually out of phase with upland erosion. This is most important because as pointed out earlier, sediment yield is usually used as a surrogate for erosion. Sediment yield must always be defined at a location within a basin and the location here is in the main valley. As agriculture expanded the rate of sediment yield to the main valley lagged behind the rate of upland erosion because sediment was being stored upstream as colluvium, and as alluvium along tributaries (figure 3.3A). As agricultural pressure increased, including grazing of woodland, overland flow from upland slopes increased, with the increased stormflow creating gullies and eroding tributary channels. Thus, many upland sediment sinks became sources. This process, which occurred rapidly once the threshold of resistance was exceeded, supplied sediment to the main valley at a rate higher than that of upland erosion (figure 3.3B). With decreased agricultural pressure and/or more protective vegetation as a result of implemented soil conservation measures, both erosion and sedimentation were decreased. The healing of gullies and other waterways reduced both the supply and the conveyance of sediment so that sediment yield became less than upland erosion, the difference again going into storage as colluvium and alluvium (figure 3.3C).

The foregoing model conceptualizes a scenario of extreme agricultural erosion. A much milder scenario is shown in figure 3.4. In this model, severe channel erosion never develops and sediment yield never exceeds slope erosion. Both the slope erosion and sedimentation functions form the hysteresis loops discussed earlier, but they may be quite disassociated. Such a model may characterize many agricultural basins of temperate zones, especially those of the eastern USA and western Europe, where sediment storage gain is the norm. This 'normal' storage gain is presented as a function of area by the sediment delivery curve (USDA, 1971) which indicates that larger areas have more storage gain, presumably because they have a greater proportional area of sediment sinks. Such stored sediment may eventually become intrinsically unstable and erode, or hydrologic changes resulting from land-use or land treatment may bring instability and further transport (Trimble, 1975, 1976, 1977).

Although it is possible to make some general statements about the relation of erosion to vegetational cover (expressed in several permutations) at the basin and regional scales, the number of variables make most predictions imprecise and unsatisfying. Because erosion is difficult to account for at that scale, sediment yield is usually its surrogate. The models presented are by no means universal, but are intended to show (1) that the relation of slope erosion to erosive landuse in a confined area is exponential and probably of a hysteretic nature and (2) erosion rates and sediment yield may be (a) greatly disparate and (b) out of phase so that one cannot be substituted for the other. Other models will be more suitable for other situations and will, it is hoped, be forthcoming.

Slope erosion and the large scale

It is at the scale of the slope or field that we get the most insight on process. I shall now discuss the role of plants and animals in controlling erosion under conditions of (a) forest, (b) cultivation and pasture, (c) orchards and vineyards, and (d) disturbed sites such as gullied land, mining, and construction. Before doing so, it will be useful to discuss prediction of erosion.

Prediction of erosion

Terrain in temperate areas is subject to rainfall erosion and often to wind erosion. To facilitate comprehension of the variables involved and particularly the vegetation-management components, predictive equations for both water and wind erosion are introduced here. Although these models are for management purposes and are stochastically, rather than deterministically, derived, they still have a strong element of physical basis. Important to this discussion is that they will allow a generalized quantitative evaluation of the vegetative components in subsequent sections.

Water erosion

The accepted prediction model for much of the world is the so-called Universal Soil Loss Equation (USLE) developed by the US Department of Agriculture (Wischmeier and Smith, 1978). It is:

A $= RKLSCP$ where:

A = estimated annual soil loss, $M_t\, ha^{-1}$
R = rainfall and runoff factor $j\, ha^{-1}$
K = soil-erodibility factor, $M_t\, j^{-1}$
LS = slope length and steepness factor, dimensionless
C = vegetative cover-management factor, dimensionless
P = erosion control management factor, dimensionless

It is the last two factors, C and P, which are of interest here and these will be used throughout the following sections. They have been derived from many years of experimental and empirical research and give a systematic, quantitative evaluation of various vegetative-management alternatives.

Wind erosion

Erosion by wind is often of considerable importance in temperate areas, especially in semi-arid regions and on cultivated fields. A long period of experimentation led to a predictive equation (Woodruff and Siddoway, 1965; Skidmore and Woodruff, 1968). It is:

E $= f\,(I', K', C', L', V)$ where:

E = predicted annual soil loss, $t\, ha^{-1}$
I' = soil-erodibility factor, $t\, ha^{-1}\, yr^{-1}$
K' = micro-meso topographic roughness factor, dimensionless
C' = climatic factor, dimensionless
L' = width of field factor, m
V = vegetative-cover factor (equivalent $Kg\, ha^{-1}$)

The last factor is of interest in this survey and will be discussed in conjunction with cultivated field, range and brush, and disturbed sites.

Forest

Forest is discussed first because it is generally the most erosion-free environment and thus becomes a standard of comparison. Although forests may not produce the most productive soils, they do tend to produce the best soils from the standpoint of hydrologic performance, especially infiltration capacity and percolation. Because the infiltration capacity of most undisturbed forest soils is greater than the intensity of any likely storms, overland flow does not occur except in limited areas. Thus, most flow from forest slopes travels by subsurface routes (Whipkey, 1965, 1969). These findings led to the revolutionary variable source area concept, or Hewlett Model of runoff (Ward, 1975). Because plant canopies and forest floor humus protect the soil from raindrop impact and with little or no overland flow, it is then not surprising that undisturbed forest has so little erosion. Following is a discussion of factors and processes which are part of, or affect forests.

Surface erosion factors

Canopy.   Conventional wisdom considers the canopy to be one of the mitigators of erosion in the forest, but the relationship is apparently more complex than first perceived. The conventional view is quantified in terms of partial C factors for forest applications of the USLE (Dissmeyer and Foster, 1980).

Table 3.1 clearly implies that erosion is a function of per cent of bare soil with canopy cover, and a canopy cover up to a height of 16 m has at least some mitigation. This effect presumably comes from shortening the fall of the drops of water, decreasing their velocity and thus reducing kinetic energy. These factors have been tested in the southeastern USA and along with other factors, have been found to give reasonably accurate predictions (Dissmeyer and Foster, 1980; Chang, *et al.*, 1982). Conflicting evidence comes from New Zealand where Mosley (1982) found more kinetic energy from raindrops under beech (*Fagus* L.) canopy than out in the open. Although the fall was shorter, the leaves allowed the drops to collect and become larger so that mass would be greater. Moreover, since terminal velocity of most raindrops is attained from falls of

Table 3.1   The canopy subfactor

Canopy height (metres)	Percentage of bare soil with canopy cover										
	0	10	20	30	40	50	60	70	80	90	100
0.5	1.00	0.91	0.83	0.74	0.66	0.58	0.49	0.41	0.32	0.24	0.16
1.0	1.00	0.93	0.86	0.79	0.72	0.65	0.58	0.51	0.44	0.37	0.30
2.0	1.00	0.95	0.90	0.85	0.80	0.75	0.70	0.65	0.60	0.55	0.50
4.0	1.00	0.97	0.95	0.92	0.90	0.87	0.84	0.82	0.79	0.76	0.74
6.0	1.00	0.98	0.97	0.96	0.94	0.93	0.92	0.90	0.89	0.87	0.85
8.0	1.00	0.99	0.98	0.97	0.96	0.95	0.95	0.94	0.93	0.93	0.92
16.0	1.00	1.00	0.99	0.99	0.98	0.98	0.98	0.97	0.97	0.96	0.96
20.0	1.00	1.00	1.00	1.00	1.00	1.00	1.00	1.00	1.00	1.00	1.00

Values in this table are represented as proportions of the amount of erosion found on bare or fallow ground with no protection.
Source: Dissmeyer and Foster, 1980

only 8–9 m, such was attained from the beech canopy. Thus the total kinetic energy was greater. The bare soils under the beech cover thus suffered more erosion than in the open.

Such conflicting evidence may possibly be reconciled by recognizing the difference of climate involved. For example, the 1 year–one hour precipitation for the southeastern USA (Caf climate) ranges from 30 to 60 mm while it is only about 8 mm for the coastal lowlands of the northwestern USA (Hershfield, 1961), a Cbf climate, and the nearest analogue to New Zealand and western Europe found in the contiguous USA. Since larger raindrops are associated with more intense precipitation (Laws and Parsons, 1943; Coleman, 1953), it follows that unaltered raindrops are larger in Caf regions than in Cbf. Hence, the enlargement of drops by vegetative canopy may be important only in marine climates. It is also in such climates that fog is collected and transformed into falling droplets, especially by needleleaf trees, in some places actually augmenting significantly the local meterological precipitation. Such magnification of raindrops and even rainfall would have little effect, however if the forest floor is intact.

Forest floor humus and roots.   Far more important than canopy in reducing erosion from forest is humus. This organic material plays two extremely important roles: first it absorbs the impact of raindrops and secondly, it has an extremely high permeability which allows water to move into the soil at the continuously high rates already mentioned. A corollary of the first factor which leads to the second is that soil particles are not detached, thus soil pores are not sealed, allowing infiltration to remain at high levels. The net result of all these factors is that some forest soils have infiltration capacities as high as 500 cm hr^{-1} which is about two orders of magnitude greater than pasture and cropland (Sidle et al., 1985). Contrary to popular belief, the organic layers do not act as a 'sponge', but at most can absorb only a few mm depth of water.

Mader et al. (1977) described the organic material under forests of the northeastern USA to be in three layers. On top was litter (undecomposed), below that was the fermentation layer (partially decomposed) and on the bottom was the humification layer (well decomposed). Below that was the A1 soil horizon which was mineral, but had 2–15 per cent organic matter by weight. The thickness and mass of this organic material is controlled primarily by age and density of stand, and can be truly amazing as figures 3.5 and 3.6 indicate.

An evaluation of the erosion-prevention quality of the organic layer was approximated by Wischmeier (1975). The relationship is shown in figure 3.7.

Not all forest produces such a fine protective organic layer. Beech trees tend to discourage understorey growth which can allow the development of a poorly-covered forest floor, as in much of the Ardenne Forest of

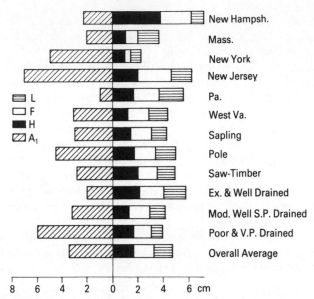

Figure 3.5 Humus layer thickness in hardwood stands in the northeastern USA by region, stand, class, soil drainage class, and overall average (from Mader *et al.*, 1977) L = litter, F = fermentation layer, H = humification layer, and $A_1 = A_1$ soil horizon

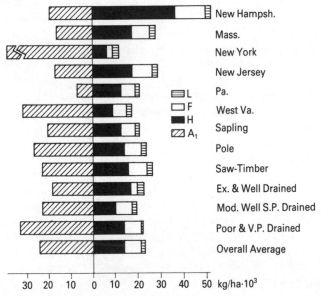

Figure 3.6 Oven-dry organic content of humus layers in hardwood stands in the northeastern US by region, stand, class, soil drainage class, and overall average (from Mader *et al.*, 1977) L = litter, F = fermentation layer, H = humification layer, $A_1 = A_1$ soil horizon

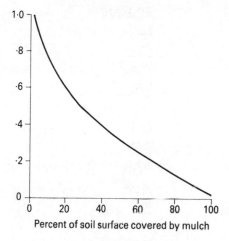

Figure 3.7 Effect of mulch, plant residues, or humus layer in preventing erosion (from Wischmeier, 1975)

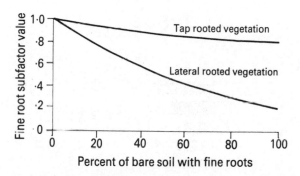

Figure 3.8 Effect of fine roots in the top 2.5–5 cm of soil in reducing erosion (from Dissmeyer and Foster, 1980)

Luxemburg and Belgium. There, partially bare soil and high canopies permit rainsplash erosion on forested slopes (Kwaad, 1977).

Some needle-leaf trees such as most pines produce especially good organic layers. Often, the needles interweave into an almost fabric-like layer, a quality which has been especially valuable in reclaiming highly-eroded abandoned farmland. (Trimble, 1974). McClurkin (1967) found that 15 years of growing shortleaf pine (*Pinus echinata*) increased the infiltration capacity from 38 to 141 mm hr^{-1}. Ursic and Dendy (1965) studied several small basins in Mississippi and found that while actively cultivated areas produced 1 to 5×10^3 t km^{-2} yr^{-1} of sediment, formerly cultivated areas with mature pine forest produced only 10 t km^{-2} yr^{-1}. Working in Texas from erosion plots, Chang *et al.* (1982) found that 40-year-old loblolly (*Pinus taeda*) and shortleaf pines permitted, over a

Table 3.2 Effect of bare soil and fine root mat of trees on untilled soils

Percentage bare soil	Percentage of bare soil with dense mat of fine roots in top 3 centimetres of soil										
	100	90	80	70	60	50	40	30	20	10	0
0	0.0000										
1	0.0004	0.0004	0.0005	0.0006	0.0007	0.0008	0.0010	0.0012	0.0014	0.0016	0.0018
2	0.0008	0.0008	0.0010	0.0012	0.0014	0.0017	0.0020	0.0023	0.0027	0.0031	0.0036
5	0.003	0.003	0.003	0.004	0.005	0.006	0.007	0.008	0.009	0.011	0.012
10	0.005	0.005	0.006	0.008	0.009	0.011	0.013	0.015	0.017	0.020	0.023
20	0.011	0.012	0.014	0.017	0.020	0.024	0.028	0.033	0.038	0.044	0.050
30	0.017	0.018	0.020	0.025	0.029	0.036	0.042	0.050	0.059	0.068	0.077
40	0.023	0.024	0.027	0.034	0.042	0.049	0.058	0.068	0.079	0.092	0.104
50	0.030	0.032	0.038	0.045	0.054	0.064	0.074	0.088	0.103	0.118	0.135
60	0.037	0.038	0.043	0.055	0.067	0.079	0.092	0.109	0.127	0.147	0.167
70	0.047	0.049	0.054	0.068	0.083	0.098	0.117	0.138	0.161	0.187	0.212
80	0.055	0.058	0.066	0.081	0.098	0.118	0.141	0.164	0.192	0.221	0.252
85	0.066	0.069	0.078	0.095	0.115	0.138	0.165	0.195	0.228	0.264	0.300
90	0.075	0.080	0.089	0.111	0.133	0.157	0.187	0.222	0.260	0.301	0.342
95	0.086	0.090	0.102	0.125	0.155	0.182	0.217	0.255	0.298	0.345	0.392
100	0.099	0.104	0.117	0.144	0.180	0.207	0.248	0.293	0.342	0.396	0.450

Values are represented as proportions of the erosion occurring on bare, fallow soil
Source: Dissmeyer and Foster, 1980

nine month period, 1.1 t km^{-2} while a similar clear-cultivated area lost 342 t km^{-2}. Much of the native hardwood areas of the USA and Europe have been replanted in needle-leaf trees, the long term result of which has been to podzolize the soil (Butzer, 1974). In southern Illinois, Rolfe and Boggess (1973) found that soils under pine plantations, as compared to local hardwood forests, had less organic material, lower pH and less exchangeable bases, but they did have lower bulk density.

Roots. Two types of roots affect erosion in the forest. The first is the fine lateral-type roots which grow in the upper part of the mineral soil. The second are large roots, discussed under the macropores section. The former, fine roots, can come from both the forest trees and from the understorey. Differentiation is made on whether plants are tap-rooted or laterally-rooted. Figure 3.8 shows the effect of both on mitigating erosion from a bare soil.

Because the forest floor is generally composed of both organic layers and fine roots, both relate to the erodibility. The combination of the two variables is given in table 3.2.

Macropores. One reason that forest soils have such a high hydraulic conductivity is the presence of macropores through which non-tension water can move, and most of these macropores have organic origins. According to Sidle *et al.* (1985, p. 43), 'temperate forests are particularly endowed with macropores because of their organic horizons, extensive rooting systems, and biotic activity'. Of greatest interest here are (a) root routes, (b) routes formed by soil fauna, and (c) structural routes (Beven and Germann, 1982; Sidle *et al.*, 1985; Aubertin, 1971).

Larger roots form efficient water routes, especially as they die and decompose (Sidle *et al.*, 1985). Tanaka (1982) found that such roots were also important routes for exfiltration of subsurface stormflow at the base of forested slopes. The erosion from this and outflow created a 'stepped' profile.

Most research done on the geomorphic impact of animals has concentrated on the more direct effects such as moving earth and these will be discussed later. Burrowing animals can create important subsurface routes for stormflow (Sidle, *et al.*, 1985; Beven and Germann, 1982). Pocket gophers appear to be active in many environments, even affecting irrigated land (Miller, 1957) and Thorn (1978) has shown that their tunnels are important for conducting stormflow, at least in alpine environments. Aside from their effects on soil morphology (discussed later), earthworms are also superb tunnel builders. Ehlers (1975) showed that earthworms created pipes 2 to 11 cm in diameter. Most of these were vertical and conducted tension-free water from the surface to as deep as 1.8 m, adding about 60 mm hr^{-1} of infiltration capacity to the soil. Because they operated only at very high rainfall intensities when tension-free water was available, they should be considered to be an auxillary

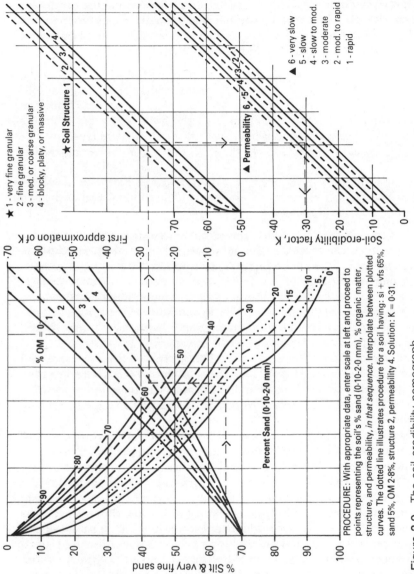

Figure 3.9 The soil erodibility nomograph
(from Wischmeier and Smith, 1978)

PROCEDURE: With appropriate data, enter scale at left and proceed to points representing the soil's % sand (0·10-2·0 mm), % organic matter, structure, and permeability, *in that sequence*. Interpolate between plotted curves. The dotted line illustrates procedure for a soil having: si + vfs 65%, sand 5%, OM 2·8%, structure 2, permeability 4. Solution: K = 0.31.

★ Soil Structure 1

1 - very fine granular
2 - fine granular
3 - med. or coarse granular
4 - blocky, platy, or massive

▲ Permeability 6

6 - very slow
5 - slow
4 - slow to mod.
3 - moderate
2 - mod. to rapid
1 - rapid

system, perhaps like an 'overflow' spillway. Well-structured soils with large aggregates are characteristic of many forest environments and such soils may have an extensive network of macropores along the faces of the peds (Sidle *et al.*, 1985). Although macropores are important in the subsurface routing of water, thus reducing or eliminating erosion from overland flow, the flow through such macropores is turbulent so that subsurface erosion can be considerable (Sidle *et al.*, 1985).

Soils.   Forest soils tend to be better aggregated than those on cultivated land (Gerrard, 1981). Moreover, the forest aggregates are much more resistant to both wetting and water drop impact (Imeson and Jungerius, 1976). The superior aggregation of forest soils is due to the presence of considerable organic material which is an important cementing agent in the formation of large water-stable aggregates (Dyrness, 1967). One study found that about 50 per cent of mean aggregate size was explained by percentage organic material (Woolridge, 1965). Quantitatively, the effect of organic material on soil erodibility is shown in figure 3.9 by the soil Erodibility Nomograph (Wischmeier and Smith, 1978). It predicts the value of the soil erodibility factor (K) used in the USLE. The nomograph considers texture, organic material (OM), structure and permeability. Figure 3.9 is marked (dotted line) to predict a K factor of 0.31 with a soil having an OM content of 2.8 per cent. Reducing the OM to 2.0 per cent raises the K factor to 0.35 or an increase of about 13 per cent. Raising the OM to 4.0 per cent decreases the k factor to 0.25, or a decrease of 19 per cent. Predictions from this nomograph, as compared to measured values, have a standard error of 0.02 so that 95 per cent of all predictions of K should be within 0.04 (Wischmeier and Smith, 1978). The nomograph was developed for agricultural and construction sites and thus does not consider OM greater than 4 per cent. For forest soils having considerably greater than 4 per cent OM, Dissmeyer and Foster (1980) suggest that the K factor be reduced by 30 per cent. However, Voroney *et al.* (1981, cited in Morgan, 1985) indicate that soil erodibility decreases linearly with percentage organic material up to 10 per cent. Morgan (1985), however, points out that very high organic material can actually increase erosion. All of this indicates that more research is needed on forest soils to consider more precisely the greater proportion of organic material.

Just as important as the presence of organic material is its transformation into organic gels by several biological processes. One of the most important of these processes is earthworm activity. Kladivko *et al.* (1986) found that earthworms increased both the number and mean weight diameter, of water-stable aggregates larger than 2 mm in diameter. The forest soil is an excellent habitat for earthworms because they prefer rich, moist, well-aerated soils with 4–5 per cent organic content (Beatty and Stone, 1986). Nielsen and Hole (1964) measured about 2.2 t of earthworms ha^{-1} in a Wisconsin forest. They are important for vertical mixing of organic material and mineral soil.

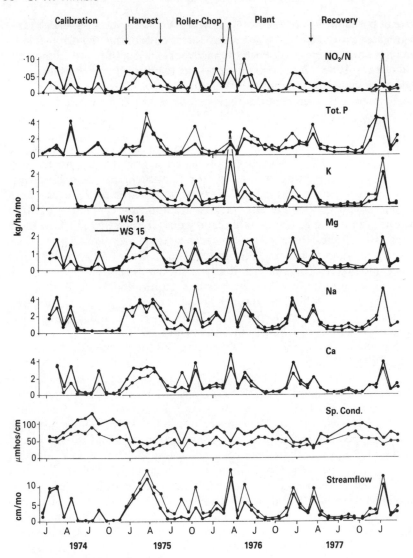

Figure 3.10 Export of minerals and water from Watershed 14 (clearcut) and Watershed 15 (control) during calibration and treatment
(from Hewlett, 1979)

It might appear that freezing of the forest soil would hinder infiltration and thus allow overland flow, but overland flow rarely occurs over forest soils even when the soil is frozen (Sartz, 1973). Forest soils do freeze although, because of their insulating cover, they are slower to do so than open or cultivated soils. Also, forest soils tend to freeze into a permeable lattice, honeycomb or granular pattern whereas open and cultivated soils tend to freeze into an impermeable (concrete) frost. Kane and Stein (1983) found that infiltration rates on frozen forest soils may exceed 20 cm d^{-1}. Because of their many voids and high permeability, forest soils are rarely saturated while compacted cultivated soils are much more likely to be saturated and thus incur concrete frost.

Subsurface and solution erosion. The subsurface routing of water through forest soils is not necessarily benign. As mentioned earlier, flow through macropores is turbulent and limited erosion can occur from the generally water-stable soil aggregates. The other component of erosion is solution, and the storage and rapid flux of water in and through forests soils make soil chemicals readily accessible for solution. However, the solute load from arable land in Luxemburg was still 50 per cent greater than from forest (Imeson and Vis, 1984). In the same area, Verstraten (1977) measured the chemical erosion rate from a small forested basin to be 13.1 t km^{-2} yr^{-1}, a value which was comparable to the sediment yield. Solution erosion was also found to be similar to particulate erosion in the Northern Rocky Mountains of the USA (Clayton and Megahan, 1986). In Somerset, England, Burt (1979) identified two distinct solutional environments related to topography. On the ridges (spurs), leaching appeared to be related entirely to infiltration while in the hollows, saturated throughflow was also important. Burt (1979) suggested that these differential processes might account for differences of slope form and development.

Solutional export is accelerated when the forest is clearcut because the trees are no longer drawing soil water so that nutrients can move out with throughflow and deep seepage. However, clearcutting has got an undeservedly bad reputation in this respect because of misreading the work of Likens *et al.* (1970). While they did report nitrate concentrations in streamflow 5 times greater than the potability standard of 10 mg l^{-1}, it has to be pointed out that not only was the New Hampshire (Dbf) forest clearcut, but the area was treated with a heavy application of herbicides for two years and this is not representative of any normal silvicultural operation.

Clearcutting of pine on the Georgia Piedmont increased concentrations of P, K, Ca, Mg, Na and especially nitrates in the form of N, which were almost doubled (Hewlett and Doss, 1984). These increases were relatively mild, however, and the effect was short term, lasting about 3 years through the periods of cutting, site preparation, and planting (Hewlett, 1979, and figure 3.10). In a wide area across Canada, however, pulpwood harvesting made no difference in solutional loss (Singh *et al.*, 1974).

Mass movement. Deep-rooted vegetation such as forest helps stabilize steep slopes by increasing the total shear strength of the soils. Ziemer (1981), for example, found that dry root biomass explained 79 per cent of the variation of soil strength on coastal sands in northern California. From 18 samples, he evolved a prediction equation:

Soil strength = $3.13 + 3.31$ biomass

Sidle et al. (1985) have written an excellent review of the effects of roots on soil strength and mass movement to which the reader is referred for a more complete discussion. Their synopsis of 11 investigations using various techniques showed that plant roots add from 1.0 to 17.5 kPa to the shear strength of the soil. Small roots appeared to be somewhat more effective than large roots and hardwood roots are generally stronger than conifer roots. Of course, the effect of roots can be no greater than the strength and stability of the lowermost earth material into which they are anchored.

When the vegetation is cut or burned, the dead roots decline in strength, sometimes exponentially with time. Meanwhile live roots increase from colonizing vegetation and some species of cut vegetation may resprout. The net result is a minimum of root strength about 3–10 years after forest removal (Sidle et al., 1985). This may pose no problem on the mild or otherwise stable slopes, but the presence of roots allows some slopes to be oversteepened beyond the capacity of the soil's inherent shear strength. In such cases there may be slope failure before new and competent vegetation can be established. An excellent example of this took place on the oversteepened slopes of the San Gabriel Mountains of southern California (Csa). There, sclerophyllous brush (Chaparral) was cut, treated with herbicide and replaced with grass in order to see if water yield would be increased. A heavy storm occurred 6 years later and there was so much slope failure that the narrow valley was buried to depths of several metres (Corbett and Rice, 1966). As this case shows and as Sidle et al. (1985) point out, the timing of slope failure may not coincide with minimum root strength because of required storm thresholds. In fact, marginally strong slopes may fail with extreme events when the vegetation has not been disturbed. Thus, Hupp (1983) attributes Quaternary mass wasting in Virginia to exceptional storms. A recent and extraordinary example was the rain from hurricane Camille in 1969, when about 60 cm of rain fell in 24 hr. Forested slopes in Virginia which had been stable for centuries failed in long slip faces (Williams and Guy, 1973). Such rare geomorphic events are apparently triggered by the additional mass, increased pore pressure, weakened soil, and perhaps lubrication of roots and slippage planes.

Finally, tree throw may be a significant form of mass movement in some forests. Disturbed areas in forests from tree falls have been reported to range from 14 to 48 per cent of the forest floor (Stephens, 1956). Lutz (1940, 1960) has demonstrated the movement of soil and rocks from uprooting and suggests that gully erosion may be instigated in some cases.

Thomas W. Small (personal communication, 1987) found that bare tree throw mounds in Pennsylvania (Daf) eroded at rates up to 120 mm y^{-1} while moss-covered mounds were effectively armoured. Types of tree throw and their subsequent morphology in New York State were described by Beatty and Stone (1986). Although there may be some net export of material from tree throw, the *in situ* landforms are often long-lived and sometimes persist for centuries.

Extrinsic disturbances and changes

The resources of the forest are set upon by man and beast, the biomass of the forest sometimes makes excellent fuel for natural or man-set fires, and humans try to undo the damage by reforestation. These points are covered in the following sections.

Meso–macro animal disturbances.   Perhaps the most obvious (but not necessarily the most significant) disturbance to the forest is that of the grazing and browsing of large domestic animals. Until recent decades, farmers allowed cattle to roam their forests, and much of the southeastern USA had open range laws for swine. Although the swine did well on the mast from the hardwood trees of the upper South, cattle did not fare so well on the limited grasses and succulent leaves within their reach. As one old farmer in Wisconsin explained it to me, getting enough to eat in the hardwood forest of that area would require a cow having 'a mouth three feet wide and moving at thirty miles an hour'. The result is that such animals are constantly on the move, with their great hoof loadings compacting the soil. On steep slopes, they can also disturb the mineral soil and displace it downslope. Nevertheless, there appears to be little evidence that such grazing causes much erosion in forests. In the upper midwestern USA Sartz (1975) contends that grazing has had little erosive effect on wooded hillsides of the Driftless Area. Although large relict gullies may be found there, they were created by runoff from upslope cultivated fields, according to Sartz. Apparently, grazing merely lowered the threshold of gullying by this upslope water. In the western USA, Packer (1953) found that even with severe trampling, high ground cover (90 per cent and over) kept overland flow and erosion to low levels.

Time is also an important factor. Given enough time, grazing and browsing animals can completely alter the nature of the forest, gradually changing it to open woodland (park) and then to open land. Less than a century of grazing in the Driftless Area changed multi-storeyed forests to open woodland. Under those conditions, the decreases of organic litter and root systems along with compaction would significantly change the character of forest soils possibly leading to overland flow. This, together with surface disturbance from trampling and rain splash from decreased canopy cover would lead to erosion. I have seen open woodland pastures in the Driftless Area which I believe to fit the above scenario.

An intriguing question is the role of large, wild animals. Early travel accounts in the eastern USA forests speak frequently of well-worn buffalo trails and stream banks being eroded at crossings, but it seems unlikely that present animals, such as deer, could do much damage. However, rooting by wild pigs in the Ardennes Forest produces local but spectacular results. In one case more than 1 t of earth was displaced over a small area in just a few days (Imeson, 1976).

Smaller animals are perhaps more efficient direct geomorphic agents than large ones. The more benign role of such animals in creating soil macropores and maintaining the highly structured and aerated forest soils has already been discussed. The animals, however, may also play destructive roles. In what is apparently the most studied locale so far, the oak–beech forest of Ardennes region in Luxemburg and Belgium, about 7 t ha^{-1} of rabbit and badger mounds were present and about 0.6 t ha^{-1} was added in a 0.5 yr period. However, downslope mass transport may be quite low at 0.02 g cm^{-1} yr^{-1} (Voslamber and Veen, 1985). Moles and voles are also active and their mounds are being lowered at a rate of about 2 mm yr^{-1} by splash erosion. This is important in creating colluvium, although the importance to stream sediment is less certain (Imeson, 1976, 1977; Imeson and Kwaad, 1976). Imeson and Kwaad (1976, p. 371) state that such areas exposed by burrowing animals 'are likely to form relatively important sediment source areas'. But since relatively undisturbed forest floor does not permit overland flow, the transport mechanism from the animal disturbance to the overland flow source areas is not apparent. Lull and Reinhart (1965) point out that sediment from logging roads is often redeposited when the flow infiltrates into the forest floor. Thus, as stated early in this essay, erosion is not the same as sediment yield. This same conveyance problem in the forest can also apply to other local disturbances such as tree throw, mass movement and logging (Burns and Hewlett, 1983).

Close by in the Keuper region of Luxemburg, again under oak–beech forest, earthworms were busy exposing the forest floor to splash erosion (Hazelhoff et al., 1981). In low damp areas, Lumbricus terrestris L. removed organic material from the surface to lower strata, thus exposing an average of 25 per cent of the forest floor to splash erosion during July and August. In better-drained areas, Allolobophora nocturna, Sav. and A. longa, Ude were bringing casts to the surface at a rate of about 15 t ha^{-1} yr^{-1}. This process bared about 20 per cent of the ground at maximum extent. However, the splash erosion from L. terrestris was much greater than from A. nocturna by a factor much greater than the relative disturbed areas would suggest. The transport rate of this material was uncertain in both cases. Since there are about 1,800 species of earthworms on earth, 220 in Europe and 165 in North America (Nielsen and Hole, 1964), and given the disparity between process and rates of the two species above, it would appear that biogeomorphologists have their work cut out for them on a global scale.

It is to be observed that the above processes largely took place under oak–beech cover where understorey was presumably sparse, as it usually is. Thus there was apparently synergism among the factors of (a) high canopy effect in a marine climate (b) exposed mineral soil from worm and mammal action and (c) a lack of protection by the low canopies and lateral root systems of understorey vegetation. To what degree would these processes be replicated in, say, the hardwood forests of the eastern USA? Are beech forests, in fact, unusual among hardwood for the lack of understorey? If so, how does this affect erosion?

Other geomorphic activity by burrowing animals has been reported in the Rocky Mountains of the USA and Canada (e.g. Thorn, 1978, 1982; Smith and Gardner, 1985), beyond the geographic range of this essay. Certainly, more is to be done in temperate zones. In the eastern hardwood forests of the USA, for example, holes and mounds of the woodchuck (groundhog) are common. Their distribution and impact could be a topic of future research.

Fire.   The body of literature seems to suggest that fire has the greatest erosional effect in exceptionally dry forest, or those areas which undergo a pronounced dry season, so that the humus layer can be burned, perhaps affecting, and certainly exposing, the mineral soil. The most extreme case of the latter occurs in Chaparral vegetation of the Csa climate. There, a hydrophobic soil often develops with ensuing disastrous erosion, a syndrome which will be discussed in the Sclerophyll forest section below. Hydrophobicity may also occur in pure forest stands. In a review of the literature, DeBano (1969) describes hydrophobic soils found throughout much of the western USA. Associated vegetation includes various pines, fire, spruce, larch, junipers, brush, eucalyptus, and oak.

The effects of fire on erosion in humid area forests, such as in the eastern USA, are also problematic. Even when dry, these forest floors often have enough damp humus to protect the mineral soil. In the hardwood forest of the north-central USA, Sartz (1975) reports that fire has little if any effect on infiltration and presumably on erosion. In the Missouri Ozarks, Arend (1941) found that continual burning in this humid climate (Caf) lowered the infiltration capacity, perhaps as a result of destroying organic material and disturbing microbiological activity. Presumably this could have caused erosion but that was not mentioned. Trimble (1974) attempted to reconstruct primeval erosion rates on the Piedmont of the southeastern USA (Caf). Regarding fire, he asked a panel of four soil scientists with a total field experience of 140 person-years to evaluate the effect of fire on erosion in the Piedmont. Three of the four believed that they have never seen erosion caused by fire while one believed that he had. However, it is now unclear whether only relatively undisturbed hardwoods were considered or if this appraisal included formerly cultivated fields, already severely eroded and reforested with pines. Ralston and Hatchell (1971,

Table 3.3 Soil losses from burned and protected woodlands in the southeastern United States

Location	Forest cover	Years of record	Annual pptn (cm)	Soil loss (t ha⁻¹ yr⁻¹)	Erosion (cm 1,000 yr⁻¹)
Holly Springs, Miss.	Scrub oak, burned	2	162	0.75	5.00
	Oak forest, protected	2	170	0.06	0.40
Guthrie, Okla	Woodland burned annually	10	78	0.25	1.70
	Virgin woodland	10	78	0.02	0.15
Statesville, NC	Hardwood, burned semi-annually	9	119	7.00	47.0
Tyler, Tex.	Hardwood, protected	9	119	0.01	0.03
	Woodland, burned annually	9	104	0.82	5.50
	Woodland, protected	9	104	0.11	0.80
East Texas	Shortleaf-loblolly, single burn	1.5	—	0.48	3.30
North Mississippi	Shortleaf-loblolly, protected	1.5	—	0.23	1.50
	Scrub oak, burned and deadened	1st	165	1.16	7.80
		2nd	103	0.045	3.10
		3rd	128	0.11	0.080
	Scrub oak, protected	1st	165	0.48	3.10
		2nd	103	0.20	1.40
		3rd	128	0.06	0.45

Source: Ralston & Hatchell, 1971, cited in Wells et al., 1978

cited in Wells *et al.* 1979) surveyed experimental work in the southeastern USA (Caf), showing that burning did affect erosion rates (Table 3.3). However, most of the soil losses after burning were still moderate.

There does not seem to be enough evidence to draw firm conclusions, and research on effects of burning seems warranted in humid areas, especially for prescribed burning, and burning of slash.

Burning also releases some nutrients for immediate use, but others are volatilized or leached away. pH often rises after a fire (Goudie, 1986).

Fire in dryer, or seasonally dryer forests seems to have more deleterious effects than in humid areas. Two good reasons for this are that such forests tend to be on steep slopes of moutainous terrain and have poor soil development. A survey of fire effects in the western USA showed a range from negligible to complete destruction of a forest with severe erosion (Wells *et al.*, 1979). In general, severe effects are spotty and are often caused by local accumulations of dry logging slash (vegetative debris such as limbs and branches). In Idaho, for example, Connaughton (1935) found that of 1,335 ha of burned virgin forest, about 28 per cent had accelerated erosion. In the same study, 1,850 ha of cutover land had 42 per cent eroded, largely because of slash. Similarly, slash burning in western Oregon caused erosion and permitted dry ravel on steep, south-facing slopes (Mersereau and Dyrness, 1972). In the Douglas fir (*Pseudotsuga menziesii*) region of the northwestern USA, Sartz (1953) measured accelerated soil erosion depths of 2–5 cm. Slope was second only to fire as the controlling variable. Laboratory tests on soils from the California coast ranges indicated that ash leachate (from Douglas fir slash) promoted aggregate dispersion and hence erosion. However, heating to 350°C reduced dispersion (Durgin, 1985). Prescribed burning in the White Mountains of Arizona significantly decreased humus and allowed some accelerated erosion (Cooper, 1961). However, in the ponderosa pine (*Pinus ponderosa*) forest of northern California, prescribed burning had little effect (Cooper, 1961). Erosion from fires in the western USA can sometimes be disastrous. Burned forest slopes upstream from Boise, Idaho were hit by several intense storms which severely eroded the slopes and dumped great amounts of sediment into Boise (Copeland, 1963). A similar scenario occurred near Verdi, Nevada.

Forest harvesting and replanting.   If trees could just be cut and removed without disturbing the forest floor, there would be little or no resulting erosion in most forests of mild slopes (Dyrness, 1967). As we have seen, it is the soil humus layer and understorey which control most erosion from forest land. In some areas, trees are cut and carried out by balloon or helicopter thus fulfilling the foregoing conditions. However, most cutting is still done with logging roads, heavy trucks, and tractors ('skidders') which drag logs to central loading areas. In the rough terrain of the eastern USA, second- and third-growth forest in good condition rarely permit sediment concentrations to exceed drinking-water standards ($11 \, \text{mg} \, \text{l}^{-1}$). This extremely low value appears to indicate that the natural factors of tree throw, burrowing animals, worms, etc. do not, in total, play an important role in producing sediment yields. However, exploitative logging increased sediment concentrations up to 1,000 times, though most of this is attributable to logging roads and skidding operations (Lull and Reinhart, 1965). In the southeastern USA, 80–90 per cent of sediment yield is caused by the construction and use of logging roads and by skidding operations (Hewlett, *et al.*, 1979; Bosche and Hewlett, 1980). In western Oregon, increasing the percentage area cut annually from 0.6 to 1.5 increased sediment yield by only 18 per cent. However, if the area occupied by roads increased from 0.1 per cent to 0.5 per cent, the sediment yield increased by 260 per cent (Anderson, 1954). A detailed discussion of logging roads is beyond the scope of this paper but two examples will demonstrate the seriousness of the problem. In the Northern Rocky Mountains of Idaho, erosion from bare road fills averaged a rate of $1,240 \, \text{t} \, \text{km}^{-2} \, \text{yr}^{-1}$ (Megahan, 1978). In the Douglas fir forest of the Cascade Mountains in Oregon, erosion from a patch-cut area with logging roads was 100 times that from an untouched control area. The erosion was primarily mass movement (Fredriksen, 1970). Cutting trees without building roads, however, increased erosion by only three times. In other places, clearcut logging on steep slopes will allow severe mass movement after the roots deteriorate, as discussed in an earlier section.

In many areas, clearcutting and replanting are done at the same time. The forest floor is normally prepared for planting by mechanical disturbance on the order of cultivation. This process alone exposes the soil and causes erosion, so that it is difficult to separate the effects of harvesting, preparation and replanting. In Arkansas (Caf), about half the soil is exposed by these operations and erosion increases significantly. However, the cover reestablishes itself within two or three years, and erosion decreases accordingly (Beasley and Granillo, 1985).

In a detailed experiment on pine regrowth in the southern Piedmont of Georgia (Caf), it was found that clearcutting, site preparation and machine replanting significantly increased streamflow peaks (Hewlett and Doss, 1984) and sediment yield (Hewlett, 1979). The mechanisms were: (a) higher antecedent soil moisture increasing the source area of saturated overland flow; and (b), as a result of (a) and surface disturbance, rills and gullies were reactivated, reducing concentration time. The effect of clearcutting on streamflow peaks is shown by the equation (Hewlett and Doss, 1984):

$$Qpt = 0.33 + 1.265 \, Qpc$$

where:

Qpt = peak discharge of clearcut basin in $\text{m}^3 \, \text{s}^{-1}$
Qpc = peak discharge of clearcut basin in $\text{m}^3 \, \text{s}^{-1}$

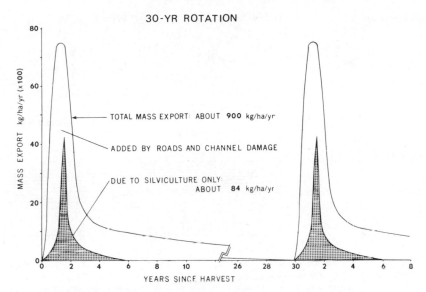

Figure 3.11   Sediment from the first order basin (32.5 ha) caused by silvicultural practices only; and by the entire operation including road and channel damage. The normal or background sediment yield of 92 kg ha^{-1} yr^{-1} is not shown (from Hewlett, 1979)

The erosive energy of streamflow from the basin was increased 55 per cent over a four-year period so that perennial stream channel erosion was possibly increased.

Such clearcutting is usually done on a basis of perhaps 30 years or so, with erosion and sediment yield declining between each harvest. For this study area, the cycle is graphically shown on a 30-year basis (figure 3.11). Note that the peaks of total mass export (sediment yield) of about 7,500 kg ha^{-1} yr^{-1} (750 t km^{-2} yr^{-1}) are almost 100 times the background level of 9.2 t km^{-2} yr^{-1}.

Sediment movement into stream may be reduced by leaving a strip buffer of forest along both sides of streams (Hewlett *et al.*, 1979). A similar practice is recommended in the USSR by Nikolayenko, 1974). Such a buffer allows sediment-laden surface water to infiltrate and deposit the sediment.

Erosion prediction procedures during harvesting and replanting using the USLE are given by Dissmeyer and Foster (1980). Practices to reduce erosion while harvesting and replanting are given by Hewlett (1979), and a model to incorporate the effect of these practices was developed by Burns and Hewlett (1983). Burns and Hewlett take strong exception to the USLE applied to forest harvesting. They point out the scattered and discontinuous nature of those vegetative and soil disturbances, and show that non-distributed or 'lumped' parameters such as the C and P factors may not

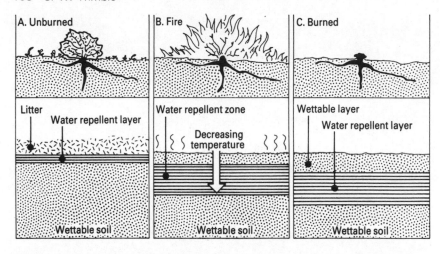

Figure 3.12 Chaparral-induced hydrophobic soil (A) before fire, hydrophobic substances accumulate in the litter layer and mineral soil immediately below it; (B) fire burns vegetation and litter layer, causing hydrophobic substances to move downward along temperature gradients; (C) after fire, water repellent layer is present below and parallel to soil surface on burned area
(from DeBano *et al.*, 1979)

be applicable. They devise a new variable which considers the discontinuous nature of such disturbances and allows prediction of sediment yield to the adjacent stream.

While afforestation is generally considered to be benign, we have just seen that it is not necessarily so in the short run. This is even more the case with recent afforestation in the uplands of the UK where boggy or peaty areas are often selected. For proper tree growth, such areas must first be drained and this is done by open ditches. These erode rapidly for the first few years increasing sediment yields in the order of 1,000 times or so, and some areas continue to erode at several times the natural rate (Battarbee *et al.*, 1985; Burt *et al.*, 1983; Painter *et al.*, 1974; Robinson and Blyth, 1982). Presumably, such areas will eventually stabilize.

Sclerophyll forest and scrub (chaparral).   This climatic–vegetative region (Csb) is discussed separately because (a) the severe dry season, (b) the highly flammable nature of the vegetation with frequent fires, (c) hydrophobic soils, and (d) steep slopes make this a highly erosive area (Howard, 1982). The vegetation here plays a key role. First, it is fairly dense, has a deep root system, and produces adequate litter, all of which are effective in holding the regolith. (Brock and DeBano, 1982). In fact, it may be argued that tectonically active mountain slopes in southern California are oversteepened in part because the deep-rooted chaparral is so effective. Disastrous mass movements partly as a result of killing the chaparral cover,

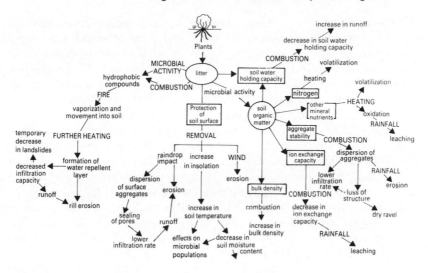

Figure 3.13   The role of organic matter in the effect of fire on soil in chaparral regions
(from Welling *et al.*, 1984)

were mentioned earlier. Plants also supply a partial canopy cover and ground litter which helps to protect against splash erosion as well as supplying organic material for chemical, biological, and structural process in the soil (Welling *et al.*, 1984). Plants and litter contain aromatic oils which (a) make the vegetation very inflammable, especially during the dry season and (b) by microbial decomposition or by fire, produce long-chain aliphatic hydrocarbons which tend to make the soil hydrophobic. In the former case, leaching carries the hydrophobic substances into the soil while in the latter case, the volitalized hydrophobic substances move down into the soil along a temperature gradient where they condense into a layer a few centimetres beneath the surface. With the frequent fires of the region, the latter process is common (figure 3.12). This permits infiltration to the hydrophobic layer only. The result in a heavy rain is a surplus of water near the surface, with high pore pressure and adequate tractive force so that the surface layer is eroded, or sometimes moves in a mass, or is cut by rills (Wells, 1981) or sometimes all three. Additionally, the intense fires often burn plants and litter down to the mineral soil, and also burn organic material in the soil, causing aggregates to break down, and thus often increase bulk density (DeBano *et al.*, 1979; Welling *et al.*, 1984). The resulting splash erosion plus the effects of the hydrophobic layer can produce erosion 30 times that of unburned areas (Rice, 1974) and the effects last for several years. Interrelationships among plants, fire, soil and erosion with special reference to chaparral areas are shown in figure 3.13.

Because of the great potential for erosion in chaparral areas, attempts have been made to rehabilitate burned slopes. One approach has been to reseed with grasses, especially annual ryegrass (*Lolium* spp.) and barley (*Hordeum* spp.) (Corbett and Green, 1965). Although the grasses help prevent surface erosion, they are too shallow-rooted to prevent soil slippage. In the longer term, moreover, grasses suppress regrowth of native vegetation so that vegetative cover after two or three years is no greater than it would have been without the grass cover. Another approach is to treat with soil with heat-shock soil fungi which helps to reaggregate the soil and reduce splash erosion (Dunn *et al.*, 1982). However, innoculations of such fungi have not always increased resistance to erosion and in some cases have actually reduced resistance (Wade G. Wells II, pers. comm.).

Animal-induced erosion has long been a problem in Mediterranean areas. Severe erosion around the Mediterranean over the past three millennia or so has been described by many including Lowdermilk (1953), Judson (1963), Vita-Finzi (1969) and Butzer (1974). While all cite deforestation and agriculture as a general cause, Lowdermilk specifically emphasizes the role of grazing and browsing animals, especially sheep and goats. In southern California, a century of sheep grazing reduced coastal sage scrub (*Artemisia californica*) and pine-oak woodland to grass or barren ground with resulting severe soil erosion (Brumbaugh *et al.*, 1982; Brumbaugh and Leishman, 1982). More detailed effects of grazing on woody vegetation, especially on Bishop pine (*Pinus muricata*) are found in Hobbs (1980), Minnich (1980), and Leishman (1981).

An especially intriguing riddle is the origin of arroyo or gully cutting in southern California and elsewhere in the southwestern USA, especially during the late nineteenth century. Two hypothetical causes offered are (a) grazing, especially by cattle, and (b) climatic change (Reeves, 1970; Cooke and Reeves, 1976). This would appear to be an especially rewarding topic in biogeomorphology.

Agriculture

It is through the aegis of agriculture that humans have made some of the most profound changes in geomorphic processes and landforms. The human role is to upset the delicate balance between geomorphic force and resistance. Principally by affecting indigenous organisms, humans reduce the ability of the landscape to resist the great natural forces of rain, running water, and wind. Moreover, the same actions which decrease resistance to geomorphic forces also may have the effect of increasing the immediate acting forces. A good example is the effect of most agricultural activities in decreasing infiltration capacity of soil (figure 3.14). Rainfall which does not infiltrate becomes overland flow, a most effective erosional agent. Human effects, then, can both reduce the resistance and increase the acting forces. Thus a relatively small expenditure of human energy allows large amounts of geomorphic work to be done.

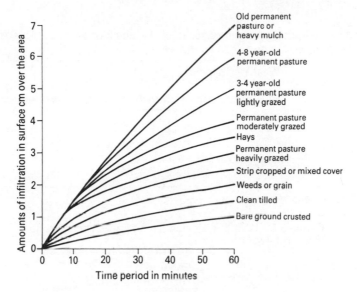

Figure 3.14   Effect of agricultural practices on soil infiltration capacities
(redrawn after Holtan and Kirkpatrick, 1950)

The amount of geomorphic work done per unit of human input is largely determined by climate. For example, it will be seen from figure 3.14 that a rainfall event of 5 cm hr^{-1} would exceed the infiltration capacities of most agricultural land uses and thus cause overland flow and erosion. That storm of 5 cm hr^{-1} is only a one-year event along the Gulf Coast of the USA (Caf), while in the Puget Sound lowlands of the northwestern USA (Cb), the closest analogue to western Europe, the 100-year–one-hour, storm is only 2.5 cm. It is not surprising then that soil erosion has not been taken seriously in much of western Europe. Rodda (1970) suggests that British farmers use with impunity practices which would be disastrous in some other areas of the world. In the spring of 1985, I observed ploughing and planting in Northern Ireland, but did not see a single field cultivated on the contour. Many were ploughed directly up and down the slope. Nevertheless, unusual events do cause perhaps isolated cases of severe soil erosion in Britain and some of these have been reported by Boardman (1983a, 1983b) and Morgan (1980). Longer-term severe erosion in western Germany (Cbf) has been reported by Hard (1971).

Conventional tillage.   Historically, the most common form of agricultural disturbance has been to clear the natural vegetation and plough or till the soil, a process known today as 'conventional tillage'. Although new and improved forms of cultivation such as 'no-till' have been developed, conventional tillage remains the standard in most temperate climates. In the

Table 3.4 Infiltration with contrasting vegetation cover on various soil series which have different depths and organic matter

Name of silt loam soil	Total infiltration in 5 hr (cm)		Difference due to land use (cm)
	Bluegrass pasture	Cornland	
Muscatine	13.7	3.4	10.3
Tama	12.8	3.8	9.0
Berwick	8.8	3.1	5.7
Clinton	7.0	5.5	1.5
Viola	4.1	3.3	0.8

Source: Holtan and Musgrave, 1947

Table 3.5 Rate of infiltration during the fifth hour for the soils in table 3.4

Silt loam soils	Bluegrass (cm hr^{-1})	Cornland (cm hr^{-1})
Muscatine	1.5	0.3
Tama	2.0	0.4
Berwick	0.9	0.3
Clinton	0.7	0.5
Viola	0.4	0.2

following discussion, conventional tillage will be covered followed by conservation tillage. Finally, the special cases of vineyards and orchards are examined.

Vegetative cover. As discussed earlier under forests, vegetative cover includes both ground cover and canopy. Actually, it is far more complex, including such factors as rooting, and long term soil effects, but those are covered later in this section. Conventional wisdom has maintained that there is a direct relationship between erosion and proportion of the ground covered by vegetation, (e.g. Heinemann and Whitaker, 1974). In general the standard of comparison is bare or fallow soil. Row crops such as maize, soybeans, or cotton would give only slight protection and small grains such as wheat, rye and barley would give substantial mitigation, while grasses would provide great protection from erosion (more exact values are given later together with time effects). While there is general agreement about the effects of ground cover, canopy cover is more problematic. Long-term experiments by the US Department of Agriculture have indicated a direct relationship between canopy cover and erosion (see table 3.1 above), but more recent laboratory experiments indicate that the relationship may be much more complex and even inverse in some cases (Noble and Morgan, 1983; Morgan *et al.*, 1986). Additionally, there may be erosional differences among plants with seemingly similar cover characteristics. Between the two row-crops of soybeans and maize, for example, soybeans allow significantly higher erosion than maize (Albers *et al.*, 1985). Another variable is the differential effects of vegetative cover on various soil types. This can be shown in part by using infiltration rates (tables 3.4 and 3.5).

Perhaps the best quantitative evaluation of the effects of vegetative cover on erosion is shown by the C values of the USLE (table 3.6). These are quite detailed, so that only excerpts from the original are shown here. Provision is made for crop, type of tillage, phase of rotation, stage of crop growth, residue management, crop yield (soil fertility) and soil amendments.

The crop stages as shown are:

Period F   *Rough Fallow.* Turn ploughing to seeding.
Period 1   *Seeding.* Seedbed preparation to 1 month after planting.
Period 2   *Establishment of crop.*
Period 3   *Growing and maturing crop.*
Period 4   *Residue or stubble.* Harvest to Period F.

The subgroups are: (1) crop residue left on soil (4L); (2) removed (4R); or (3) left plus a winter cover crop (4L + WC).

The number for each crop entry under each crop stage is a percentage, being the ratio of soil loss from cropland to the corresponding loss from continuous clean-tilled fallow. The latter is the standard or index cover factor and has a value of 1 (or 100 per cent). For agriculture all other land uses are less erosive than fallow. Breaking the C factors into crop stages has two important functions: first, it allows proper description of the crop cover value at that growth stage and secondly, it allows the proper partial allocation of average annual rainfall erosive energy. The full application of the equation is quite complex and for complete details, the reader is directed to Wischmeier and Smith (1965), and Wischmeier and Smith (1978). The latter is more complex than the former, better reflecting the many permutations and combinations of management in modern agriculture.

The effects of different vegetative covers can be seen by comparing lines 15 (maize), with 93 (small grain). Soil-loss ratios are less for all stages with small grains but period 3 (growing and maturing crop) is most indicative. Then, small grain permits only 23 per cent (5 per cent ÷ 22 per cent) as much erosion as maize.

Soil condition.   The physical condition of a soil is a primary determinant of erosion rates and a major factor of soil condition is organic material (OM) and biological activity (McCalla, 1942). Wischmeier (1966) found that increasing OM from 1 per cent to 4 per cent decreased overland flow by 50 per cent. As shown earlier (figure 3.9), the resistance of a soil to

Table 3.6 Vegetative cover factors

Line no.	Cover, sequence, and management	Productivity		Soil-loss ratio for crop-stage period						
		Hay field (t ha⁻¹)	Corn field (t ha⁻¹)	F (%)	1 (%)	2 (%)	3 (%)	4L Residue left (%)	4R Residue removed (%)	4L + WC Winter cover (%)
	Corn rotation									
	First-year maize after meadow									
2	Spring turn ploughed, conv. tillage	4–7	5.0	10	28	19	12	18	40	11
5	Spring turn ploughed, conv. tillage	2–4	2.5–4.0	15	32	30	19	30	50	15
9	Spring turn ploughed, min. tillage	4–7	5.0	—	10	10	7	18	40	10
	Second-year maize after meadow									
15	Spring turn ploughed, conv. tillage	4–7	5.0	32	51	41	22	26	—	15
20	Spring turn ploughed, min. tillage	4–7	5.0	—	32	32	13	26	60	15
	Third or fourth-year maize after meadow or second-year maize after grain or clover									
36	Spring turn ploughed, conv. tillage	7–11	5.0	6	63	50	26	30	—	—

Line										
	Cotton in rotation									
61	First-year cotton after meadow	7–11	—	8	25	30	20	22	—	15
	Small grain in rotation									
	With meadow seed after disked									
93	Second or third-year maize after meadow	7–11	5.0	—	32	19	5	3	—	—
	Without meadow seed after disked									
93+ 115	Second or third-year maize	7–11	5.0	—	32	19	10	10	20	—
120	Grass and legume meadow	7+	—	—	—	—	—	4	—	—

The soil-loss ratios are presented as percentages, and are the ratios of soil loss from cropland to the corresponding loss from continuous clean-tilled fallow. Line numbers are from the original source
Source: Adapted from Wischmeier and Smith, 1965

erosion is a function in part of percentage OM which promotes aggregate stability. Long-term cultivation without proper measures will reduce the OM in soil: sixty years of cereal production in Sweden reduced OM from about 6 per cent to about 4.5 per cent (cited in Morgan, 1985). Greenland *et al.* (1975) suggest that soils with less than 3.4 per cent OM may have unstable aggregates. Organic material in soil may be increased and soil condition improved by (1) applying organic material such as manure from external sources, (2) use of green manure, (3) management of crop residues, and (4) crop rotations.

Organic material as available from external sources may be added to soils. For example, the recommended application rate of manure to non-irrigated land in southern Alberta, Canada (BSk) is 22–27 t ha^{-1} yr^{-1}. Such applications decreased bulk density and increased the number of aggregates larger than 1 mm diameter which would decrease both water and wind erosion (Sommerfeldt and Chang, 1985). In Wisconsin (Daf), manure spread on the soil surface generally reduced both runoff and soil-loss (Mueller *et al.*, 1984). Because so many farms no longer have livestock, farmyard manure is increasingly difficult to procure. Some farmers in the southeastern USA have used chicken manure from commercial operations with some success. Sewage sludge was once considered a potential source but the presence of heavy metals had made that source problematic. Other materials have been assigned organic values relative to farmyard manure (Kolenbrander, 1974, cited in Morgan, 1985). These organic equivalents are:

farmyard manure	1.00
green manure	0.35
cereal straw	0.45
roots of crops	0.55
deciduous tree litter	1.40
coniferous tree litter	1.60
peat moss	2.50

Organic farming has often been advocated as a means to cure the ills of modern agriculture, including soil erosion. The foregoing discussion has indicated some of the attributes of large organic amendments to soil, but there appears to be very little well-designed data which would show that organic farming *per se* significantly reduces soil erosion. One study in the midwestern USA (Daf) showed that soil loss was about one-third less on organic farms, but this was attributable to different rotations and not to the organic farming itself. However, an average of 9 yrs of organic farming imparted 6 per cent greater organic soil material than on conventional farms (Lockeretz *et al.*, 1978), so the long-term effects may be important. Crop yields on organic farms are considerably less than on conventional farms (Lockeretz *et al.*, 1978). Thus, if there were a general shift to organic farms, the land base would have to expand, at least in the short run. Because this would include more sloping land, total soil erosion could possibly be increased.

Green manure is a young crop, preferably a legume, ploughed under. Although it has a low organic equivalent, as we have just seen, Troeh *et al.* (1980) praise this practice highly, pointing out that the fresh organic material decomposes rapidly and the population of soil microbes multiples, producing the cementing agents which promote soil aggregation.

Crop residue management is the returning to the soil of the less-commercial parts of the plants such as leaves and stalks. Note that many crop residues have alternative commercial uses such as livestock feed, energy sources, and raw material for cellulose products, so that using them for erosion control is a conscious and somewhat costly proposition. The best practice is to leave the chopped residue on the surface, a process called stubble-mulching, but Cogo *et al.* (1984) found that partial incorporation into the soil was more effective in controlling erosion. The immediate value is surface mulch, and erosion is inversely proportional to percentage ground cover (figure 3.7). Not only are water and wind erosion immediately reduced, but soil water is conserved and there may be longer-term improvements to soil condition, especially to structure and fertility (Siddoway, 1963; Ramig and Mazurak, 1964; Taylor *et al.*, 1964; Unger, 1968; Black, 1973; Pikul and Allmaras 1986; Skidmore *et al.*, 1986). A major problem is that the decomposing microbes require considerable nitrogen and take it from the soil (Troeh *et al.* 1980) so that nitrogen must often be added. Residue management is also used on irrigated land. Although the mulch effect is significant in reducing erosion (Miller and Aarstad, 1983), one study found that there were no long-term improvements in soil condition (Skidmore *et al.*, 1986).

Stubble-mulching has other problems. It is difficult to plant through the residue and implements become clogged. The residue may act as a harbour for disease and pests. Finally, the residues tend to insulate the ground, slowing its warming in spring, and thus retard the emergence of crops, make it possible that a full erosion-resistant stand will not develop until later in the season. This can be a severe problem at higher latitudes.

Crop rotations.   Crop rotation is the practice of alternating soil conserving vegetation with crops which deplete the soil and/or permit higher rates of erosion. Not only is erosion reduced during the conservatory periods, but soil condition is often improved so that there is usually a residual carryover which reduces the erosion rate while the erosive crop is growing. This is clearly shown by long-term experiments in the southeastern USA (Caf) where cotton in rotation was compared with continuous cotton (figure 3.15). Not only was the average rate of erosion greatly reduced, the annual soil loss from cotton in rotation was only about 20 per cent that of continuous cotton.

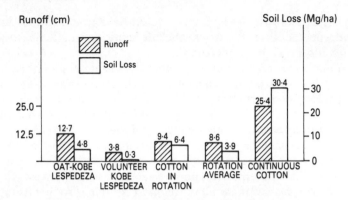

Figure 3.15   Average annual runoff and soil loss from 22 m long, 7 per cent slope plots of sandy clay loam cropped to a three-year rotation of (1) oats–Kobe lespedeza, (2) volunteer lespedeza, and (3) cotton; compared with continuous cotton, 1945–1952, Watkinsville, Georgia. Annual rainfall averaged 116 cm (from Hendrickson *et al.*, 1963)

The conservatory crop is usually a forage, sometimes a legume, which may be left in place several years and then turned under for green manure before an erosive crop such as maize is planted. The longer the conservatory crop remains in place, the greater are the benefits to soil condition and hence to mitigating erosion. In the north-central USA (Daf–Dbf), under otherwise identical conditions, when rotation with only 1 year of grass cover lost an annual average of $2,200\,t\,km^{-2}$, rotation with 2 years of grass cover lost only $1,300\,t\,km^{-2}$. At the same time, a rotation with 3 years of grass cover lost $800\,t\,km^{-2}$, and a rotation with 4 years of grass cover lost only $700\,t\,km^{-2}$ (Hays, *et al.*, 1949). In the Driftless Area of that region, Trimble and Lund (1982) attributed much of a significant decrease of upland erosion to longer periods of forage. Alfalfa (lucerne) is an especially important forage crop in this respect because it is leguminous, very deeply rooted (up to 30 m), and its longevity is up to 7 years. There can also be important differences among erosive crops. For example, soybeans (*Glycine max* L.) allows more erosion than other row crops, especially maize (*Zea mays* L.) (Laflen and Moldenhauer, 1979; Bathke and Blake, 1984; Van Doren *et al.*, 1984). Soybeans apparently produce less residue and permit more soil disaggregation.

Soil fertility helps to decrease erosion of cultivated land by insuring prompt and vigorous plant growth. Wischmeier (1966) found that when maize yields were increased from $1.3\,t\,ha^{-1}$ to $7.5\,t\,ha^{-1}$, the overland flow decreased from about 65 per cent of that experienced on fallow ground to about 10 per cent. In Iowa (Daf), large applications of nitrogen on maize fields greatly reduced erosion, overshadowing even percentage organic material and aggregate stability (Moldenhauer, *et al.*, 1967).

Soil crusting is a thin, less-permeable surface layer formed on bare soil by raindrops disintegrating structural particles and sealing the surface pores. There is also evidence that raindrop impact helps form this layer by compaction. Its effects are to (1) retard infiltration and thus increase erosion and (2) retard plant emergence resulting in a late and thin crop stand so that erosion is increased. Crusting can be reduced by increasing aggregate stability. Kladivco *et al.* (1986) report that earthworm (*Lumbricus rubellus*) activity strongly decreases crusting and its effects. Infiltration capacities were phenomenally increased from only 8 cm hr^{-1} to as much as 118 cm hr^{-1}.

Weed control.   Weeds tend to discourage the use of such important conservation practices as steep-backslope terraces, forage-strip cropping, and grassed waterways because such areas are harbours for weed pests (Brock, 1982). In Kansas (Daf), windbreaks have been removed for weed control (Sorensen and Marotz, 1977). One reason that stubble mulching never reached full potential in the USA is the weed problem. Attempts to cultivate (1) clog implements with the residue, and (2) bury much of the residue, reducing its effectiveness. Of course, weeds can be controlled by herbicides, but these can increase runoff and erosion and reduce crop yields. Experiments in Missouri (Daf) showed that herbicide treatments increased runoff up to 2–4 times, soil loss up to 4–8 times, and reduced maize yields up to 30 per cent (Whitaker *et al.*, 1973), but the ubiquity of this phenomenon is not yet known.

Soil frost.   As discussed earlier in the section on forests, wet, unprotected agricultural soils tend to freeze into a nearly impermeable 'concrete' frost which promotes accelerated erosion. Such erosion is a particularly severe problem in the wheat area of eastern Washington and Oregon (BSk). There, Pikul *et al.* (1986) found that crop residue management had the added benefit of reducing soil frost depth. A problem in less severe climates is 'needle' or 'spew' frost whereby the bare soil surface freezes overnight, raising soil particles up to several centimetres on ice projections. When the ice melts the next day, the loose soil particle can move downslope. Since this is often a diurnal event, the amount of material loosened and moved becomes considerable. The loosened material is especially susceptible to erosion by rainfall and overland flow. Ireland *et al.* (1939) found this mechanism to be important in gully erosion in South Carolina (Caf). Needle frost would presumably be inoperative if the soil had vegetative cover, and this would appear to be an important theme of biogeomorphological research.

Microbial and animal activity.   The role of microbes in creating organic gels and other cementing substances, and their role in increasing aggregate stability, has already been mentioned. A considerable body of literature exists, but most of it concerns the effect on aggregates, with most studies

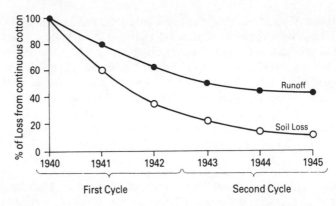

Figure 3.16 Annual runoff and soil loss from cotton plots shown in figure 3.15 under a three-year rotation of (1) oats–Kobe lespedeza, (2) volunteer lespedeza, and (3) cotton; compared with those from continuous cotton, 1940–1945, Watkinsville, Georgia
(from Hendrickson *et al.*, 1963)

finding that fungi are the most effective of the microbes (e.g., Greenland *et al.*, 1962). Few studies have directly studied the role of microbes on erosion. Two of these (Peele and Beale, 1940, 1941) showed that runoff and erosion were reduced by microbes, especially fungi. Gasperi-Mago and Troeh (1979), using a sterile control, showed that microbes had a significant effect on resistance to erosion, with fungi having the greatest. As indicated before, earthworms have a profound effect on agricultural soil. In addition to creating vertical tunnels which greatly enhance infiltration capacity (Ehlers, 1975), they also improve soil structure and reduce soil crusting (Kladivco *et al.*, 1986). Ehlers (1975) found that earthworm activity was twice as much on no-till land as compared to tilled land.

Little appears to be known about erosional effects, if any, of wildlife on agricultural land. Large animals such as deer often eat significant areas of crops and burrowing animals are sometimes found in pasture, but their erosional effects are uncertain. The increase of no-till agriculture in the USA presents new research questions in this respect because the soil is not greatly disturbed for long periods. In Iowa (Daf) for example, Clark and Young (1986) found that although field rodents and ground squirrels accounted for some seedling mortality on no-till maize, their effect was only a fraction of that from insects and weather. Rodents, in fact, helped control insects, and animal control was deemed to be unwarranted.

Time. Most of the soil conservation techniques mentioned in this essay have both immediate and long-term effects. For example, a good stand of forage immediately reduced erosion by protecting and anchoring the surface. The long-term effects, however, might expand to include increased organic material, improved structure, decreased bulk density, increased fertility, increased infiltration capacity, and increased resistance to erosion. Such improvement may continue for several years. For example, the improvements from the soil rotation shown in figure 3.15 continued over at least a six-year period (figure 3.16). For similar results, see Borst *et al.* (1945) and Harold *et al.* (1962). Longer periods may be even more significant. Fifty years of pasture created a soil condition which permitted a constant infiltration rate over seven times greater than an adjacent cultivated field (Skidmore *et al.*, 1975).

The above process is also reversible. That is, when a soil in excellent condition is put into an erosive crop, the degenerative process takes several years. Agronomists refer to this phenomena as the residual effect. The idea is to precede an erosive crop by one or more years of a soil-building land use, the best of which is sod or forest (Wischmeier and Smith, 1978). The effects can be demonstrated quite readily by the change of C values for various crops under different rotations or management arrangements. In table 3.6, for example, lines 2, 15 and 36 compare the erosion from a crop of maize directly after the field has been in meadow (line 2), 2 years after (line 15), and 3 and 4 years after (line 36). The most diagnostic crop period is 1 (rough fallow) when the soil is almost bare so that it is the resistance of the soil to erosion that is measured. By the second year after meadow, the soil-loss has increased to over 300 per cent that of the first year after meadow. By the third and fourth year (line 36), there are still greater increases. Forty years of cultivation on a prairie soil in Kansas (Daf) reduced aggregation in the plough layer by 80 per cent (Olmstead, 1946).

The lag demonstrated in the foregoing section constitutes part of the response lag discussed early in this essay accounting for part of the hysterectic effect shown in figures 3.2, 3.3 and 3.4. The downslope erosional momentum of overland flow cascading from deteriorated soils, or the *lack* of downslope erosional momentum caused by overland flow *not* being produced by soils still in good condition, enhances the hysterectic effect. To that may be added the presence, or lack of, temporary channels such as rills and gullies which add to the momentum and aid in its downslope transfer.

Conservation tillage.   There are degrees of conservation tillage ranging from stubble-mulching to no-till. Stubble-mulching was discussed earlier and it remains here to discuss minimum and no-till and their profound effects in reducing erosion. Both methods were developed to decrease soil compaction by eliminating many passes with heavy machinery, and to decrease erosion by having the bare soil exposed over shorter periods or not at all. Reduced or minimum tillage usually involves seed bed preparation and planting in one integrated step. With no-tillage the soil is never ploughed; seeds are drilled into standing vegetation or stubble. With both types, weeds are controlled by herbicides, although some ploughing may occur with minimum-tillage. Even with no-till, residue management is still important, especially with row crops like maize, soybeans, or cotton. Decreases of erosion using no-till have been truly spectacular. Langdale *et al.* (1978) report that in Georgia (Caf), soil losses under no-till were about 0.5 per cent of those from conventional tillage. Similar results from a wide range of studies are reported in an excellent review by Phillips *et al.* (1980). Other reviews are Crosson (1981), and Schnepf (1983). More recent studies generally support the superiority of no-till over conventional tillage in preventing erosion, although the factor of superiority seems to have decreased considerably over earlier reports (Shelton *et al.*, 1983; Andraski *et al.*, 1985; Wendt and Burwell, 1985). However, a surprising study in the Blackland Prairie (Black Belt) of Mississippi (Caf) showed that while minimum- and no-till soil-losses were about 60–90 per cent that of conventional tillage, the soybean yields under no-till were only about 70 per cent of conventional yields (Hairston *et al.*, 1984). Another surprising recent finding is that up to 8 years of no-till made no significant difference in soil bulk density (Hill and Cruse, 1985). In Kentucky, 10 years of continuous no-till also produced no difference in bulk density, but it did double the organic matter in the upper 5 cm and increased the saturated hydraulic conductivity (Blevins *et al.*, 1983).

In general, crops grown under no-till are no more susceptible to disease and insects than with conventional agriculture. Because it saves time, fuel, and soil, no-till has become popular in the USA over the past two decades, but the current agricultural depression and cheap fuels make that continued growth uncertain in the immediate future. Its utility in western Europe remains uncertain.

Orchards and vineyards.   For several reasons, orchards and vineyards tend to be on steep slopes. In areas of adequate rainfall and milder slopes, there is sometimes intertilling of grass or small grains but in areas of sparse rainfall and/or extremely steep slopes, the ground is kept relatively bare, creating a severe erosion potential and this is especially important in continental Europe (Tropeano, 1984a, 1984b). Some of the most spectacularly steep vineyards are along the gorges of the Moselle and Rhine Rivers in Germany (Cbf). Richter and Negendank (1977) have measured soil-loss rates from those slopes up to 4.4 cm 1,000 yr^{-1} (57.2 t km^{-2} yr^{-1} at 1.3 g cm^{-3}), a very modest rate indeed. They state that most of this loss comes from 'catastrophic rainstorms' (p. 278). However, the most extreme event they cite (p. 269) is 8.4 mm 10 min^{-1}. In the USA, a mere 2 yr–10 min event is about 15 mm in the north-central states (Daf), and along the Gulf Coast (Caf) is 21 cm. My point is that soil losses from these steep, almost bare areas are modest because the climate is so mild. A 100 yr–10 min rainfall along the US Gulf Coast is 40 cm and that could truly be termed catastrophic. The Moselle–Rhine

Table 3.7 USLE 'C' factors for deciduous, non-tilled orchards–almond, peach, walnut etc.

Stage of development	Raised canopy cover[1]	% Ground cover[2] 0	20	40	60	80	95+
Clearing to 2 years old[3]	0[a]	0.45	0.20	0.10	0.042	0.013	0.003
	0[b]	1.00	0.44	0.22	0.092	0.029	0.007
2 to 5 years old	25	0.40	0.18	0.093	0.040	0.013	0.003
5 to 10 years old	50	0.34	0.16	0.085	0.038	0.012	0.003
10 or more years old	75	0.28	0.14	0.080	0.036	0.012	0.003

[1]Portion of total-area surface hidden by canopy from 'bird's eye view'
[2]Ground cover might be accomplished through use of a mulch or cover crop, which is especially important in winter months
[3]Tillage for clearing may occur in grove development. Two alternatives are presented here, i.e.
[a]Hand or light mechanical clearing which leaves at least the root network from previous plants;
[b]Clearing by bulldozer or other equipment, removing all surface residues including previous plant root network
No tillage operations are done after the establishment year
Source: USDA, SCS, 1985

vineyards could never exist in a Daf or Caf climate. In Italy, Tropeano (1984a) states the erosion from the vineyards of the central Piedmont (Caf) is much higher than other European sites. He measured soil loss up to 4,700 t km^{-2} yr^{-1} which he attributes to deep ploughing and heavy herbicide treatment. One wonders about the role of Caf climate relative to Cbf and Csb found in most other wine regions of Europe. The wine region of Hungary (Dbf) is also one with significant erosion problems (Pinczes, 1982).

In California (Csa), C factors for the USLE have been estimated for various management and growth stages of vineyards and orchards (table 3.7).

Bare areas under both orchards and vineyards are highly susceptible to erosion. One common organic management practice in orchards is to use pruned branches as a mulch. This appears to be effective, but it appears possible from the work of Jamison (1969) that citrus cuttings may enhance soil hydrophobicity and thus potentially increase erosion.

Grazing of forage.   Grazing by large animals tends to thin the vegetative stand and compact the soil, especially when wet. Sheep tend to graze very close to the ground, thus weakening and thinning the forage stand. With heavy grazing, bare soil may be visible and this puts high stress on the

Table 3.8 Factor C for permanent pasture, range, and idle land[1]

Vegetative canopy			Cover that contacts the soil surface					
Type and height[2]	Percentage cover[3]	Type[4]	Percent ground cover					
			0	20	40	60	80	95+
No appreciable		G	0.45	0.20	0.10	0.042	0.013	0.003
canopy		W	0.45	0.24	0.15	0.091	0.043	0.011
Tall weeds or	25	G	0.36	0.17	0.09	0.038	0.013	0.003
short brush		W	0.36	0.20	0.13	0.083	0.041	0.011
with average	50	G	0.26	0.13	0.07	0.035	0.012	0.003
drop fall		W	0.26	0.16	0.11	0.076	0.039	0.011
height of	75	G	0.17	0.10	0.06	0.032	0.011	0.003
50 cm		W	0.17	0.12	0.09	0.068	0.038	0.011
Appreciable	25	G	0.40	0.18	0.09	0.040	0.013	0.003
brush or		W	0.40	0.22	0.14	0.087	0.042	0.011
bushes, with	50	G	0.34	0.16	0.08	0.038	0.012	0.003
average drop		W	0.34	0.19	0.13	0.082	0.041	0.011
fall height	75	G	0.28	0.14	0.08	0.036	0.012	0.003
of 2 m		W	0.28	0.17	0.12	0.078	0.040	0.011
Trees, but no	25	G	0.42	0.19	0.10	0.041	0.013	0.003
appreciable		W	0.42	0.23	0.14	0.089	0.042	0.011
low brush.	50	G	0.39	0.18	0.09	0.040	0.013	0.003
Average drop		W	0.39	0.21	0.14	0.087	0.042	0.011
fall height	75	G	0.36	0.17	0.09	0.039	0.012	0.003
of 4 m		W	0.36	0.20	0.13	0.084	0.041	0.011

[1]The listed C values assume that the vegetation and mulch are randomly distributed over the entire area

[2]Canopy height is measured as the average fall height of water drops falling from the canopy to the ground. Canopy effect is inversely proportional to drop fall height and is negligible if fall height exceeds 10 m

[3]Portion of total-area surface that would be hidden from view by canopy in a vertical projection (a bird's eye view)

[4]G: cover at surface is grass, grasslike plants, decaying compacted duff, or litter at least 5 cm deep.

W: cover at surface is mostly broadleaf herbaceous plants (as weeds with little lateral-root network near the surface) or undecayed residues or both

Source: Wischmeier and Smith, 1978

forage so that replanting or interplanting should be done every few years or when necessary. Additionally, soil condition should be closely monitored and amendments applied as needed. While cattle do not graze so closely, they have high unit hoof loadings and compact the soil. Especially in negotiating hillsides, cattle can exert tremendous force through the small area of one or two hooves. With non-grazed forest and heavily grazed pasture in juxtaposition along a fenceline, it is not uncommon to find the pasture lying several cm lower than the forest floor. The general reduction in infiltration from increasing grazing intensity is shown in figure 3.14. It indicates that heavy grazing can reduce infiltration by about 60 per cent. More specifically, Gifford and Hawkins (1978) report that heavy grazing reduced infiltration in Louisiana (Caf) from 4.6 to 1.8 cm hr^{-1} and 3.3 to 2.0 cm hr^{-1} in Kansas (Daf). Moving from infiltration to erosion, Wischmeier and Smith (1978) give C values for pasture and range with varying ground cover and canopy (table 3.8).

By far the most erosive of animals are swine when confined to a relatively small area. In these conditions I have observed soil truncations of over 30 cm on areas of 1–2 ha. Swine are usually enclosed on slopes to facilitate washing away of wastes (into streams!) and this increases the erosion potential. High densities of cattle, such as in barnyards, can also greatly accelerate erosion.

Wind erosion.   Wind erosion continues to be a problem in some humid and semi-humid agricultural areas (Lyles, 1977). This is especially true in the Great Plains and midwestern USA (Daf and Dbf). Wind erosion may occur at any time of year when the soil is bare and friable. An especially vulnerable time is winter with fallow soils and no snow cover. Then, moisture can be removed from the upper few millimetres by sublimation, leaving a loose surface mulch of disaggregated soil which is highly susceptible to wind erosion. This can be mitigated by residue management (Skidmore *et al.*, 1966; Bilbro and Fryear, 1985) cover crops, or manure (Woodruff *et al.*, 1974), but the ubiquity of dirty snowbanks downwind from midwestern fields indicate that many farmers do not protect against this problem. Graphs showing the effectiveness of selected crop residues (in various densities and heights) for reducing wind erosion are given by Skidmore and Woodruff (1968). Windbreaks have long been used on the Great Plains (Davis, 1976). They are effective, but they unfortunately occupy productive space, shade crops, and offer a haven for weed and animal pests (Sorensen and Marotz, 1977). Trees are the most common windbreak and a single row of evergreens is probably the most effective in terms of the space it occupies (Troeh *et al.*, 1980). Suedkamp (1976) lists trees and shrubs suitable for windbreaks on the Northern Great Plains, and grass barriers are helpful in some cases (Aase *et al.*, 1985). Charts showing the mitigation of wind erosion by windbreaks of varying heights and porousness are given by Hagen (1976, cited in Troeh *et al.*, 1980).

Disturbed areas

Disturbed areas include those naturally disturbed, eroded agricultural land, construction areas, surface mining, chemically denuded land, recreational areas, and roadside ditches and banks. The common bond among these categories is sparse vegetation and active erosion.

Naturally disturbed.   Badlands can occur in marginally humid areas where parent material and edaphic conditions do not allow the natural establishment of vegetation. Well-known examples are found in South Dakota and Alberta (Bsk). The latter has been studied by Campbell (1977) who found the badlands to have much higher erosion rates than the surrounding grasslands so that their relatively small area supplies most of the sediment yield of the Red Deer River. Landslides are another type of naturally disturbed area. In the Blue Ridge Mountains of northern Virginia (Caf), Hupp (1983) found that landslides were colonized by the surrounding forest, but the vegetation could not stabilize the slope.

Eroded agricultural land. Abused agricultural land is often sparsely vegetated with much of the topsoil truncated, leaving a finer, less-fertile, often shallow subsoil, usually with little or no organic material nor biotic activity. Infiltration and retention of water is also greatly reduced. The resulting overland flow and erosion commonly gully the slopes. Original causes of erosion may have been tillage, or grazing, or some combination of the two. A commonly perceived example is the area surrounding the Mediterranean Sea (see section above on sclerophyll forest). Butzer (1974), however, contends that these areas are no more eroded than the comparable areas in USA. Reclamation of the more eroded areas has been partially successful by first temporarily stabilizing slopes with contouring, stone terraces, and check-dams, especially of living material such as willow (*Salix* spp.) and alder (*Alnus* spp.). Concurrently, other types of vegetation such as various grasses and black locust (*Robinia pseudoacacia*) are planted. The idea is to rebuild an 'organic soil mantle' (Margaropoulos, 1967). *Robinia* is used to stabilize disturbed areas in all temperate climates because (a) it is leguminous and therefore provides its own nitrogen, (b) it establishes easily, (c) it tolerates a wide range of pH and moisture conditions, and (d) it is vegetatively reproductive with the roots sending out adventitious sprouts. Moreover, the wood is long-lived and has commercial value so that badlands can be transformed into productive areas.

Highly eroded areas of the eastern USA have long been the target of stabilization schemes (Trimble, 1985). The south-east (Caf), in particular has had hundreds of km^2 devastated since European occupance. Stabilization has been effected by a combination of engineering methods such as check dams, and vegetative controls (McClurkin, 1967; Trimble, 1974). Pines have proven to be a particular godsend because they can colonize and survive in severe edaphic conditions. They also have considerable commercial value. Among the several pines, McClurkin (1967) argues that the loblolly (*Pinus taeda*) is superior to all in controlling erosion because it establishes better, grows faster, and produces more litter. While about 1 cm of pine litter will halt most erosion, a 10-year-old stand of loblolly will have a net accumulation of more than 2.5 cm. In one example, where abandoned fields allowed $41\, t\, km^{-2}\, yr^{-1}$, only $4\, t\, km^{-2}\, yr^{-1}$ of erosion occurred under loblolly cover (McClurkin,

1967). The best non-tree covers are Kudzu (*Pueraria thunbergiana*), sericea lespedcza (*Lespedeza cuneata*) and lovegrass (*Eragrostis curvala*).

Abused and deteriorated areas in semi-humid areas can also be restored. The procedures are much the same as already described but the establishment of vegetation is more difficult (Heede, 1979).

Construction.   In addition to baring soil, construction also temporarily accelerates erosion by steepening slopes, loosening some soil areas and compacting others, and brings less-permeable subsoil to the surface. The erosional effects of urbanization have been documented by Vice *et al.* (1969), who measured short-term rates of erosion from construction to be about ten times those of cultivated land. Wolman (1967), in a well-known curve, showed the erosional effects of urbanization to be a high but short-term blip on the historical sediment budget.

Relative erodibility for construction areas has been described using C factors of the USLE (table 3.9). These C values should be modified to reflect physical practices. For example, compacted and smooth up-and-down slope from bulldozing would require a factor of 1.3 while leaving the surface loose and rough with 30 cm depth or more reduced the factor to 0.8 (USDA, Soil Conservation Service, 1985). To reduce erosion during the short period of construction, several vegetative and biotechnical techniques are used aside from the engineering practices of paving and building holding-ponds. Other than seeding grass or laying sod, the area may be mulched in one or more of several ways. In addition to the more common mulches listed in table 3.9, many commercially-prepared chemical and organic soil binders are available (Maryland Department of Water Resources *et al.*, 1972). These stabilize the soil until vegetation can take over. A common technique is hydroseeding, whereby a slurry composed of water, grass seeds, fertilizer, mulch, and soil binder is sprayed onto bare slopes.

Surface mining.   Surface mining spoil, when regraded, has essentially all the same physical problems already encountered in construction areas (Collier, 1970; Toy, 1984). An important additional problem, however, is that acid or toxic mineral materials may be incorporated into the surface, thus inhibiting plant growth and prolonging erosion. Pyrites, in particular, oxidize to sulphuric acid and can lower pH below threshold values for many plants. Liming and other soil amendments may ameliorate the condition, perhaps temporarily, and at great cost. The best course is to bury such acid or toxic materials and cover the area with the original topsoil which has been set aside and stored. This is often a critical step for effective revegetation and 40 to 60 cm of soil is usually needed (Schuman *et al.*, 1985).

For initial establishment of vegetation, hydroseeding and soil stabilizers are used to control erosion (Kay, 1977). Several trees including *Robinia* and *Alnus* are suitable for mine spoil in humid regions (Linstrom, 1960;

Table 3.9   Cover index factor C for construction sites

Type of cover		Factor C	Percentage[1]
None (fallow ground)		1.0	0.0
Temporary seedings (90% stand):			
Ryegrass (perennial type)		0.05	95
Ryegrass (annuals)		0.1	90
Small grain		0.05	95
Millet or sudan grass		0.05	95
Field bromegrass		0.03	97
Permanent seedings (90% stand)		0.01	99
Sod (laid immediately)			
Mulch:			
rate of application ($t\,ha^{-1}$):			
Hay	1.1	0.25	75
	2.3	0.13	87
	3.4	0.07	93
	4.5	0.02	98
Small grain straw	4.5	0.02	98
Wood chips	13.6	0.06	94
Wood cellulose	4.0	0.1	90
Fibreglass	1.1	0.05	95
Asphalt emulsion ($12,000\,l\,ha^{-1}$)		0.02	98

Fibre matting, excelsior, gravel and stone may also be used as protective cover.
[1]Percentage soil-loss reduction as compared with fallow ground.
Source: USDA, Soil Conservation Service, 1985

May and Striffler, 1967). Grasses include lovegrass and orchardgrass (Troeh *et al.* 1980). Vegetation requirements in more arid areas are complex because sodicity and salinity as well as acidity and moisture conditions must be considered (Hassell, 1977; Plummer, 1977). Even with restoration and revegetation, it is difficult to reduce erosion rates to the natural level. In Wyoming (Bsk), erosion from revegetated mine spoils was several times the rate from undisturbed areas (Lusby and Toy, 1976). However, the reclaimed spoils were steeper and contained more clay so the results were not definitive.

Chemically denuded land.   This category is land which has been exposed to chemicals which have not only killed the vegetation, allowing disastrous erosion, but have also made the soil acid or toxic thus discouraging natural revegetation. Two famous examples are Ducktown, Tennessee (Caf) and Sudbury, Ontario (Dbf). Both areas are bandlands created after the surrounding vegetation was killed by dilute sulphuric acid fumes from the smelting of copper. In many cases the soil has been eroded down to weathered bedrock and even that has been acidified. Little appears to be known about the lag time between the demise of biotic life and the onset of severe erosion. These two regions are perhaps the most eloquent statement of the role of organisms in preventing erosion.

In Tennessee, revegetation efforts since the 1930s have reduced about 125 km² of largely denuded area to about 35 km² although many areas had to be replanted several times (Tyre and Barton, 1986). Parts of the remaining area are being replanted to the previously described loblolly pine by first ploughing deeply and then treating with sewage sludge and fertilizer. Some success was also encountered using sericea lespedeza, ryegrass, and love grass.

Recreational areas. Hiking and camping disturb vegetation and compact the soil. Lutz (1945) found that soils of picnic grounds had about 28 per cent pore volume while adjacent unused areas had 40 per cent. In the southeastern USA, Lockaby and Dunn (1984) found that soil in camping areas had a bulk density of $1.4 \, g \, cm^{-2}$ while control areas had a mean of $0.8 \, g \, cm^{-2}$. Several investigators found much less vegetative litter on recreational areas as compared to nearby control areas (Young, 1978; Dawson *et al.*, 1978). Soil OM averaged 6 per cent on campgrounds in Iowa (Daf) but 13 per cent in adjacent forest, presumably because less litter was available to be incorporated into the soil. Also, soil compaction with lessened moisture and air content resulted in some crown dieback of upland trees (Dawson *et al.*, 1978). Considering all these factors, it is not surprising that erosion has been observed (La Page, 1962; Settergren and Cole, 1970). Off-road recreational vehicles have a similar effect to the foregoing except that the tyres are more effective in disturbing the surface.

Roadside ditches and banks. These areas are highly susceptible to erosion because the roadways furnish ample stormflow which tends to erode the roadside ditches which, in turn, undermines the roadbanks. In Idaho, Megahan *et al.* (1983) found unvegetated granitic roadbanks to be retreating at about $1.1 \, cm \, yr^{-1}$. A problem is that roadbanks tend to be steep and formed on subsoil or weathered bedrock. Some of the cure may be found in engineering, but most erosion prevention is by means of vegetation. Seeding and mulching should be done as soon as construction is completed (Massie and Bubenzer, 1974). Hydroseeding and chemical soil binders are commonly used. Many grasses are used but lovegrass and sericea lespedeza are common in the southeastern USA.

Conclusions

Much is known about the effects of organisms on erosion in temperate regions, although most of this knowledge has been accumulated during the past 25 years and a great deal has evolved over the past decade. Yet it is clear that what we do not know exceeds what we do by several orders of magnitude. By indicating some research areas I hope this essay will encourage work on this important topic.

Acknowledgements

I thank A. P. Barnett (USDA), John Hewlett (Georgia), Cliff Hupp (USGS), and Wade Wells (USFS) for contributions to this paper. Claire Bourne, Ray Doss, Chase Langford and Mike Murphy kindly drew the diagrams. Any errors remain my own.

References

Aase, J. K., Siddoway, F. H. and Black, A. L. 1985: Effectiveness of grass barriers for reducing wind erosiveness. *Journal of Soil and Water Conservation*, 40, 354–57.

Albers, E. E., Wendt, R. C., and Burwell, R. E. 1985: Corn and soybean cropping effects on soil losses and C factors. *Soil Science Society of American Journal*, 49, 721–8.

Anderson, H. W. 1954: Suspended sediment discharge as related to streamflow, topography, soil and land use. *Transactions, American Geophysical Union*, 35, 268–81.

Andraski, B. J., Miller, D. H. and Daniel, T. C. 1985: Effects of tillage and rainfall simulation date on water and soil losses. *Soil Science Society of America Journal*, 49, 1512–17.

Arend, J. L. 1941: Infiltration rates of forest soils in the Missouri Ozarks as affected by wood burning and litter removal. *Journal of Forestry*, 39, 726–8.

Aubertin, G. M. 1971: Nature and extent of macropores in forest soils and their influence on subsurface water movement. *Research Paper NE-192*, United States Department of Agriculture, Forest Service.

Bathke, G. R. and Blake, G. R. 1984: Effects of soybeans on soil properties related to soil erodibility. *Soil Science Society of America Journal*, 48, 1398–1401.

Battarbee, R. W., Appleby, P. G., Odell, K. and Flower, R. J. 1985: 210Pb dating of Scottish lake sediments, afforestation and accelerated soil erosion. *Earth Surface Processes and Landforms*, 10, 137–42.

Beasley, R. S. and Granillo, A. B. 1985: Soil protection by natural vegetation on clearcut forest land in Arkansas. *Journal of Soil and Water Conservation*, 40, 379–82.

Beatty, S. W. and Stone, E. L. 1986: The variety of soil microsites created by tree falls. *Canadian Journal of Forest Resources*, 16, 539–48.

Beven, K. and Germann, P. 1982: Macropores and water flow in soils. *Water Resources Research*, 118, 1311–25.

Bilbro, J. D. and Fryrear, D. W. 1985: Effectiveness of residues from six crops for reducing wind erosion in a semiarid region. *Journal of Soil and Water Conservation*, 40, 358–9.

Black, A. L. 1973: Soil property changes associated with crop residue management in a wheat-fallow rotation. *Soil Science Society of America Proceedings*, 37, 943–6.

Blevins, R. L., Smith, M. S., Thomas, G. W., and Frye, W. W. 1983: Influence of conservation tillage on soil properties. *Journal of Soil and Water Conservation*, 38, 301–5.

Boardman, J. 1983a: Soil erosion at Albourne, West Sussex, England. *Applied Geography*, 3, 317–29.
—— 1983b: Soil erosion on the lower Greensands near Hascombe, Surrey, 1982–1983. *Journal Farnham Geological Society*, 1, 2–8.
Borst, H. L., McCall, A. G. and Bell, F. G. 1945: Investigations in erosion control and the reclamation of eroded land at the Northwest Appalachian Conservation Experiment Station, Zanesville, Ohio, 1934–1942. United States Department of Agriculture *Technical Bulletin* 888.
Bosch, J. M. and Hewlett, J. D. 1980: Sediment control in South African forests and mountain catchments. *Suid-Afrikaanse Bosboutydskrif*, 115, 50–5.
Brock, B. G. 1982: Weed control versus soil erosion control. *Journal of Soil and Water Conservation*, 37, 73–8.
Brock, J. H. and DeBano, L. F. 1982: Runoff and sedimentation potentials influenced by litter and slope on a chaparral community in central Arizona. In C. E. Conrad and W. C. Oechel (eds), *Dynamics and Management of Mediterranean-Type Ecosystems*. United States Forest Service *General Technical Report* PSW-58.
Brumbaugh, R. W. and Leishman, N. J. 1982: Vegetation change on Santa Cruz Island, California: The effect of feral animals. In C. E. Conrad and W. C. Oechel (eds), *Dynamics and Management of Mediterranean-Type Ecosystems*. United States Forest Service *General Technical Report* PSW-58.
——, Renwick, W. H., and Loeher, L. L. 1982: Effects of vegetation change on shallow landsliding: Santa Cruz Island, California. In C. E. Conrad and W. C. Oechel (eds), *Dynamics and Management of Mediterranean-Type Ecosystems*. United States Forest Service *General Technical Report* PSW-58.
Brune, G. M. 1948: Rates of sediment production in midwestern United States. United States Soil Conservation Service, Publication No. SCS-TP-65.
Büdel, J. 1982: *Climatic Geomorphology*. Princeton, New Jersey: Princeton University Press.
Burns, R. G. and Hewlett, J. D. 1983: A decision model to predict sediment yield from forest practices. *Water Resources Bulletin*, 19, 9–14.
Burt, T. P. 1979: The relationship between throughflow generation and the solute concentration of soil and stream water. *Earth Surface Processes*, 4, 257–66.
——, Donohoe, M. A. and Vann, A. R. 1983: The effect of forestry drainage operations on upland sediment yields: the results of a storm-based study. *Earth Surface Processes and Landforms*, 8, 339–46.
Butzer, K. W. 1974: Accelerated soil erosion: A problem of man–land relationships. In I. R. Manners and M. W. Mikesell (eds), *Perspectives on Environment*, pp. 57–78. Washington, DC: Association of American Geographers.
Campbell, I. A. 1977: Stream discharge, suspended sediment and erosion rates in the Red Deer River basin, Alberta, Canada. International Association of Hydrological Sciences, *Publication* 122, 244–59.
Chang, M., Poth, F. A. II and Hunt, Ellis V. Jr. 1982: Sediment production under various forest-site conditions. International Association of Hydrological Sciences, *Publication* 137, 13–22.
Clark, W. R. and Young, R. E. 1986: Crop damage by small mammals in no-till cornfields. *Journal of Soil and Water Conservation*, 41, 338–41.
Clayton, J. L. and Megahan, W. F. 1986: Erosional and chemical denudation rates in the southwestern Idaho batholith. *Earth Surface Processes and Landforms*, 11, 389–400.

Cogo, N. P., Moldenhauer, W. C., and Foster, G. R. 1984: Soil loss reductions from conservation tillage practices. *Soil Science Society of America Journal*, 48; 368–73.

Coleman, E. A. 1953: *Vegetation and Watershed Management*. New York: The Ronald Press Company.

Collier, C. R., 1964: Influences of strip mining on the hydrologic environment of parts of Beaver Creek basin, Kentucky, 1955–59. US Geological Survey *Professional Paper* 427-B.

Connaughton, C. A. 1935: Forest fires and accelerated erosion. *Journal of Forestry*, 33, 751–2.

Cooke, R. U. and Reeves, R. W. 1976: *Arroyos and Environmental Change*. Oxford: Oxford University Press.

Cooper, C. F. 1961: Controlled burning and watershed conditions in the White Mountains of Arizona. *Journal of Forestry*, 59, 438–42.

Copeland, O. L., Jr. 1963: Land use and ecological factors in relation to sediment yields. *Federal Inter-Agency Sedimentation Conference Proceedings, Jackson, Mississippi, January 28–31, 1963.*

Copley, T. L., Forrest, L. A., McCall, A. G., and Bell, F. G. 1944: Investigations in erosion control and reclamation of eroded land at the central piedmont conservation experiment station, Statesville, N.C., 1930–40. United States Department of Agriculture, Soil Conservation Service *Technical Bulletin* No. 873.

Corbett, E. S. and Green, L. R. 1965: Emergency revegetation to rehabilitate burned watersheds in southern California. USDA Forest Service *Research Paper* PSW-22.

—— and Rice, R., 1966: Soil slippage increased by brush conversion. USDA Forest Service *Research Note* PSW-128.

Crosson, P. 1981: *Conservation tillage and conventional tillage: A comparative assessment*. Soil Conservation Society of America, Ankeny, Iowa.

Davis, R. M. 1976: Great Plains windbreak history: An overview. In *Shelterbelts on the Great Plains*. Publication No. 78, 8–11. Lincoln: Great Plains Agricultural Council.

Dawson, J. O., Countryman, D. W. and Fittin, R. R. 1978: Soil and vegetative patterns in northeastern Iowa campgrounds. *Journal of Soil and Water Conservation* 33, 39–41.

DeBano, L. F. 1969: Observations on water-repellent soils in Western United States. In DeBano, L. F. and Letey, J. (eds), *Water-Repellent Soils*, University of California, Riverside, 17–28.

——, Rice, R. M., and Conrad, C. E. 1979: Soil heating in chaparral fires: effects on soil properties, plant nutrients, erosion, and runoff. United States Department of Agriculture, Forest Service, *Research Paper* PSW-145.

Derbyshire, E. (ed.) 1976: *Geomorphology and Climate*. New York: John Wiley and Sons.

Dissmeyer, G. E. and Foster, G. R. 1980: A guide for predicting sheet and till erosion on forest land. United States Department of Agriculture, Forest Service, *Technical Publication* SA-TP11.

Douglas, I. 1976: Erosion rates and climate: Geomorphological implications. In E. Derbyshire (ed.), *Geomorphology and Climate*, pp. 269–87. New York: John Wiley and Sons.

Dunn, P. H., Wells, W. G., II, Dickey, J. and Wohlgemuth, P. M. 1982: Role of fungi in postfire stabilization of chaparral ash beds. In C. E. Conrad and W. C. Oechel (eds), *Dynamics and Management of Mediterranean-Type Ecosystems*. United States Forest Service General Technical Report PSW-58.

Durgin, P. B. 1985: Burning changes the erodibility of forest soils. *Journal of Soil and Water Conservation*, 40, 299–301.

Dyrness, C. T. 1967: Erodibility and erosion potential of forest watersheds. In Sopper, W. E. and Lull, H. E., *International Symposium on Forest Hydrology*, New York: Pergamon.

Ehlers, W. 1975: Observations on earthworm channels and infiltration on tilled and untilled loess soil. *Soil Science*, 119, 242–9.

Fournier, F. 1982: Climatic Factors in Soil Erosion. In J. B. Laronne and M. P. Mosley (eds), *Erosion and Sediment Yield*, pp. 175–80. Stroudsburg, Pennsylvania: Hutchinson Ross Publishing Company.

Fredriksen, R. L. 1970: Erosion and sedimentation following road construction and timber harvest on unstable soils in three small western Oregon watersheds. USDA Forest Service Research Paper PNW 104.

Gasperi-Mago, R. R. and Troeh, F. R. 1979: Microbial effects on soil erodibility. *Soil Science Society of America Journal*, 43, 765–8.

Gerrard, A. J. 1981: *Soils and landforms: an integration of geomorphology and pedology*. London: George Allen & Unwin.

Gifford, G. F. and Hawkins, R. H. 1978: Hydrologic impact of grazing on infiltration: a critical review. *Water Resources Research*, 14, 305–13.

Goudie, A. 1986: *The Human Impact on the Natural Environment* (2nd edn). Oxford: Basil Blackwell; Cambridge, Mass: MIT Press.

Greenland, D. J., Histron, G. R. and Quirk, J. P. 1962: Organic materials which stabilize natural soil aggregates. *Soil Sciences Society of America Proceedings*, 26, 366–71.

——, Rimmer, D., and Payne, D. 1975: Determination of the structural stability class of English and Welsh soils using a water coherence test. *Journal of Soil Science*, 26, 294–303.

Hadley, R. F. (ed.) 1986 *Drainage Basin Sediment Delivery*. International Association of Hydrological Sciences *Publication* No. 159.

Hagen, L. J. 1976: Windbreak design for optimum wind erosion control. *Shelterbelts on the Great Plains, Proceedings of the Symposium*. Great Plains Agricultural Council, Publication No. 78.

Hairston, J. E., Sanford, J. O., Hayes, J. C. and Reinschmiedt, L. L. 1984: Crop yield, soil erosion, and net returns from live tillage systems in the Mississippi Blackland Prairie, *Journal of Soil and Water Conservation*, 39, 391–5.

Hard, G. 1971: Excessive bodenerosion um und nach 1800. *Erdkunde*, 24, 290–308.

Harold, L. L., Brakensiek, D. L., McGuinness, J. L., Amerman, C. R., and Dreibelbis, F. R. 1962: Influence of land use and treatment on the hydrology of small watersheds at Coshoton, Ohio, 1938–1957. United States Department of Agriculture *Technical Bulletin* 1256.

Hassell, W. G. 1977: Plant species for critical areas. In J. L. Thames (ed.), *Reclamation and Use of Disturbed Land in the Southwest*, pp. 340–52. Tucson, Arizona: The University of Arizona Press.

Hays, O. E., McCall, A. G., and Bell, F. G. 1949: Investigations in erosion control and the reclamation of eroded land at the Upper Mississippi Valley Conservation Experiment Station near la Crosse, Wisconsin, 1933–1943. United States Department of Agriculture *Technical Bulletin* 973.

Hazelhoff, L., Vantloof, P., Imeson, A. C., and Kwaad, F. J. P. M. 1981: The exposure of forest soil to erosion by earthworms. *Earth Surface Processes and Landforms*, 6, 235–50.

Heede, B. H. 1979: Deteriorated watersheds can be restored: A case study. *Environmental Management*, 3, 271–81.

Heinemann, H. G. and Whitaker, F. D. 1974: Soil cover governs soil loss on United States claypan soils. International Association of Hydrological Sciences Pub. 113, 109–113.

Hendrickson, B. H., Barnett, A. P., Carreker, J. R. and Adams, W. E. 1963: Runoff and erosion control studies on cecil soil in the Southern Piedmont. United States Department of Agriculture *Technical Bulletin* No. 1261.

Hershfield, D. M. 1961: Rainfall frequency atlas of the United States for durations from 30 minutes to 24 hours and return periods from 1 to 100 years. United States Department of Agriculture. Soil Conservation Service, *Technical Paper* no. 40.

Hewlett, J. D. 1979: *Forest Water Quality: An Experiment In Harvesting And Regenerating Piedmont Forest*. Athens, Georgia: School of Forest Resources, University of Georgia.

—— and Doss, R. 1984: Forests, floods and erosion: a watershed experiment in the southeastern piedmont. *Forest Science*, 30, 424–34.

——, Thompson, W. P., and Brightwell, N. 1979: Erosion control on forest land in Georgia. School of Forest Resources, University of Georgia.

Hill, R. L. and Cruse, R. M. 1985: Tillage effects on bulk density and soil strength of two Mollisols. *Soil Science Society of America Journal*, 49, 1270–3.

Hobbs, E. R. 1980: The Effects of Feral Sheep Grazing on Bishop Pine (*Pinus muricata*) Forests, Santa Cruz Island, California. Unpublished master's thesis, University of California, Los Angeles.

Holtan, H. N. and Kirkpatrick, M. H. 1950: Rainfall, infiltration, and hydraulics of flow in run-off computation. American Geophysical Union, *Transactions* 31 771–9.

—— and Musgrave, G. W. 1947: Soil water and its disposal under corn and under bluegrass. United States Department of Agriculture, Soil Conservation Service. *Technical Pamphlet* 68.

Howard, R. B. 1982: Erosion and sedimentation as part of the natural system. In C. E. Conrad and W. C. Oechel (eds), *Dynamics and Management of Mediterranean-Type Ecosystems*. United States Forest Service General Technical Report PSW-58.

Hupp, C. R. 1983: Geo-botanical evidence of late quaternary mass wasting in block field areas of Virginia. *Earth Surface Processes and Landforms*, 8, 439–50.

—— 1983: Seedling establishment on a landslide site. *Castanea*, 48, 89–98.

Imeson, A. C. 1976: Some effects of burrowing animals on slope processes in the Luxembourg Ardennes, Part 1. *Geografiska Annaler*, 58, 115–25.

—— 1977: Splash erosion, animal activity and sediment supply in a small forested Luxembourg catchment. *Earth Surfaces Processes*, 2, 153–60.

—— and Jungerius, P. D. 1976: Aggregate stability and colluviation in the Luxembourg Ardennes; an experimental and micromorphological study. *Earth Surface Processes*, 1, 259–71.

—— and Kwaad, F. J. P. M. 1976: Some effects of burrowing animals on slope processes in the Luxembourg Ardennes, Part 2. *Geografiska Annaler*, 58, 317–28.

—— and Vis, M. 1984: The output of sediments and solutes from forested and cultivated clayey drainage basins in Luxembourg. *Earth Surface Processes and Landforms*, 9, 585–94.

Ireland, H. A., Sharpe, C. F. S., and Eargle, D. H. 1939: Principles of gully erosion in the piedmont of South Carolina. United States Department of Agriculture, Technical Bulletin No. 633.

Jamison, V. C. 1969: Wetting resistance under citrus trees in Florida. In DeBano, L. F. and Letey, J. (eds), *Water-Repellant Soils*, pp. 9–15. University of California, Riverside.

Jansen, J. M. L. and Painter, R. B. 1974: Predicting sediment yield from climate and topography. *Journal of Hydrology*, 21, 371–80.

Judson, S. 1963: Erosion and deposition of Italian stream valleys during historic time. *Science*, 140, 898–9.

Kane, D. L. and Stein, J. 1983: Water movement in seasonally frozen soils. *Water Resources Research*, 19, 1547–57.

Kay, B. L. 1977: Hydroseeding and erosion control chemicals. In J. L. Thames (ed.), *Reclamation and Use of Disturbed Land in the Southwest*, pp. 238–47. Tucson, Arizona: The University of Arizona Press.

Kirkby, M. J. 1980: The problem. In M. J. Kirkby and R. P. C. Morgan (eds), *Soil Erosion*. New York: John Wiley and Sons.

Kladivko, E. J., Mackay, A. D. and Bradford, J. M. 1986: Earthworms as a factor in the reduction of soil crusting. *Soil Science Society of America Journal*, 50, 191–6.

Kolenbrander, G. J. 1974: Efficiency of organic manure in increasing soil organic matter content. *Transactions of the 10th International Congress on Soil Science*, vol. II, 129–36.

Kwaad, F. J. P. M. 1977: Measurements of rainsplash erosion and the formation of colluvium beneath deciduous woodland in the Luxembourg Ardennes. *Earth Surface Processes*, 2, 161–73.

Laflen, J. M. and Moldenhauer, W. C. 1979: Soil and water losses from corn–soybean rotations. *Soil Science Society of America Journal*, 43, 1213–15.

Langbein, W. B. and Schumm, S. A. 1958: Yield of sediment in relation to mean annual precipitation. American Geophysical Union *Transactions*, 39, 1076–84.

Langdale, G. W., Barnett, A. P., and Box, J. E., Jr. 1978: Conservation tillage systems and their control of water erosion in the Southern Piedmont. In J. T. Touchton and D. G. Cummins (eds), *Proceedings, First Annual Southeastern No-Till Systems Conference*. Special Publication no. 5, Georgia Experimental Station, Athens, 22–29.

La Page, W. F., 1962: Recreation and the forest site. *Journal of Forestry*, 60, 319–21.

Laws, J. O. and Parsons, D. A. 1943: The relation of rain drop size to intensity. American Geophysical Union *Transactions*, 24, 452–9.

Leishman, N. J. 1981: Effects of Feral Animals on Woody Vegetation: Santa Cruz Island, California. Unpublished MA thesis, University of California Los Angeles.

Leopold, L. B. 1956: Land use and sediment yield. In W. L. Thomas, Jr. (ed.), *Man's Role in Changing the Face of the Earth*, pp. 639–47. Chicago: University of Chicago Press.

Linstrom, G. A. 1960: Forestation of strip-mined land in the Central States. United States Department of Agriculture, *Agriculture Handbook* no. 166.

Lockaby, B. G. and Dunn, B. A. 1984: Camping effects on selected soil and vegetative properties. *Journal of Soil and Water Conservation*, 39, 215–16.

Lockeretz, W., Shearer, G., Klepper, R. and Sweeney, S. 1978: Field crop production on organic farms in the Midwest. *Journal of Soil and Water Conservation*, 33, 130–4.

Lowdermilk, W. C. 1953: Conquest of the land through 7,000 years. United States Department of Agriculture, Soil Conservation Service, *Agriculture Information Bulletin* No. 99.

Lull, H. W. and Reinhart, K. G. 1965: Logging and erosion on rough terrain in the east. Proceedings of the Federal Inter-Agency Sedimentation Conference, 1963. United States Department of Agriculture, *Miscellaneous Publication* No. 970, 43–7.

Lusby, G. L. and Toy, T. J. 1976: An evaluation of surface-mine spoils area restoration in Wyoming using rainfall simulation. *Earth Surface Process*, 1, 375–86.

Lutz, H. J. 1940: Disturbance of forest soil resulting from the uprooting of trees. *Bulletin of Yale University*, Sch. No. 45.

—— 1945: Soil conditions in picnic grounds in public forest parks. *Journal of Forestry*, 43, 121–7.

—— 1960: Movement of rock by uprooting of forest trees. *American Journal of Science*, 258, 752–6.

Lyles, L. 1977: Wind erosion: processes and effects on soil productivity. *Transactions, American Society of Agricultural Engineers*, 20, 880–4.

McCalla, T. M. 1942: Influence of biological products on soil structure and infiltration. *Soil Science Society of America Proceedings*, 7, 209–14.

McClurkin, D. C. 1967: Vegetation for erosion control in the southern coastal plain of the United States. In Sopper, W. E. and Lull, H. E. (eds), *International Symposium on Forest Hydrology*. New York: Pergamon.

Mader, D. L., Lull, H. W., Swenson, E. T. 1977: Humus accumulations in hardwood stands in the northeast. *Massachusetts Agricultural Experiment Station Research Bulletin*, No. 648.

Margaropoulos, P. 1967: Woody revegetation as a pioneer action towards restoring of totally eroded slopes in mountainous watersheds. In Sopper, W. E. and Lull, H. E. (eds), *International Symposium on Forest Hydrology*. New York: Pergamon.

Maryland Department of Water Resources, Becker, B. C. and Mills, T. R. 1972: Guidelines for erosion and sediment control planning and implementation. United States Environmental Protection Agency, Pub. No. EPA-R2-72-015.

Massie, L. R. and Bubenzer, G. D. 1974: Improving road bank erosion control. *Journal of Soil and Water Conservation*, 29, 176–8.

May, R. F. and Striffler, W. D. 1967: Watershed aspects of stabilization and restoration of strip-mined areas. In Sopper, W. E. and Lull, H. E. (eds), *International Symposium on Forest Hydrology*. New York: Pergamon.

Meade, R. H. 1969: Errors in using modern stream-load data to estimate natural rates of denudation. *Geological Society of America Bulletin*, 80, 1265–74.

—— 1982: Sources, sinks and storage of river sediment in the Atlantic drainage of the United States. *Journal of Geology*, 90, 235–52.

Megahan, W. F. 1978: Erosion processes on sleep granitic road fills in central Idaho. *Soil Science Society of America Proceedings*, 42, 350–7.

——, Seyedbagheri, K. A., and Dodson, P. C. 1983: Long-term erosion on granitic road cuts based on exposed tree roots. *Earth Surface Processes and Landforms*, 8, 19–28.

Mersereau, R. C. and Dyrness, C. T. 1972: Accelerated mass washing after logging and slash burning in western Oregon. *Journal of Soil and Water Conservation*, 27, 112–14.

Miller, D. E. and Aarstad, J. S. 1983: Residue management to reduce furrow erosion. *Journal of Soil and Water Conservation*, 38, 366–70.

Miller, I. A. 1957: Burrows of the Sacramento Valley pocket gopher in flood-irrigated alfalfa fields. *Hilgardia*, 26, 431–52.

Minnich, R. A. 1980: Vegetation of Santa Cruz and Santa Catalina Islands. In Power, D. (ed.), *A Multidisciplinary Symposium on the California Islands*, pp. 123–37. Santa Barbara, California: Santa Barbara Museum of Natural History.

Moldenhauer, W. C., Wischmeier, W. H., and Parker, D. T. 1967: The influence of crop management on runoff, erosion, and soil properties of a Marshall silty clay loam. *Soil Science Society of America Proceedings*, 31, 541–6.

Morgan, R. P. C. 1980: Soil erosion and conservation in Britain. *Progress in Physical Geography*, 4, 24–47.

—— 1985: Soil degradation and erosion as a result of agricultural practice. In K. S. Richards, R. R. Arnett, and S. Ellis (eds), *Geomorphology and Soils*, pp. 379–95. London: George Allen & Unwin.

——, Finney, H. J., Lavee, H., Merritt, E., and Noble, C. A. 1986: Plant cover effects on hillslope runoff and erosion: evidence from two laboratory experiments. In A. D. Abrahams (ed.), *Hillslope Processes*, pp. 77–96. Boston: Allen & Unwin.

Mosley, M. P. 1982: The effect of a New Zealand beech forest canopy on the kinetic energy of water drops and on surface erosion. *Earth Surface Processes and Landforms*, 7, 103–7.

Mueller, D. H., Wendt, R. C., and Daniel, T. C. 1984: Soil and water loss as affected by tillage and manure application. *Soil Science Society of America, Journal*, 48, 896–900.

Nielsen, G. A. and Hole, F. D. 1964: Earthworms and the development of coprogenous A1 horizons in forest soils of Wisconsin. *Soil Science Society of America, Proceedings*, 28, 426–30.

Nikolayenko, V. T., 1974: The role of forest stands in the control of erosion processes and other negative natural phenomena. International Association of Scientific Hydrology, *Publication* 113.

Noble, C. A. and Morgan, R. P. C. 1983: Rainfall interception and splash detachment with a brussels sprouts plant: a laboratory simulation. *Earth Surface Processes and Landforms*, 8, 569–77.

Olmstead, L. B. 1946: The effect of long-time cropping systems and tillage practices upon soil aggregation at Hays, Kansas. *Soil Science Society of America Proceedings*, 11, 89–92.

Packer, P. E. 1953: Effects of trampling disturbance on watershed condition, runoff, and erosion. *Journal of Forestry*, 51, 28–31.

Painter, R. B., Blyth, K., Mosedale, T. C. and Kelly, M. 1974: The effect of afforestation on erosion processes and sediment yield. International Association of Hydrological Sciences, *Publication* 113, 62–7.

Peele, T. C. and Beale, O. W. 1940: Influence of microbial activity upon aggregation and erodibility of lateritic soils. *Soil Science Society of America Proceedings*, 5, 33–5.

—— 1941: Effect on runoff and erosion of improved aggregation resulting from stimulation of microbial activity. *Soil Science Society of America Proceedings*, 6, 176–82.

Phillips, R. E., Blevins, R. L., Thomas, G. W., Frye, W. W. and Phillips, S. H. 1980: No-tillage agriculture. *Science*, 208, 1108–13.

Pikul, J. L., Jr. and Allmaras, R. R. 1986: Physical and chemical properties of a Haploxeroll after fifty years of residue management. *Soil Science Society of America Journal*, 50, 214–9.

——, Zuzel, J. F., and Greenwalt, R. N. 1986: Formation of soil frost as influenced by tillage and residue management. *Journal of Soil and Water Conservation*, 41, 196–9.

Pinczes, Z. 1982: Variations in runoff and erosion under various methods of protection. In R. F. Hadley (ed.), *Recent Developments in the Explanation and prediction of Erosion and Sediment Yield*, International Association of Hydrological Sciences, Publication 137.

Plummer, A. P., 1977: Revegetation of disturbed intermountain area sites. In J. L. Thames (ed.), *Reclamation and Use of Disturbed Land in the Southwest*, pp. 302–339. Tucson, Arizona: The University of Arizona Press.

Ralston, C. W. and Hatchell, G. E. 1971: Effects of prescribed burning on physical properties of soil. In *Prescribed Burning Symposium Proceedings*, United States Department of Agriculture, Forest Services Southeast Forest Experimental Station, Asheville, North Carolina, 68–85.

Ramig, R. E. and Mazurak, A. P. 1964: Wheat stubble management: I. Influence on some physical properties of a Chernozem soil. *Soil Science Society of America Proceedings*, 28, 554–7.

Reeves, R. W. 1970: Modern Channel Entrenchment in the Coastal Ranges of Central and Southern California. Unpublished Ph.D. thesis, University of California, Los Angeles.

Rice, R. M. 1974: The hydrology of chaparral watersheds. In M. Rosenthal (ed.), *Symposium on Living with the Chaparral, Proceedings*, San Francisco: Sierra Club, 27–34.

Richter, G. and Negendank, J. F. W. 1977: Soil erosion processes and their measurement in the German areas of the Moselle River. *Earth Surface Processes*, 2, 261–78.

Robinson, M. and Blyth, K. 1982: The effect of forestry drainage operations on upland sediment yields: a case study. *Earth Surface Processes and Landforms*, 7, 85–90.

Rodda, J. C. 1970: Rainfall excesses in the United Kingdom. *Transactions of the Institute of British Geographers*, 49, 49–60.

Rolfe, G. L. and Boggess, W. R. 1973: Soil conditions under old field and forest cover in southern Illinois. *Soil Science Society of America Proceedings*, 37, 314–18.

Sartz, R. S. 1953: Soil erosion on a fire-denuded forest area in the douglas-fir region. *Journal of Soil and Water Conservation*, 8, 279–81.

—— 1973: Effect of forest cover removal on depth of soil freezing and overland flow. *Soil Science Society of America Proceedings*, 37, 774–7.

—— 1975: Controlling runoff in the driftless area. *Journal of Soil and Water Conservation*, 30, 92–3.

Schnepf, M. (ed.) 1983: Conservation tillage: special issue of *Journal of Soil and Water Conservation*, 38, 134–319.

Schuman, G. E., Taylor, E. M., Jr., Ravzi, F. and Pinchak, B. A. 1985: Revegetation of mined land: Influence of topsoil depth and mulching method. *Journal of Soil and Water Conservation*, 40, 249–52.

Schumm, S. A. 1965: Quaternary paleohydrology. In Wright, H. E., Jr. and Frey, D. G. (eds), *The Quaternary of the United States*, pp. 783–94. Princeton: Princeton University Press.

Sedimentation and Erosion Sub-Task Force of the Federal Interdepartmental Task Force 1967: *Report on the Potomac: Hyattsville, Maryland*. United States Department of Agriculture, Soil Conservation Service.

Settergren, C. D. and Cole, D. M. 1970: Recreation effects on soil and vegetation in the Missouri Ozarks. *Journal of Forestry*, 68, 231–3.

Shelton, C. H., Tompkins, F. D., and Tyler, D. D. 1983: Soil erosion from five soybean tillage systems. *Journal of Soil and Water Conservation*, 38, 425–8.

Siddoway, F. H. 1963: Effects of cropping and tillage methods on dry aggregate soil structure. *Soil Science Society of America Proceedings*, 27, 452–454.

Sidle, R. C., Pearce, A. J. and O'Loughlin, C. L. 1985: *Hillslope Stability and Land Use*. Washington, DC: American Geophysical Union.

Singh, T., Kalra, Y. P., Hillman, G. R. 1974: Effects of pulpwood harvesting on the quality of stream waters of forest catchments representing a large area in Western Alberta, Canada. International Association of Hydrological Sciences, *Publication* 113, 21–7.

Skidmore, E. L. and Woodruff, N. P. 1968: Wind erosion forces in the United States and their use in predicting soil loss. United States Department of Agriculture, Agricultural Research Service, *Agriculture Handbook* No. 346.

——, Carstenson, W. A., and Banbury, E. E. 1975: Soil changes resulting from cropping. *Soil Science Society of America Proceedings*, 39, 964–5.

——, Layton, J. B., Armbrust, D. V., and Hooker, M. L. 1986: Soil physical properties as influenced by cropping and residue management. *Soil Science Society of America Journal*, 50, 415–19.

——, Nossaman, N. L., and Woodruff, N. P. 1966: Wind erosion as influenced by row spacing, row direction, and grain sorghum population. *Soil Science Society of America Proceedings*, 30, 505–9.

Smith, D. J. and Gardner, J. S. 1985: Geomorphic effects of ground squirrels in the Mount Rae area, Canadian Rocky Mountains. *Arctic and Alpine Research*, 17, 205–10.

Sommerfeldt, T. G. and Chang, C. 1985: Changes in soil properties under annual applications of feed lot manure and different tillage practices. *Soil Science Society of America Journal*, 49, 983–7.

Sorensen, C. J. and Marotz, G. A. 1977: Changes in shelter belt mileage statistics over four decades in Kansas. *Journal of Soil and Water Conservation*, 32, 276–81.

Stephens, E. P. 1956: The uprooting of trees: a forest process. *Soil Science Society of America, Proceedings*, 20, 113–16.

Suedkamp, J. F. 1976: Tree and shrub species for conservation use in the northern Great Plains. *Shelter belts on the Great Plains, Proceedings of the Symposium*. Great Plains Agricultural Council, Publication No. 78, 130–3.

Swanson, F. J., Tanda, R. J., Dunne, T. and Swanston, D. N. 1982: Sediment budgets and routing in forested drainage basins. United States Department of Agriculture, Forest Service, *General Technical Report* PNW-141.

Tanaka, T. 1982: The role of subsurface water exfiltration in soil erosion processes. International Association of Hydrological Sciences, Publication 137, 73–80.

Taylor, R. E., Hays, O. E., Bay, L. E., and Dixon, R. M. 1964: Corn stover mulch for control of runoff and erosion on land planted to corn after corn. *Soil Science Society of America Journal*, 28, 123–5.

Thorn, C. E. 1978: A preliminary assessment of the geomorphic role of pocket gophers in the alpine zone of the Colorado Front Range. *Geografiska Annaler*, 60, 181–7.

—— 1982: Gopher disturbance: its variability by Braun–Blanquet vegetation units in the Niwot Ridge alpine tundra zone, Colorado Front Range, USA. *Arctic and Alpine Research*, 14, 45–51.

Toy, T. J. 1984: Geomorphology of surface-mined lands in the Western United States. In Costa, J. E. and Fleisher, P. J. (eds), *Developments and Applications of Geomorphology*, pp. 134–70. Berlin: Springer-Verlag.

Trimble, S. W. 1974: *Man-Induced Soil Erosion on the Southern Piedmont, 1700–1970*. Soil Conservation Society of America.

—— 1975: Denudation studies: Can we assume stream steady state? *Science*, 188, 1207–8.

—— 1976: Unsteady state denudation. *Science*, 191, 871.

—— 1977: The fallacy of stream equilibrium in contemporary denudation studies. *American Journal of Science*, 277, 876–87.

—— 1983: A sediment budget for Coon Creek basin in the Driftless Area, Wisconsin, 1853–1977. *American Journal of Science*, 283, 454–74.

—— 1985. Perspectives on the history of soil erosion control in the eastern United States. *Agricultural History*. 59, 162–80.

—— and Lund, S. W. 1982: Soil conservation and the reduction of erosion and sedimentation in the Coon Creek basin, Wisconsin. US Geological Survey *Professional Paper* 1234.

Troeh, F. R., Hobbs, J. A. and Donahue, R. L. 1980: *Soil and Water Conservation for Productivity and Environmental Protection*. Englewood Cliffs, New Jersey: Prentice-Hall, Inc.

Tropeano, D. 1984a: Rate of soil erosion processes on vineyards in Central Piedmont (NW Italy). *Earth Surface Processes and Landforms*, 9, 253–66.

—— 1984b: Soil loss and sediment yield from a small basin in the Langhe area (Piedmont, NW Italy): a first report. *Geologia Applicata e Idrogeologia*, 19, 269–87.

Tyre, G. L. and Barton, R. G. 1986: Treating critical areas in the Tennessee copper basin. *Journal of Soil and Water Conservation*, 41, 381–2.

Unger, P. W. 1968: Soil organic matter and nitrogen changes during 24 years of dryland wheat tillage and cropping practices. *Soil Science Society of America Proceedings*, 32, 427–9.

United States Department of Agriculture, Soil Conservation Service 1971: Sedimentation. In *National Engineering Handbook*, section 3, 1–19.

United States Department of Agriculture, Soil Conservation Service 1985: *Guides for Erosion and Sediment Control*. Davis, California.

Ursic, S. J. and Dendy, F. E. 1965: Sediment yields from small watersheds under various land uses and forest covers. Proceedings of the Federal Inter-Agency Sedimentation Conference, United States Department of Agriculture, *Miscellaneous Publication* no. 970, 47–52.

Van Doren, D. M., Jr., Moldenhauer, W. C. and Triplett, G. B. Jr. 1984: Influence of long-term tillage and crop rotation on water erosion. *Soil Science Society of America Journal* 48, 636–40.

Verstraten, J. M. 1977: Chemical erosion in a forested watershed in the Oesling, Luxembourg. *Earth Surface Processes*, 2, 175–84.

Vice, R. B., Guy, H. P., and Ferguson, G. E., 1969: Sediment movement in an area of suburban highway construction, Scott Run Basin, Fairfax County, Virginia, 1961–64. U.S. Geological Survey *Water Supply Paper* 1591-E.

Vita-Finzi, C. 1969: *The Mediterranean Valleys: Geological Changes in Historical Times*. Cambridge: Cambridge University Press.

Voroney, R. P., vanVeen, J. A. and Paul, E. A. 1981: Organic carbon dynamics in grassland soils. II. Model validation and simulation of the long-term effects of cultivation and rainfall erosion. *Canadian Journal of Soil Science*, 61, 211–24.

Voslamber, B. and Veen, A. W. L. 1985: Digging by badgers and rabbits on some wooded slopes in Belgium. *Earth Surface Processes and Landforms*, 10, 79–82.

Walling, D. E. 1983: The sediment delivery problem. *Journal of Hydrology*, 69, 209–37.

Ward, R. C. 1975: *Principles of Hydrology*. New York: McGraw-Hill.

Welling, R., Singer, M., and Dunn, P. 1984: Effects of fire on shrubland soils. In J. J. DeVries (ed.), *Shrublands in California: Literature Review and Research Needed for Management*, pp. 42–50. Berkeley: California Water Resources Center.

Wells, C. G., Campbell, R. E., DeBano, L. F., Lewis, C. E., Fredriksen, R. L., Franklin, E. C., Froelich, R. C. and Dunn, P. H. 1979: Effects of fire on soil: A state-of-knowledge review. United States Department of Agriculture, Forest Service, *General Technical Report*, WO-7.

Wells, W. G., II. 1981: Some effect of bushfires on erosion processes in coastal southern California. International Association of Hydrological Sciences, *Publication* 132, 305–42.

Wendt, R. C. and Burwell, R. E. 1985: Runoff and soil losses for conventional, reduced, and no-till corn. *Journal of Soil and Water Conservation*, 40, 450–4.

Whipkey, R. Z. 1965: Subsurface stormflow from forested slopes. International Association of Scientific Hydrology, *Bulletin*, 10, 74–85.

—— 1969: Storm runoff from forested catchments by subsurface routes. *Floods and Their Computation, UNESCO–IASH–WMO Studies and Reports in Hydrology*, No. 3, 773–9.

Whitaker, F. D., Heinemann, H. G., and Wischmeier, W. H. 1973: Chemical weed controls affect runoff, erosion, and corn yields. *Journal of Soil and Water Conservation*, 28, 174–6.

Williams, G. P. and Guy, H. P. 1973: Erosional and depositional aspects of Hurricane Camille in Virginia, 1969. U.S. Geological Survey *Professional Paper* 804.

Wilson, L. 1973: Variation in mean annual sediment yield as a function of mean annual precipitation. *American Journal of Science*, 273, 335–49.

Wischmeier, W. H. 1966: Relation of field-plot runoff to management and physical factors. *Soil Science Society of America Proceedings*, 30, 272–7.

—— 1975: Estimating the soil loss equation's cover and management factor for undisturbed areas. In *Present and Prospective Technology for Predicting Sediment Yields and Sources*, United States Department of Agriculture, Agricultural Research Service, No. ARS-S-40.

—— and Mannering, J. V. 1965: Effect of organic matter content of the soil on infiltration. *Journal of Soil and Water Conservation*, 20, 150–2.

—— and Smith, D. D. 1965: Predicting rainfall-erosion losses from cropland east of the Rocky Mountains. United States Department of Agriculture, *Agricultural Handbook*, No. 282.

—— and Smith, D. D. 1978: Predicting rainfall losses – a guide to conservation planning. United States Department of Agriculture, *Agriculture Handbook* No. 537.

Wolman, M. G. 1967: A cycle of sedimentation and erosion in urban river channels. *Geografiska Annaler*, 49A, 385–95.

Woodruff, N. P., Lyles, L., Dickerson, J. D. and Armbrust, D. V. 1974: Using cattle feedlot manure to control wind erosion. *Journal of Soil and Water Conservation* 29, 127–9.

—— and Siddoway, F. H. 1965: A wind erosion equation. *Soil Science Society of America, Proceedings*, 29, 602–8.

Woolridge, D. D. 1965: Soil properties related to erosion of wild-land soils in central Washington. In *Forest–Soil Relationships in North America*. Corvallis: Oregon State University Press, 141–52.

Young, R. A., 1978: Camping intensity effects on vegetative ground cover in Illinois campgrounds. *Journal of Soil and Water Conservation*, 33, 36–9.

Ziemer, R. R. 1981: Roots and the stability of forested slopes. International Association of Hydrological Sciences, *Publication* 132, 343–57.

Part II
Tropical, Arid and Periglacial Environments

4 The tropical rain forest landscape

Carl F. Jordan

Introduction

A spectrum of biogeomorphological processes may extend along global or regional gradients. At one end of the spectrum, landscapes and geomorphology strongly influence plant and animal communities, but the reverse influence is weak. For example, desert and tundra ecosystems are highly stressed by physical factors, but organisms in these communities have a small effect on the landscape (but see chapters 6 and 7 of this volume). Tropical rain forests may be on the other end of the spectrum. This chapter explores the idea that tropical rain forests may strongly influence the inorganic landscape, but this landscape and its associated climate may have relatively little effect on the rain forest ecosystem.

Figure 4.1 A nineteenth-century view of rain forest vegetation in the Amazon rain forest
(from Bates, 1892)

Figure 4.2 A nineteenth-century view of birdlife within the Amazon rain forest (from Bates, 1892)

Rain forest impact on geomorphological processes

Why should rain forests have a relatively large influence on geomorphological processes in the wet tropics? The answer is related to (1) the annual amount of biological activity in the tropical humid landscape, and (2) the way this biological activity influences the landscape. A large amount of information has been collected upon these, and other, facets of the tropical rain forest environment in recent years, which has radically changed ideas on ecology in the wet tropics. Earlier perceptions of the wet tropical environment were strongly influenced by reports from nineteenth-century explorers and their accompanying drawings, as illustrated by figures 4.1 and 4.2.

Biological activity

Geographically, the tropics are between 23.5° N. and S. latitudes, but a thermal definition of the tropics is more useful here. The tropics can refer to the area where the mean annual isotherm is 20°C or above (Tricart, 1972). This definition restricts the tropics to lowland areas. Average temperatures in mountain ranges within the geographical tropics may be well below 20°C or even below the freezing point.

The rate of enzymatic reactions approximately doubles for each 10°C increase in temperature, between 0° and about 38°C. Thus metabolic activity of higher plants, and decomposers such as bacteria, is greater where temperatures are higher. Where temperatures are high throughout the year, total annual metabolism expressed in grams of carbon photosynthesized or respired per unit of landscape per year is potentially high. The high potential is realized, however, only if there is adequate moisture for living organisms. Although species vary widely in their ability to tolerate drought, 100 mm of rainfall per month is often used as an index of the occurrence of potential moisture stress in the tropics (Walter, 1979).

Moisture regimes vary greatly within the tropics. Gradients between areas having only a few months of the year with any rainfall, to areas with all 12 months of the year having greater than 100 mm per month occur in the Americas, Africa, and South East Asia. Tropical regions with all months receiving more than 100 mm of rainfall would be expected to have the world's highest annual totals of photosynthesis and respiration. Such regions when undisturbed by man usually support an evergreen rain forest.

Evidence for effects of temperature and moisture

Is there any evidence that photosynthesis and respiration in warm wet regions are relatively high on an annual basis? Comparisons of biological rates in different ecosystems are often difficult, because of structural and functional differences between species. Nevertheless, if comparisons are restricted to the same general types of organisms and tissues, global patterns often can be seen. For example, table 4.1 compares leaf and fine litter production of late successional to mature broad-leaved forests along a latitudinal gradient on sites not severely limited by low rainfall. There is a very marked trend of increasing leaf productivity with decreasing latitude.

If primary productivity is higher in tropical latitudes than in higher zones, rates of decomposition also must be higher in the tropics. If they were not, there would be a continual build-up of organic matter on the forest floor in the tropics. Table 4.2 clearly shows that this is not the case. In fact, this table shows that accumulation of organic matter in the soil is less in the tropics than at higher latitudes, for any given moisture regime.

Why are stocks of soil organic matter less in the tropics, when productivity there is higher? The amount of organic matter in and on the soil represents the balance between inputs from the death of plants, and outputs due to decomposers. If decomposers and producers responded in exactly the same way to changes in temperature, a doubling of productivity would be accompanied by a doubling of decomposition, and the stock of organic matter on the forest floor would remain constant. However, this is not the case. As temperature increases, photosynthesis increases less rapidly than respiration, especially above 20°C. An increase in temperature causes a relatively greater increase in respiration than in photosynthesis.

Table 4.1 Leaf and fine litter production in broad-leaved mesic forests in five latitudinal zones

Zone	Tonnes ha^{-1} yr^{-1} Average	1 std devn	No. sites studied
Boreal	2.8	1.3	23
N. temperate	3.3	1.1	24
S. temperate	3.7	1.2	35
Sub-tropical	6.4	2.0	20
Tropical	9.6	3.6	33

Source: Jordan, 1983

Table 4.2 Soil organic carbon (kg per surface cubic metre) as a function of latitudinal belt and rainfall

Latitudinal belt	Precipitation range (mm yr^{-1}) 250-500	500-1000	1000-2000	2000-4000
Tropical	1-3	3-7	7-10	10-14
Sub-tropical and warm temperate	3-8	6-9	8-10	12-17
Cool temperate	8-11	9-14	12-17	18-22
Boreal	11-15	15-20	18-33	—

Source: Zinke *et al.*, 1984

Therefore, along a latitudinal gradient from boreal to tropical regions, annual totals of photosynthesis increase, but respiration increases even faster, and consequently stocks of soil organic matter decrease.

Moisture also has a differential effect on producers and decomposers. Along a gradient from a wet to a dry environment at any latitude, net primary productivity falls off faster than decomposition. As a result, the standing stock of organic matter on the forest floor is lower in dry than in wet ecosystems for a given latitudinal belt (table 4.2).

Biological activity and the landscape

Due to high annual rates of photosynthesis and decomposition, total respiration is greater in a tropical rain forest environment than in regions that are cooler or drier. Much of the respiration in forest ecosystems occurs in the soil or near the soil surface, because of roots and decomposers. An end product of respiration is carbon dioxide, and this gas is released into the soil by roots and decomposers. If the soil is moist, as it is in tropical rain forests, the carbon dioxide reacts with water to form carbonic acid.

Carbonic acid readily dissociates into positively charged hydrogen ions and negatively charged biocarbonate ions (Johnson et al., 1977). The hydrogen ions replace nutrients such as potassium within the crystal structure of soil minerals. This is part of the process of 'weathering'. Once released into the soil solution, nutrient ions are rapidly leached away into ground water or nearby drainage streams. It is the continual year-round weathering of minerals in the parent rock, and in the clay fraction of the soils that results in the very deep, but nutrient-poor, quality of most of the soils of the humid tropics (Sanchez, 1976).

From a landscape perspective, the effect of the deep weathering on the parent rock beneath the soil is critical. The continual intense chemical weathering penetrates deeply into the parent rock. The weathering not only releases nutrient ions, it also breaks down the primary minerals which comprise the rock, and softens them into clay.

Effects of deep weathering

In the alpine zones of tropical mountains such as the Andes, the mountains are jagged and angular, just as are the mountains in higher latitudes. But below about 1,000 m, the low mountains and hills of the tropics have a very different appearance than the lower mountains in higher latitudes. The tropical hills are much more rounded, and less angular.

The deep chemical weathering due to year-round soil respiration is the cause of this landform in the humid tropics. Chemical weathering there can penetrate to a depth of six metres (Jenny, 1980), and soften the material between the active soil horizon and the underlying rock. This highly weathered rock, or 'saprolite', often looks as if it has the consistency of the underlying rock, but it is soft, and the blade of a knife is easily pushed into it.

This soft subsoil is relatively unresistant to physical forces such as landslides. Large landslides are not uncommon in the tropics. In 1976, two earthquakes struck near the southeastern coast of Panama, and denuded about 54 square kilometres of steep terrain originally covered by rain forest (Garwood et al., 1979). Garwood et al. also mentioned accounts of an earthquake in New Guinea in 1935 that had carried away 130 square kilometres of tropical rain forest.

Landslides in the humid tropics frequently occur during or after heavy rains, when the soil becomes saturated, and the flow of water through or over the saprolite loosens it, and breaks it from the underlying rock, especially in forests where the rooting is shallow. For example, extremely heavy rainfall in Puerto Rico in 1979 during hurricane 'David' (Lugo et al., 1983) saturated the subsoil on a slope in the Luquillo mountains, causing a landslide that carried away several square kilometres of the mountainside, and left a chasm 10 metres deep and half a kilometre across where there previously had been a road. Five years before the landslide, Lewis (1974) had predicted that landslides were likely in that forest during heavy rains, because of the shallow rooting of the trees.

Because of the soft nature of the subsoil, the edges of the land forms remaining after the landslides are rounded. The soft contours of crumbling saprolite dominate the landscape. In contrast, where landscape formation is dominated more by winds and the action of freezing and thawing, the landscape has a much more rugged appearance.

Laterization

The intense chemical weathering that occurs in the humid tropics leaches not only nutrient elements from the soil horizon, it also leaches silica. This process is called laterization, and it leaves soil material which is almost pure iron and aluminium oxide (Jenny, 1980). When this material is subjected to extreme drying, it forms a hard durable material. In 1800, in the red earth country of India's Malabar Coast, F. Buchanan, MD, observed Indians cutting slabs out of soft red and ivory mottled clay strata, air-drying them to a hard rock, and using them as building stones. He called the brick and hardened soil mantles 'laterite' from the Latin word 'later' meaning brick.

A number of years ago when rates of tropical deforestation began to increase, the fear was expressed that the exposure of tropical soils to the sun would cause the tropical soils to harden into a brick-like pavement (McNeil, 1964). In most cases, however, this did not happen. The reason may be that an area of rain forest which is cut often quickly regrows with new vegetation. If the deforested area is not planted to crops or pasture, natural vegetation quickly establishes. The shading of the soil surface by the vegetation prevents the soil surface from drying out and becoming hard.

If the vegetation is continually and deliberately removed, as it is where the subsoil is used to build roads and air strips, then a hard and durable crust does develop.

Impact of the humid tropical environment on rain forests

Ecosystems under stress usually have low biomass, small structure, and low productivity (Woodwell, 1970; Odum, 1985; Rapport *et al.*, 1985). Deserts show such symptoms, and thus are ecosystems strongly influenced by the environment. When a stress such as low water availability has existed for long periods, as it has in deserts, some species have adapted to the conditions, and it may not be correct to say that they are stressed. For example, Solbrig and Orians (1977) reviewed some of the adaptations of desert plants to drought conditions. Some plants are phreatophytes with long deep roots that are able to tap underground sources of water. Some plants such as perennial evergreen shrubs have small leathery leaves specialized to withstand water stresses. A third type are succulents such as cacti, usually without leaves, an ability to store water, and very low photosynthetic rates. A fourth type, ephemerals which evade drought, have adapted by having a life span short enough to complete a life cycle during periods of moisture availability.

Since these species are adapted to drought, they cannot be said to experience drought stress under usual desert conditions. The term drought stress is more applicable to a wilting corn plant in the desert. Nevertheless, the low biomass, low productivity, and small structure of native desert vegetation suggest that it has been influenced very strongly by the environment, if not at present, then during the evolutionary past.

In contrast to desert ecosystems, tropical rain forests do not immediately appear to be negatively affected by the environment. Rain forests have large structure, biomass, and high primary productivity. The hypothesis that the tropical humid landscape has little influence on the rain forest ecosystem appears to be at least superficially true.

However, there is another indicator of stress which might be more meaningful for ecosystems with soils low in nutrients and in low nutrient-holding capacity, as are most rain forest soils. The cycling index, that is the efficiency with which nutrients are recycled, may be relatively low in ecosystems stressed by forces such as leaching and weathering (Odum, 1985). A forest with a leaky cycle would have a low index (Finn, 1976, 1978). A forest with a tight cycle for some or all its nutrients has a high cycling index. If humid tropical environments have little influence on rain forests as hypothesized, tropical rain forests should have a high nutrient cycling index. On the other hand, if the continual forces of leaching and weathering somehow disrupt the functioning of rain forests, they should have a low index.

The cycling index can be calculated as that proportion of the total amount of material entering a compartment which is recycled and eventually returns to that compartment (Finn, 1978). For example, suppose that 10 kg ha^{-1} yr^{-1} of a nutrient enter the soil by throughfall, stemflow, litter-fall and tree-fall, that 2 kg ha^{-1} yr^{-1} are leached from the soil, and the stock in the soil remains constant. This means that 8 kg ha^{-1} yr^{-1} are taken up by the roots of trees and eventually return again to the soil. The cycling index for the soil would be 0.8.

Cycling index case study

As part of a rain forest ecosystem study on an Oxisol low in nutrients in the upper reaches of the Rio Negro near San Carlos, Venezuela, cycling indices were determined for several nutrients. The models on which the indices for calcium and potassium are based are shown in figures 4.3 and 4.4. The cycling indices are calculated in table 4.3.

The rate of nutrient movement into the humus mat compartment on the forest floor is the sum of throughfall, stemflow, litter-fall and tree-fall (row *a*, table 4.3). The rate of nutrient uptake from the mat (row *c*) is the difference between row *a* and the total leached from the mat (row *b*).

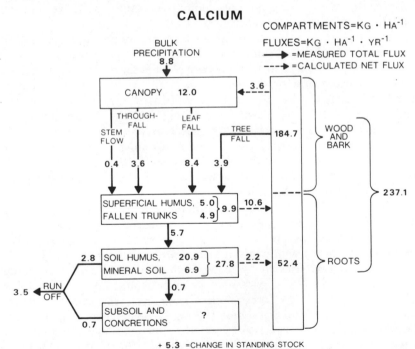

Figure 4.3 Stocks and flows of calcium in a rain forest ecosystem on Oxisol, near San Carlos de Rio Negro, Venezuela

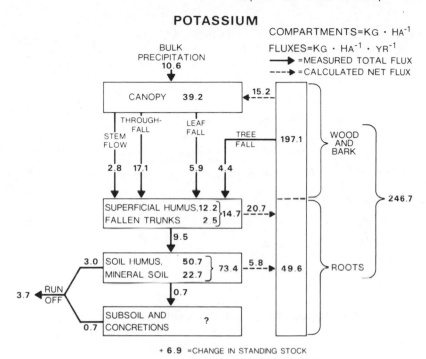

Figure 4.4   Stocks and flows of potassium in a rain forest ecosystem on Oxisol, near San Carlos de Rio Negro, Venezuela

Since all the nutrients taken up by the vegetation eventually return again to the humus, row *c* divided by row *a* is the cycling index for the humus compartment (row *d*).

The same procedure is used to calculate the cycling index for the mineral soil alone. The difference between rates of nutrient movement into and out of the mineral soil (row *f*) is the rate of nutrient uptake from mineral soil to vegetation. Row *f* divided by the amount entering the mineral soil (row *b*) is the cycling index for the mineral soil (row *g*).

The cycling index for the humus mat plus the mineral soil (row *h*) is calculated as the total recycling rates (rows *c* and *f*) divided by total entering the forest floor (row *a*).

The data in table 4.3 quantify the proportion of recycled nutrients that are cycled directly in the San Carlos forest, when 'direct nutrient cycling' is defined as the movement of nutrients from decomposing litter to roots without passage through the mineral soil. Direct nutrient cycling is important in rain forest ecosystems, because nutrients maintained in the organic portion of the ecosystem are less susceptible to losses than those simply exchanged on the surface of mineral clays. The proportion of total recycled nutrients which is recycled 'directly' (row *i*) is uptake from the

Table 4.3 Cycling indices for cations in the humus mat and total soil compartments in the Oxisol site. Data from figures 4.3 and 4.4

	Ca	*K*
a Total entering forest floor*	16.3	30.2
b Total leached through mat*	5.7	9.5
c Total cycled from mat $(a-b)$*	10.6	20.7
d Cycling index, mat compartment $\left(\dfrac{c}{a}\right)$	0.65	0.69
e Total leached out of soil*	3.5	3.7
f Total cycled from soil $(b-e)$*	2.2	5.8
g Cycling index, soil alone $\left(\dfrac{f}{b}\right)$	0.39	0.61
h Cycling index, mat+soil $\left(\dfrac{c+f}{a}\right)$ or $\left(\dfrac{a-e}{a}\right)$	0.79	0.88
i Proportion of total cycling that occurs 'directly $\left(\dfrac{c}{c+f}\right)$ or $\left(\dfrac{c}{a-e}\right)$	0.83	0.78

*Rates in kg ha^{-1} yr^{-1}

humus (row *c*) divided by total uptake. Direct nutrient cycling clearly is important for nutrient cations in the Oxisol forest at San Carlos.

Comparison of cycling indices for San Carlos with other ecosystems is difficult because of few comparable data. Although watershed studies often include nutrient cycling data (Likens *et al.*, 1977), nutrient runoff as determined in watershed studies may not be suitable for the calculation of a recycling index, because an unknown proportion of the nutrients in the runoff could be moving directly from weathered rock to drainage streams, without cycling through the vegetation.

Table 4.4 lists three sites besides the San Carlos Oxisol plot, for which enough data are available for comparisons. One is a lower montane forest in Puerto Rico on basaltic derived soils, the second is a deciduous, mixed mesophytic forest on calcium-rich soil in Tennessee, and the third is a Douglas fir forest in Washington. The cycling index of calcium in the Puerto Rican forest is much lower than that for San Carlos, and in the Tennessee forest it is slightly lower. There was almost no difference in the nutrient cycling index for potassium between the sites for which data were available.

The potential for nutrient leaching should be greater at the San Carlos site than the other sites. Annual rainfall at San Carlos averages about

Table 4.4 Comparison of cycling indices of calcium and potassium

	a Amount entering forest floor (litter-fall + throughfall, kg ha^{-1} yr^{-1})	b Leaching from rooting zone (kg ha^{-1} yr^{-1})	c Cycling Index $\left(\dfrac{a-b}{a}\right)$
Calcium			
Oxisol			
San Carlos	16.3	3.5	0.79
Montane forest Puerto Rico			
(Jordan *et al.*, 1972)*	62.9	43.1	0.31
Mesophytic forest on calcium			
rich soil Tennessee			
(Shugart *et al.*, 1976)	84.1	27.4	0.67
Douglas fir forest			
Washington (Cole *et al.*,			
1967)	18.5	4.5	0.76
Potassium			
San Carlos	30.2	3.7	0.88
Puerto Rico	138.9	20.8	0.85
Washington	15.8	1.0	0.94

*Tree-fell data lacking

3,600 mm, while it is about 2,800 mm at the Puerto Rico site, 1,265 at the Tennessee site, and 1,360 at the Washington site. Further, the growing season at San Carlos is longer than in the temperate zone sites, and therefore the potential for carbonic acid production and consequent leaching is lower in the temperate sites. Despite the higher potential for nutrient leaching at San Carlos, comparisons of cycling indices give no evidence that the continually hot and humid environment in Amazonia results in less efficient recycling of nutrients than in other forest ecosystems.

Reasons for efficient recycling

Although it might seem that the high leaching potential of the wet tropics could stress the rain forest ecosystem and the species that make up the ecosystem, cycling indices gave no evidence of stress. Thus far, the original hypothesis that the humid tropics has relatively little influence on the ecosystems that exist there seems to hold true.

If there is a high potential for nutrient leaching in the humid tropics, why isn't the potential realized? The reason is that nutrient conserving mechanisms have evolved in rain forest species, and these mechanisms allow the species to survive and the ecosystem to function without apparent symptoms of stress.

Figure 4.5 An important nutrient conserving mechanism in many tropical forests is the concentration of roots on top of, or near, the soil surface (scale in cm)

During the course of the San Carlos study, at least 19 mechanisms were identified which appear to play a role in nutrient conservation. The mechanisms, and their mode of operation are as follows:

1 *Large root biomass.* A relatively large root biomass occupies more fully the volume of soil where nutrients are held after release from decomposition, and increases the probability of nutrient uptake before leaching losses can occur (Chapin, 1980).

2 *Root concentration near surface,* as shown in figure 4.5. Trees with roots near the soil surface may be better able to compete for nutrients, since nutrient mineralization occurs at or near the soil surface (Stark and Jordan, 1978).

3 *Aerial roots.* Adventitious roots growing on tree branches are able to exploit nutrients captured by bryophytes, lichens, mosses, bromeliads, ferns, and other epiphytes that live on the tree (Nadkarni, 1981).

4 *Mycorrhizae.* Due to the infertility of many rain forest soils, especially the lack of readily soluble phosphorus, mycorrhizae appear to play an important role in rain forest nutrition (Janos, 1983; St. John and Coleman, 1983).

5 *Below ground communities.* A complex and large below ground community of micro flora and fauna may play an important role in the nutrition of tropical rain forest ecosystems (Jordan, 1985).

6 *Low nutrient uptake kinetics.* Many rain forest species have roots with relatively low nutrient uptake kinetics, an advantage where nutrients are scarce (Olson *et al.*, 1981; Chapin, 1983).

7 *Long life span.* The long life span of most rain forest trees enables individuals to take up nutrients beyond their immediate need and store them for later use (Chapin, 1980).

8 *Leaf longevity.* The long life of leaves on most evergreen rain forest trees (Mabberly, 1983) reduces the necessity for high nutrient uptake to replace leaves which are shed.

9 *Efficiency of nutrient use.* The relatively low nutrient concentrations in tissues of many rain forest species suggests that they are able to produce a unit of biomass with a low amount of nutrients (Vitousek, 1982, 1984).

10 *Reproduction.* Many tree species in infertile habitats do not produce a large seed crop every year, thereby conserving nutrients (Janzen, 1974).

11 *Tolerance of acid soils.* Native tropical rain forest species have a relatively high tolerance for the high levels of acid and aluminium rich soils in the humid tropics (Baker, 1976).

12 *Species deversity.* The high species diversity of most rain forests may be a defence against herbivores which are often mono-specific, and this therefore may be a defence against nutrient loss to herbivores (Orians *et al.*, 1974).

13 *Allelopathy.* Rain forest plants synthesize substantial quantities of secondary compounds that can repel or inhibit herbivores (Whittaker and Feeny, 1971), thereby conserving nutrients.

14 *Scleromorphic leaves.* Thick and leathery leaves often found on species of nutrient-poor soils of the tropics (Klinge and medina, 1979) protect the plant from nutrient loss to herbivory.

15 *Nutrient translocation.* Translocation of mobile elements out of leaves before they fall (Charley and Richards, 1983) may be a common mechanism to conserve nutrients in tropical rain forests.

16 *Epiphylls*. Epiphylls such as algae, mosses and lichens are common in the wet tropics, and can fix nitrogen (Forman, 1975), and scavenge nutrients from the rainwater (Witkamp, 1970).

17 *High silicate*. The high concentration of silicates in roots of many native rain forest species may be a mechanism through which scarce phosphates in surface soils are mobilized, when the roots are shed (Jordan, 1985).

18 *Thick bark*. The thick bark of many species in nutrient poor rain forests may be a defence against consumers such as termites which remove nutrients from living trees (Jordan and Uhl, 1978).

19 *Drip tips*. Drip tips on tropical leaves may reduce the residence time of water on the leaves, and thus the leaching potential (Dean and Smith, 1978).

The evaluation of the impact of tropical environment on the rain forest

The naturally occurring rain forest vegetation may not be under stress from the low nutrient environment, in the sense that individuals are debilitated or diseased. The rain forest, as it is examined today, shows little evidence of environmental stress. It has a large structure, and a high rate of organic matter production. It recycles nutrients efficiently. It seems to be able to maintain structure, productivity and nutrient cycling because mechanisms have evolved that prevent nutrient loss. However, the evidence for their role in nutrient conservation is circumstantial.

Direct evidence for their importance comes from studies of changes in nutrient cycling and primary productivity when the naturally occurring rain forest is cut and replaced with crop species. Native rain forest species, and economically important crop species, are structurally and functionally very different. While native species are well adapted to the nutrient-poor soils, economically important species which are planted following removal of the rain forest usually have been bred to grow rapidly and to produce fruit, grains, or tubers, at the expense of nutrient conserving mechanisms. Instead of conserving nutrients through efficient recycling, most crop plants obtain their nutrients by depleting stocks held in the mineral soil.

The effect of replacement of rain forest by crops

There is a general pattern of nutrient dynamics and primary productivity which follows replacement of the native rain forest with crop species. The immediate result of deforestation (figure 4.6) is a short-term increase in soil fertility. The increase may come about through the decomposition of slash lying on top of the soil, or through the ash, when the downed trees are burned. It is this pulse of nutrients which permits growth of crops or pasture for a few years. Almost immediately, however, processes of nutrient loss begin.

There are three important nutrient loss processes. One is volatilization, which affects nitrogen (Delwiche, 1970) and sulphur (Ewel *et al.*, 1981). If the slash on the soil is burned after it dries, a significant proportion of these nutrients may be lost during the burn. Regardless of whether the slash is burned, the increased soil temperatures and increased carbon availability result in increasing activity of denitrifying bacteria, with a consequent continuous loss of nitrogen (Delwiche, 1977).

A second type of loss is through leaching (Vitousek *et al.*, 1979). As the organic material on the soil surface decomposes, nutrient cations such as calcium, potassium, and magnesium are carried down into the mineral soil by rainwater. However, the capacity of many tropical soils to retain the cations is very limited, because of their highly weathered condition, and the cations are leached out into drainage streams. The monovalent cations such as potassium are leached most readily, and potassium loss is often an early problem in tropical agriculture. After a few years, leaching losses of divalent cations such as calcium and magnesium become important, and eventually, even trace elements such as zinc and copper are significantly reduced. Leaching also results in important losses of nitrate, sulphate and phosphate.

A third type of loss, fixation by clays and by soluble iron and aluminium, affects phosphorus. Although tropical soils frequently contain relatively large stocks of phosphorus, most of it is bound in insoluble forms which cannot be taken up by most crop plants. Phosphorus availability is frequently the most important problem in tropical agriculture (Olson and Englestad, 1972). Immediately after a forest is cut and burned, available phosphorus in the soil increases, possibly due to pH-mediated changes in the soil resulting from the ash. As the soil pH decreases again within a few years, phosphorus moves back into a primarily insoluble state (Brady, 1974).

Productivity

As a result of nutrient losses during shifting cultivation in the humid tropics, crop productivity declines rapidly (Jordan, 1987). Native secondary successional or 'weed' species are much less affected by nutrient loss, because of their adaptations to the nutrient-poor environment. Competition by these weeds further reduces the productivity of crop species. Production dynamics of a shifting cultivation site near San Carlos in the upper Rio Negro of Venezuela are shown in figure 4.7. Throughout the three-year cropping period, total productivity of the principal crop

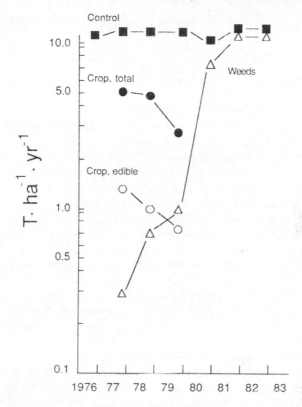

Figure 4.7 Net primary productivity in an experimental slash and burn site on an Oxisol near San Carlos de Rio Negro, Venezuela
Data are given for total crop (principally manioc) productivity, edible manioc tubers, secondary successional vegetation ('weeds'), and forest vegetation in adjacent control plot
(from Jordan, 1987)

manioc (*Manihot esculenta*) drops, and productivity of edible tubers decines to half the initial rate. Meanwhile, net primary productivity of the control plot in the undisturbed rain forest remains high. Productivity of successional species (weeds) increases rapidly, and within two years after the plot has been abandoned, the rate equals that of the control forest.

Thus the evidence from crop systems which have few if any of the suggested nutrient conserving mechanisms is that if these mechanisms are not present, nutrient losses can increase, and primary productivity will decline. Both direct and indirect evidence suggests that these mechanisms

Figure 4.6 Nutrient conserving mechanisms in tropical forests are destroyed on cutting of the forest.

are in fact adaptations to the nutrient leaching potential of the hot, humid tropical environment.

Is the hypothesis true or false?

This chapter began with the hypothesis that in the humid tropics, the vegetation had a strong influence on the landscape, especially the soils and rocks, but the influence of the landscape on the vegetation was low. The first part of the hypothesis is true. The intense deep weathering brought about by the year-round respiration of soil organisms results in soils that are very deep, and that form smooth rounded landforms after erosion following hurricanes and landslides.

Evidence from studies of the structure and function of naturally occurring tropical ecosystems suggested that the second part of the hypothesis was also true: that is, the landscape and its climate have little effect on ecosystem function. The problem with this conclusion is that while there seems to be little current effect, the presence of nutrient conserving mechanisms strongly suggests that the environment has had a strong modifying influence on the rain forest in the past. At some time during the course of evolution, the native rain forest species must have been subject to environmental stress. Otherwise they should not have evolved nutrient conserving mechanisms.

If there has been an environmental stress on rain forest ecosystems in the past that has led to the evolution of nutrient conserving mechanisms, could not that same stress still be influencing the course of rain forest evolution? It would seem that the answer is more likely to be yes than no. Evolution is an ongoing process, not just an event of the past.

Conclusion

The tropical rain forest has an important influence upon the tropical landscape. A consideration of mechanisms that appear to conserve nutrients in the present day rain forest suggests that the tropical humid rain landscape has also strongly influenced the rain forest. The influence is not readily apparent as long as the system is undisturbed, because the influence is manifest in relatively subtle nutrient conserving mechanisms. Only when the native forest is cut and the conserving mechanisms are destroyed does the impact of the environment on the tropical ecosystem become apparent.

Acknowledgement

Data presented in this chapter are from a study sponsored by the Ecosystems Studies Program of the US National Science Foundation.

References

Baker, D. E. 1976: Soil chemical constraints in tailoring plants to fit problem soils. 1. Acid soils. In M. J. Wright (ed.), *Plant adaptation to mineral stress in problem soils, Proceedings of a workshop*, pp. 127–40. Beltsville, Maryland, Nov. 1976: Agency for International Development, Washington, DC.

Bates, H. W. 1892: *The Naturalist on the River Amazon*. London: John Murray.

Brady, N. C. 1974: *The nature and properties of soils*. New York: Macmillan.

Chapin, F. S. 1980: The mineral nutrition of wild plants. *Annual Review of Ecology and Systematics* 11, 233–50.

—— 1983: Patterns of nutrient absorption and use by plants from natural and man modified environments. In H. A. Mooney and M. Godron (eds), *Disturbance and ecosystems: components of response*, pp. 175–87. Berlin: Springer-Verlag.

Charley, J. L., and Richards, B. N. 1983: Nutrient allocation in plant communities: mineral cycling in terrestrial ecosystems. In O. L. Lange, P. S. Nobel, C. B. Osmond, and H. Ziegler (eds), *Physiological plant ecology. Vol. IV. Ecosystem processes: mineral cycling, productivity and man's influence*, pp. 5–45. Berlin: Springer-Verlag.

Cole, D. W., Gessell, S. P. and Dice, S. F. 1967: Distribution and cycling of nitrogen, phosphorus, potassium, and calcium in a second growth Douglas fir ecosystem. In H. E. Young (ed.), *Symposium on primary productivity and mineral cycling in natural ecosystems*, pp. 197–232. Orono: College of Life Sciences and Agriculture, University of Maine.

Dean, J. M. and Smith, A. P. 1978: Behavioral and morphological adaptations of a tropical plant to high rainfall. *Biotropica* 10, 152–4.

Delwiche, C. C. 1970: The nitrogen cycle. *Scientific American* 223 (3), 136–46.

—— 1977: Energy relations in the global nitrogen cycle. *Ambio* 6, 106–11.

Ewel, J., Berish, C., Brown, B., Price, N. and Raich, J. 1981: Slash and burn impacts on a Costa Rican wet forest site. *Ecology* 62, 816–29.

Finn, J. T. 1976: Measures of ecosystem structure and function derived from analysis of flows. *Journal of Theoretical Biology* 56, 363–80.

—— 1978: Cycling index: a general definition for cycling in compartment models. In D. C. Adriano and I. L. Brisbin (eds), *Environmental chemistry and cycling processes*, pp. 138–64. CONF-760429. Technical Information Center, US Dept. of Energy, Washington, DC.

Forman, R. T. 1975: Canopy lichens with blue-green algae: a nitrogen source in a Colombian rain forest. *Ecology* 56, 1176–84.

Garwood, N. C., Janos, D. P. and Brokow, N. 1979: Earthquake-caused landslides: a major disturbance to tropical forests. *Science* 205, 997–9.

Janos, D. P. 1983: Tropical mycorrhizas, nutrient cycles, and plant growth. In S. L. Sutton, T. C. Whitmore and A. C. Chadwick (eds), *Tropical rain forest: ecology and management*. pp. 327–45. Oxford: Blackwell Scientific Publications.

Janzen, D. H. 1974: Tropical blackwater rivers, animals, and mast fruiting by the Dipterocarpaceae. *Biotropica* 6, 69–103.

Jenny, H. 1980: *The soil resource. Origin and Behavior*. NY: Springer-Verlag.

Johnson, D. W., Cole, D. W., Gessell, S. P., Singer, M. J. and Minden, R. V. 1977: Carbonic acid leaching in a tropical, temperate, subalpine, and northern forest soil. *Arctic and Alpine Research* 9, 329–43.

Jordan, C. F. 1983: Productivity of tropical rain forest ecosystems and the implications for their use as future wood and energy sources. In F. B. Golley (ed.) *Ecosystems of the World 14A: Tropical rain forest ecosystems*, pp. 117–36. Amsterdam: Elsevier.

—— 1985: *Nutrient cycling in tropical forest ecosystems*. Chichester, John Wiley.

—— 1987: Shifting cultivation. In C. F. Jordan (*ed.*), *Amazon rain forests: disturbance and recovery*, pp. 9–23. NY: Springer-Verlag.

—— and Uhl, C. 1978: Biomass of a 'tierra firme' forest of the Amazon Basin. *Oecologia Plantarum* 13, 387–400.

—— Kline, J. R. and Sasscer, D. S. 1972: Relative stability of mineral cycles in forest ecosystems. *American Naturalist* 106, 237–53.

Klinge, H. and Medina E. 1979: Rio Negro caatingas and campinas, Amazonas states of Venezuela and Brazil. In R. L. Specht (ed.), *Ecosystems of the world, vol. 9A: Heathlands and related shrublands*, pp. 483–8. Amsterdam: Elsevier.

Lewis, L. A. 1974: Slow movement of earth under tropical rain forest conditions. *Geology* 2, 9–10.

Likens, G. E., Bormann, F. H., Pierce, R. S., Eaton, J. S. and Johnson, N. M. 1977: *Biogeochemistry of a forested ecosystem*. New York: Springer-Verlag.

Lugo, A. E., Applefield, M., Pool, D. J. and McDonald, R. B. 1983: The impact of hurricane David on the forests of Dominica. *Canadian Journal of Forest Research* 13, 201–11.

Mabberly, D. J. 1983: *Tropical rain forest ecology*. Glasgow: Blackie.

McNeil, M. 1964: Laterite soils. *Scientific American* 211, 68–73.

Nadkarni, N. 1981: Canopy roots: convergent evolution in rainforest nutrient cycles. *Science* 214, 1023–4.

Odum, E. P. 1985: Trends expected in stressed ecosystems. *BioScience* 35, 419–22.

Olson, R. A. and Engelstad, O. P. 1972: Soil phosphorus and sulfur. In *Soils of the humid tropics*, pp. 82–101. Washington, DC. National Academy of Sciences.

Olson, R. A., Clark, R. B. and Bennet, J. H. 1981: The enhancement of soil fertility by plant roots. *American Scientist* 69, 378–84.

Orians, G., and 10 others 1974: Tropical population ecology. In E. Farnworth and F. Golley (eds), *Fragile ecosystems*, pp. 5–65. NY: Springer-Verlag.

Rapport, D. J., Regier, H. A. and Hutchinson, T. C. 1985: Ecosystem behavior under stress. *American Naturalist* 125, 617–40.

Saldarriaga, J. G. 1987: Recovery following shifting cultivation. In C. F. Jordan (*ed.*), *Amazonian rain forests: disturbance and recovery*, pp. 24–33. NY: Springer-Verlag.

Sanchez, P. A. 1976: *Properties and management of soils in the tropics*. New York: John Wiley.

Shugart, H. H., Reichle, D. E., Edwards, N. T. and Kercher, J. R. 1976: A model of calcium-cycling in an east-Tennessee Liriodendron forest: model structure parameters and frequency response analysis. *Ecology* 57, 99–109.

Solbrig, O. T. and Orians, G. H. 1977: The adaptive characteristics of desert plants. *American Scientist* 65, 412–21.

Stark, N. M. and Jordan, C. F. 1978: Nutrient retention by the root mat of an Amazonian rain forest. *Ecology* 59, 434–7.

St. John, T. V. and Coleman, D. C. 1983: The role of mycorrhizae in plant ecology. *Canadian Journal of Botany* 61, 1005–14.

Tricart, J. 1972: *The landforms of the humid tropics, forests, and savannas*. London: Longman.

Vitousek, P. 1982: Nutrient cycling and nutrient use efficiency. *American Naturalist* 119, 553–72.

—— 1984: Litterfall, nutrient cycling, and nutrient limitation in tropical forests. *Ecology* 65, 285–98.

——, Gosz, J. R., Grier, C. C., Melillo, J. M., Reiners, W. A. and Todd, R. L. 1979: Nitrate losses from disturbed ecosystems. *Science* 204, 469–74.

Walter, H. 1979: *Vegetation of the earth*. Berlin: Springer-Verlag.

Whittaker, R. H. and Feeny, P. P. 1971: Allelochemics: chemical interactions between species. *Science* 171, 757–70.

Witkamp, M. 1970: Mineral retention by epiphyllic organisms. In H. T. Odum and R. F. Pigeon (eds), *A tropical rain forest*, pp. H-177–H-179. Washington DC: Div. Technical Info., US Atomic Energy Commission.

Woodwell, G. M. 1970: Effects of pollution on the structure and physiology of ecosystems. *Science* 168, 429–32.

Zinke, P. J., Stangenberger, A. G., Post, W. M., Emanuel, W. R. and Olson, J. S. 1984: Worldwide organic soil carbon and nitrogen data. Oak Ridge National Laboratory, Environmental Sciences Division No. 2212, US Department of Energy, Oak Ridge, Tennessee.

5 The geomorphological role of termites and earthworms in the tropics

A. S. Goudie

In recent years, some major international research projects have been mounted in tropical environments, including savannas and forests, in an attempt to assess the ecological, pedological and geomorphological importance of certain abundant and energetic organisms. Particular attention has been paid to the role of termites and worms, and the purpose of this chapter is to review some of the geomorphological implications of recent work.

Termites

Termites, of which there are several thousand species, are members of the Isoptera order, and about four-fifths of the known species belong to the Termitidae family (Harris, 1961). They vary in size according to their species, from the large African *Macrotermes*, with a length of around 20 mm and a wing span of 90 mm, down to the Middle Eastern *Microcerotermes* which is only around 6 mm long with a wing span of 12 mm. Major recent taxonomic and ecological surveys include those of Brian (1978) and Krishna and Weesner (1970), while Lee and Wood (1971a) provide a detailed study of the effects of termites on soils. Snyder (1956) provides a bibliography of earlier work on termites.

Termites, though 'fierce, sinister and often repulsive' (Maeterlinck, 1927), are remarkable for having been adapted to living in highly organized communities for as long as 150–200 million years (Skaife, 1955), and much of their success is due to their development of elaborate architectural, behavioural, morphological and chemical strategies for colony defence (Prestwich, 1984). They occur in great numbers – 2.3 million ha^{-1} in Senegal and 9.1 million ha^{-1} in the Ivory Coast (UNESCO, UNEP, FAO, 1979). Maeterlinck (1927) regarded them as 'the most tenacious, the most deeply rooted, the most formidable, of all the occupants and conquerors of this globe'.

The vast majority of termite species are found in the tropics, though as figure 5.1 shows their distribution is wider than this; they extend to 45–48°N and to 45°S.

Termite numbers decrease rapidly with increasing altitude and latitude. So, for example, in Sarawak 58 species occur in lowland forest, 10 in montane forest above 1,000 m, and none above 1,860 m (Collins, 1983).

The geomorphological importance of termites became apparent from the late eighteenth century onwards (Smeathman, 1781) with the increased tempo of European exploration in low latitudes. Branner (1896), for example, reported how along the upper Paraguay river in the Matto Grosso of Brazil he had 'seen places where the nests are so close together that one could almost walk upon them for several hundred yards at a time, while over many areas of ground no one of the nests was more than 10 feet from another'. However, the most significant early treatment of termite activity was that of Drummond (1888, p. 146) who worked in Africa. He reported that their heaps and mounds were 'so conspicuous that they may be seen for miles, and so numerous are they and so useful as cover to the sportsman, that without them in certain districts hunting would be impossible'. He went on to suggest that 'the soil of the tropics is in a state of perpetual motion . . . there is so to speak, a constant circulation of earth in the tropics, a ploughing and harrowing . . .' (p. 154). He postulated that while Egypt was the gift of the Nile, that river's sediments resulted from 'the labours of the humble termites in the forest slopes about Victoria Nyanza' (p. 158).

It comes as something of a surprise, therefore, that many recent works on the geomorphology of low latitudes (e.g. Thomas, 1974; Faniran and Jeje, 1983; Büdel, 1982) pay scant attention to the role of termites. A notable exception to this is Tricart (1972), who reviews some of the available French literature on mound construction, soil and cuirasse modification, and sediment entrainment.

Termite constructions

Termite mounds and hills are the most striking manifestations of termite activity, and have a large range of sizes and morphologies (see figure 5.2 for an example). Branner (1896) likened their round mounds to enormous Irish potatoes and the tall slender forms to 'stone friars', while Maeterlinck (1927) described such forms as 'wrinkled hillocks, battered sugar-loafs, gigantic stalagmites, mushrooms, portentous sponges, ricks of storm-tossed hay or corn'. Not all termites build large termitaria, however, and Gay (1970) indicates that only about one fifth of all Australian species do.

The heights of termite constructions vary considerably according to species (table 5.1). There are records in the literature of mounds attaining heights in excess of 9 m, though most are less than this. Among the species that create the tallest mounds are *Bellicositermes bellicosus, Bellicositermes*

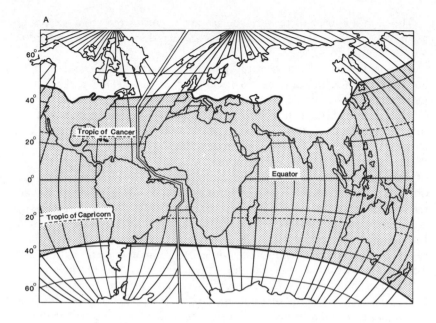

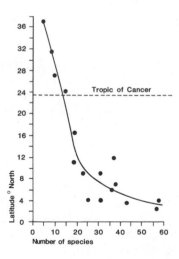

Figure 5.1 The world distribution of termites
(A) The general global pattern
(after Harris, 1961)
(B) The number of termite species at world-wide latitudes north of the equator
(after Collins, 1983)

natalensis, Macrotermes bellicosus, Macrotermes falciger and *Nasutitermes triodiae*.

Substantial quantities of data are now also available on mound densities per unit area (table 5.2). In general the densities vary considerably according to both environmental conditions (e.g. soil type) and termite species. So, for example, the density of the very large mounds produced by *Macrotermes bellicosus, Macrotermes subhyalinus, Macrotermes falciger, Bellicositermes bellicosus,* and *Nasutitermes triodiae* tends to be less (often around 2–10 ha^{-1}) than those for the smaller types of mound (often around 200–1,000 ha^{-1}). The large mound building Macroterminitinae are especially important in savanna environments rather than in rainforest (Buxton, 1981a), though the tolerances of different species varies. Pomeroy (1978) suggests that while minimum mean annual values of rainfall are 700 mm for *Macrotermes bellicosus,* those for *Macrotermes subhyalinus* are only 300 mm.

Figure 5.2 An active occupied *Macrotermes* mound and its associated wash slope from *Acacia* savanna in Namibia (this example is approximately 2.5 m high)

Table 5.1   Heights of termite mounds

Dominant termite species	Location	Height (m)	Source
Amitermes laurensis	N. E. Australia	0.5–1.37	Spain et al., 1983
Anacanthotermes ahngerianus	Tajik SSR	0.6	Vallachmedov, 1981
Bellicositermes bellicosus	Central African Republic	3.6	Boyer, 1973
Bellicositermes natalensis	Central African Republic	3.4	Boyer, 1973
Cubitermes spp.	Ivory Coast	0.5	López-Hernández and Febres, 1984
Discupiditermes nemorosus	W. Malaysia	0.18–0.24	Matsumoto, 1976
Drepanotermes rubriceps	Northern Territory, Australia	0.005	Williams, 1978
Hommalotermes foraminifer	W. Malaysia	0.08	Matsumoto, 1976
Macrotermes bellicosus	Ivory Coast	4.6	López-Hernández and Febres, 1984
Macrotermes falciger	Shaba, Zaire	7.6	Malaisse, 1978
Nasutitermes triodiae	Australia	7	Gay, 1970
Nasutitermes triodiae	Northern Territory, Australia	0.5	Williams, 1978
Odontotermes obesus	Delhi, India	0.04–0.08	Mohindra and Mukerji, 1982
Trinervitermes geminatus	Ivory Coast	0.5	López-Hernández and Febres, 1984
Tumulitermes hastilis	Northern Territory, Australia	0.08	Williams, 1978
Tumulitermes pastinator	Northern Territory, Australia	0.05	Williams, 1978

The distribution of large mounds is an intriguing problem, for as Pullan (1979, p. 269) has remarked, 'termite hills are absent from environments which would appear to be similar in all respects to those which contain them'. Pullan stresses the importance of intensive activity by man and animals as an important distributional control in Africa, but stresses that over huge areas information on distribution patterns remains fragmentary. He also sees such activities as major controls on mound form and presents an evolutionary model of development.

Undoubtedly soil characteristics are also an important control of mound formation. Mounds tend to be rare on sands (where there is insufficient binding material), on deeply cracking self-mulching clays (which are unstable), or on shallow soils (where there is a shortage of building

Table 5.2 Density of termite mounds

Dominant termite species	Location	Density (mounds ha^{-1})	Source
Amitermes laurensis	N. E. Australia	70–643	Spain et al., 1983
Amitermes laurensis	Savanna, N. Australia	28–210	Lee and Wood, 1971a
Amitermes vitiosus	N. Queensland, Australia	283	Holt et al., 1980
Anacanthotermes anhgerianus	Steppe, Central Asia	162–570	Lee and Wood, 1971a
Bellicositermes bellicosus rex	Central African Republic	10–12	Boyer, 1973
Bellicositermes natalensis	Central African Republic	Up to 30	Boyer, 1973
Coptotermes lacteus	S. Australia	1–2	} Lee and Wood, 1971a
Cubitermes exiguus	Steppe savanna, Zaire	0–652	
Cubitermes fungifaber	Rainforest, Zaire	875	
Cubitermes sankurensis	Steppe savanna, Zaire	8–850	
Discupiditermes nemorosus and Homallotermes foraminifer	West Malaysia	236–385	Matsumoto, 1976
Drepanotermes rubriceps	Tropical Australia	200–450	Spain et al., 1983
Macrotermes spp.	Ivory Coast	86	Bodot, 1964
Macrotermes bellicosus	Uganda	0.75–9.0	Pomeroy, 1977
Macrotermes bellicosus	N. Sierra Leone	5	Miedema and Van Vuure, 1977
Macrotermes bellicosus	Ivory Coast	4–41.7	Lepage, 1984
Macrotermes falciger	Shaba, Zaire	2.7–4.9	Goffinet, 1976
Macrotermes subhyalinus	Uganda	0.75–13.25	Pomeroy, 1977
Nasutitermes exitiosus	S. Australia	4–9	Lee and Wood, 1971a
Nasutitermes longipennis	Tropical Australia	425	Spain et al., 1983
Nasutitermes magnus	E. Australia	61	} Lee and Wood, 1971a
Nasutitermes triodiae	N. Australia	3–7	
Odontotermes sp.	Kenya	5–7	
Pseudoacanthotermes	Uganda	0.37–7.32	Pomeroy, 1977
Trinervitermes trinervoides	South Africa	534	Lee and Wood, 1971a
Trinervitermes spp.	Zaria, Nigeria	74–148	Sands, 1965
Tumulitermes hastilis	Tropical Australia	245–1108	Spain et al., 1983

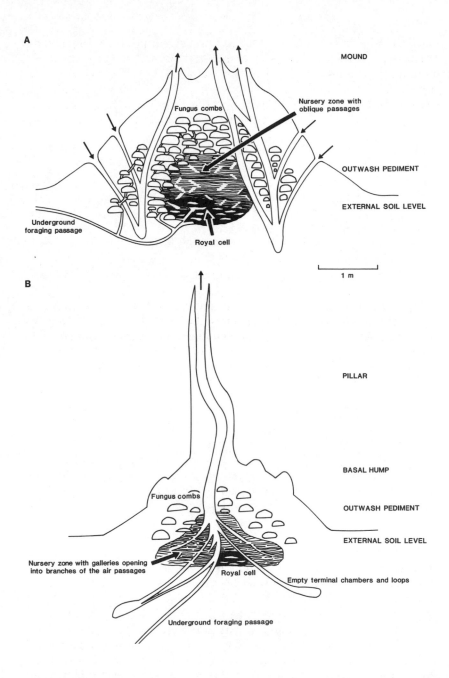

Figure 5.3 Diagrammatic vertical section through a mound of *Macrotermes subhyalinus* of the Bissel type (A), and of the Marigat type (B) (after Darlington, 1984)

material) (Lee and Wood, 1971a). Soil drainage may also be important. For example, Grigg (1973) discusses the distribution of oriented 'magnetic' meridional mounds in tropical Australia and shows that whereas *Amitermes meridionalis* always builds meridional mounds, two other species, *Amitermes laurensis* and *Amitermes vitiosus* only build them in some particular combination of circumstances, particularly in low-lying ill-drained areas.

Likewise, the particular form of mound produced by any individual species may vary according to environmental conditions. For example, Darlington (1984) has identified two distinct types of mound in Kenya, both of which have been produced by *Macrotermes subhyalinus*: the Bissel type (figure 5.3A) and the Marigat type (figure 5.3B). The prime difference lies in their ventilation systems which serve to cool a large, highly centralized nest. She believes that the former type may be most appropriate on a compact, clay-rich soil, while the latter would be most effective in a porous, unconsolidated soil such as sand or volcanic ash.

Although in most cases it is obvious that particular mounds have been produced by particular species of termites, there have been some arguments about the origin of some mounds found in the tropics and elsewhere (Ojani, 1968). For example, Cox and Gakaha (1983) have argued that certain mounds in Kenya are created by the mole-rat, *Tachyoryctes splendens*, whereas Darlington (1985) argues that the same mounds are produced by *Odontotermes*.

Termites and soil properties

Whether termites live in hidden subsoil chambers or in the conspicuous mounds just discussed, they have a significant effect on soils, partly because of their mechanical activities and partly because of their feeding habits. As yet there are few quantitative data on their effects on geomorphologically important soil properties such as infiltration capacity, but for other variables, such as acidity, more data are available.

When one compares the acidity of mound materials with the acidity of adjacent soils (table 5.3) no very clear picture emerges. Out of the 26 examples given ten have mounds which are relatively less acidic and sixteen have mounds which are relatively more acidic.

The role of termites in mobilizing soil elements can be illustrated by a consideration of the calcium contents of termite mounds in comparison with neighbouring soils (table 5.4). The picture here is relatively consistent, for in most examples, the mounds have markedly higher calcium contents. Out of the 34 pairs there are only three cases where the mounds are depleted in calcium, and there are eight cases where there is an augmentation of calcium of greater than fivefold. Taking all 34 pairs, the mean calcium augmentation is 3.42 times.

Table 5.3 pH values of termite mounds compared to surrounding soil

Termite	Location	Source	Mound	Soil
Amitermes evuncifer	Nigeria	Omo Malaka (1977)	4.6	4.2
Amitermes laurensis			6.04	6.34
Amitermes meridionalis	Australia	Lee and Wood, 1971a	5.53	5.37
Amitermes vitiosus			5.9	5.95
Bellicositermes bellicosus	Central African Republic	Boyer, 1973	7.19	5.46
Bellicositermes natalensis			6.79	5.46
Coptotermes acinaciformis	Australia	Lee and Wood, 1971a	4.93	5.85
Coptotermes lacteus			4.75	5.67
Cubitermes spp.	Nigeria	Omo Malaka, 1977	4.7	4.4
Cubitermes spp.	Ivory Coast	López-Hernández and Febres, 1984	6.31	5.76
Cubitermes spp.	Nigeria	Anderson and Wood, 1984	5.7	5.2
Drepanotermes rubriceps	Australia	Lee and Wood, 1971a	6.10	6.64
Macrotermes bellicosus	Nigeria	Omo Malaka, 1977	4.5	4.7
Macrotermes bellicosus	Ivory Coast	López-Hernández and Febres, 1984	5.60	5.16
Macrotermes michaelseni	Kenya	Arshad, 1981	6.1	6.3
Macrotermes spp.	Zimbabwe	Watson, 1975	6.0	5.5
Macrotermes subhyalinus	Kenya	Arshad, 1981	6.0	6.3
Nasutitermes exitiosus			4.67	5.02
Nasutitermes longipennis	Australia	Lee and Wood, 1971a	5.7	6.15
Nasutitermes magnus			6.1	6.4
Nasutitermes triodiae			5.77	6.14
Procubitermes spp.	Nigeria	Anderson and Wood, 1984	5.5	4.9
Schedorhinotermes spp.	Australia	Lee and Wood, 1971a	5.8	6.3
Trinervitermes geminatus	Nigeria	Omo Malaka, 1977	5.1	4.7
Trinervitermes geminatus	Ivory Coast	López-Hernández and Febres, 1984	7.24	6.11
Tumulitermes hastilis	Australia	Lee and Wood, 1971a	5.38	5.83
Tumulitermes pastinator			5.5	6.0

Table 5.4 Calcium contents of termite mounds compared to surrounding soil

Termite type	Location	Source	Calcium augmentation
Amitermes evuncifer	Nigeria	Omo Malaka, 1977	×5.08
Amitermes laurensis	Australia	Lee and Wood, 1971a	×2.8 to ×6
Amitermes meridionalis			×2
Amitermes vitiosusus			×1.44
Bellicositermes bellicosus	Central African Republic	Boyer, 1973	×5.44
Bellicositermes natalensis			×1.88
Coptotermes acinaciformis	Australia	Lee and Wood, 1971a	×2 to ×3
Coptotermes lacteus			×0.6
Cubitermes spp.	Nigeria	Omo Malaka, 1977	×1.82
Drepanotermes rubriceps	Australia	Lee and Wood, 1971a	×2 to ×3
Macrotermes bellicosus	Nigeria	Omo Malaka, 1977	×0.33
Nasutitermes longipennis			×4
Nasutitermes magnus	Australia	Lee and Wood, 1971a	×2.5
Nasutitermes triodiae			×2 to ×12
Odontotermes obesus	Delhi, India	Mohindra and Mukerji, 1982	×3
Scherorhinotermes intermedius actuosus	Australia	Lee and Wood, 1971a	×3.2
Trinervitermes geminatus	Ivory Coast	López-Hernández and Febres, 1984	×1.5 to ×3
Trinervitermes geminatus	Nigeria	Omo Malaka, 1977	×1.04

There are cases reported in the literature where calcium contents of mounds are sufficiently elevated for calcium carbonate nodules and concretions to develop (e.g. Pendleton, 1942, in Thailand; and Watson, 1974, in Zimbabwe). There are two possible mechanisms that have been postulated to account for this situation. On the one hand some termite species use partially digested organic material in construction and such material may have a high calcium content (Trapnell et al., 1976; Omo Malaka, 1977). On the other, calcium carbonate may be precipitated in the alkaline environment of some termite mounds, such as those of Macrotermes in Malawi (Crossley, 1984), as capillary rise brings solutions from a high ground water table. The mounds act in effect as a wick from which evaporation takes place, with the process being accelerated by transpiring plants on the mounds. Confirmation of this hypothesis came from East Africa, where Hesse (1955) found that calcareous mounds tended to occur close to waterlogged areas, so that they were prone to periodic inundation which brought calcium charged ground water to their bases.

A more contentious issue is the study of laterites and the role that termites play in their development. The vesicular and vermicular structures of some laterites have been seen as resembling the interior structures of small mounds constructed by species such as Cubitermes and Nasutitermes. This was, for example, the view of Erhart (1951) who referred to 'cuirasses termitiques'. Likewise, Taltasse (1957) believed that consolidation of the Cabeças de Jacaré crusts on the Chapadas plateaux of North East Brazil and the development of their scoriaceous structure was the product of termite activity.

By contrast, Tricart (1957) believed that the burrowing activities of termites and other soil fauna would tend to cause the dismantling and subsidence of cuirasses rather than their consolidation. In any event the soil structure must be modified by the ramifying systems of passages and storage pits created by the termites. Darlington's (1982) study of a Macrotermes michaelseni mound in Kenya revealed the presence of 6 km of passage, and 72,000 storage pits in an area of just 8,000 m^2.

Grassé and Noirot (1959) dismissed the importance of termites in laterite formation in West Africa on various grounds: the volume of individual masses of vesicular laterite is often much greater than that of termitaria; the vesicles of laterite are not arranged in ways that even vaguely resemble any part of termite nests; and neither termitaria nor surrounding gallery networks have a form resembling that of lateritic crusts.

Termites have a considerable impact on nutrient cycling and on soil organic characteristics and litter cover. The impression that Drummond (1888, p. 141) gained that the lack of litter and fallen branches in the African bush compared to the forests of the Rocky Mountains was a result of termites, has been confirmed by various recent ecological studies. In Tsavo, Kenya, a relatively arid area, Buxton (1981b) found that almost

Table 5.5   Litter decomposition in the Southern Guinea savanna of Nigeria

	Annual production (g m^{-2})	Annual consumption by termites (g m^{-2})	Percentage consumed by termites
Wood	139	83.5	60.1
Leaves	239	6.8	2.9
Grass	162	98.0	60.5
Total	540	188.3	34.9

Source: Collins, 1981

all dead wood is removed by termites, of which 90 per cent is achieved by the fungus-cultivating sub-family of the Macrotermitinae. In the wetter southern Guinea savanna of Nigeria, Collins (1981) calculated the annual consumption of wood, leaves and grass by termites (table 5.5) and found that 34.9 per cent of all litter decomposition was brought about by termites (60.1 per cent of all wood decomposition). In the wetter rainforest environment the role of termites appears to be relatively less, only achieving 0.9–16.3 per cent of litter consumption in Malaysia (Collins, 1983).

Material granulometry

Termites play an important role in modifying the granulometric characteristics of sediments through preferential appropriation of fine material and by the crushing of minerals like micas by worker termites (Boyer, 1975b). A clear manifestation of this is provided when one considers the clay contents of mounds (table 5.6). In all the sixteen examples cited, the clay content of mounds is elevated in comparison with neighbouring soils. The mean clay content of mounds is 36.47 per cent, while that of the soils is 24.91 per cent.

This ability of termites to mobilize particular soil fractions may be important in the development of a prominent feature of tropical soils – the stone-line (de Ploey, 1964). It can also cause the relative downward movement of coarse particles such as human artefacts, thereby creating the potential for confusion if the artefacts are used for stratigraphic dating purposes (Moeyersons, 1978). Likewise over a long time period the accumulation of fine materials washed off decaying mounds may produce a stone-free upper layer of loamy sandy cover deposits (Pomeroy, 1976; de Dapper, 1978; Komanda, 1978). In relatively arid areas such as Tajikistan in Soviet Central Asia it has been postulated that the translocation of clay and salt by Anacanthotermes ahngèrianus may contribute to the development of salty clay plains called takyrs (Vallachmedov, 1981). Another consequence of soil sorting by termites may be the obliteration of pre-existing sedimentary structures in deposits like alluvium or dune sand.

Table 5.6 Clay contents of termite mounds compared to surrounding soil

Termite	Location	Source	Percentage clay	
			mound	soil
Amitermes evuncifer	Nigeria	Omo Malaka, 1977	28.8	24.9
Amitermes meridionalis	Australia	Lee and Wood, 1971a	19.5	6.0
Bellicositermes bellicosus	Central African Republic	Boyer, 1973	55.64	29.35
Bellicositermes natalensis			56.50	29.35
Coptotermes acinaformis	Australia	Lee and Wood, 1971a	31	18.5
Cubitermes sp.	Nigeria	Omo Malaka, 1977	35.2	17.2
Drepanotermes rubriceps	Australia	Lee and Wood, 1971a	15.7	14.6
Macrotermes bellicosus	Nigeria	Omo Malaka, 1977	36.6	22.5
Macrotermes michaelseni	Kenya	Arshad, 1981	54.28	39.71
Macrotermes spp.	East Africa	Hesse, 1955	26.33	25.83
Macrotermes spp.	Zimbabwe	Watson, 1975	49.66	47.33
Macrotermes subhyalinus	Kenya	Arshad, 1981	51.52	39.71
Nasutitermes exitiosus	Australia	Lee and Wood, 1971a	42.0	26.0
Nasutitermes triodiae			25.7	13.3
Trinervitermes geminatus	Nigeria	Omo Malaka, 1977	23.2	18.2
Tumulitermes hastilis	Australia	Lee and Wood, 1971a	30.3	26
Amitermes vitiosus				

The depths to which termite activity proceed are variable. Boyer (1975a, 1975b) found that *Bellicositermes bellicosus* obtained material from depths up to 12 m and *Bellicositermes natalensis* from depths up to 2 m. These values may be greater than the norm, and Hesse (1955) suggested that *Macrotermes* collected subsoil from a depth of 60–100 cm.

Termites and soil erosion

Lee and Wood (1971a) identify three main ways in which termites can contribute to accelerated rates of soil denudation:

(i) by removing the plant cover;
(ii) by digesting or removing organic matter which would otherwise be incorporated into the soil, and thus making the soil more susceptible to erosion;
(iii) by bringing to the surface fine grained materials for subsequent wash and creep action.

The huge numbers of termites and their large total biomass in favoured localities ensures that these three mechanisms are important. As table 5.7(a) shows, the live weight biomass of termites can be substantial and is comparable to the live weight biomass of large mammalian herbivores in tropical areas (table 5.7(b)).

Some attempts have been made to quantify the speed with which termite mounds are constructed and destroyed. Skaife (1955), found that in the Cape area of South Africa mounds grew at a rate of 25 mm per year (*Amitermes atlanticus*) and that mound growth ceased after around 25 years. Lepage (1984) noted that *Macrotermes* mounds in the Ivory Coast reached a height of 1 m between 2 to 3 years after their first appearance above ground, and that a 3 m mound will be 8–10 years old. Mounds are then abandoned and observations by Holt *et al.* (1980) in Queensland, Australia, suggest that *Amitermes vitiosus* mounds are inhabited for 20 to 40 years. In Senegal Lepage (1974) showed that mounds of *Bellicositermes bellicosus* were occupied for rather longer and continued to grow for at least 75–80 years, by which time they had a volume of 50 m^3. In Australia, Spain *et al.* (1983) calculated that the generation time of mounds was of the order of 30 years.

It would probably be misleading, however, to give the impression that mounds are always short-lived. As Darlington (1985) has pointed out, once mounds are established they provide the best sites for new nests, so that in some cases they will persist over long periods of the order of centuries by repeated recolonization of the same sites. Mounds can also possess high compressive strengths that may provide them with some resistance to erosion. Schmidt Hammer tests on termite mounds in Nigeria (Adepegba and Adegoke, 1974) showed that they had compressive strengths in kg cm^{-2} of 17.4–48.5 compared to 1.4–3.8 for neighbouring unstabilized soils.

Table 5.7 Live weight biomass of termites and other organisms

Species	Location	$g\,m^{-2}$	Source
(a) *Live weight biomass of termites*			
Gnathanitermes tubiformans	Semi-arid, S. America	5.2	Wood and Sands, 1978
Nasutitermes costalis	Rainforest, S. America	0.1	
All	Sahel, W. Africa	1.0	Lepage, 1974
All	South Guinea	1.7–11.1	
All	Semi-deciduous forest, W. Africa	8.0	
All	Riverine forest, C. Africa	11.0	Wood and Sands, 1978
All	Rainforest, Malaysia	3.4	
All	Grazed land, W. Africa	2.8	
All	Maize, W. Africa	1.7–18.9	
Cubitermes exiguus	Steppe savanna, Congo	1.2–1.9	Lee and Wood, 1971a
Nasutitermes exitiosus	Dry sclerophyll forest, Australia	3.0	
Trinervitermes geminatus	Nigeria	10.0	
All	Fété Olé, Senegal	9.7	UNESCO, UNEP, FAO 1979
All	Lamto, Ivory Coast	19.5	
All	Sarawak	0.5–3.6	Collins (1983)
(b) *Live weight biomass of other organisms*			
Uganda Kob	Savanna, Uganda	2.2	
Wild ungulates	Wooded savanna, Tanzania	12.3–17.5	Lee and Wood, 1971a
Wild ungulates and cattle	Open plains, Kenya	5.2–12.6	
Wild ungulates	Bush savanna, Zimbabwe	4.4	
Red kangaroo	Semi-arid woodland, Australia	0.013–0.100	

Once the mounds have been abandoned they are subjected to erosion. Lepage (1974) estimated that a 1.75 cm high mound can be completely eroded in about 50 years, and that 20 to 25 years is required to erode a 8 m³ mound (Lepage, 1984), whereas Williams (1968) working in the Northern Territory of Australia, estimated that abandoned mounds of *Tumulitermes hastilis* are eroded to near ground level in three years and those of *Nasutitermes triodiae* in about ten years. As the organic matter incorporated in the mound decays, the aggregates that give the mounds their strength and hardness when occupied are destroyed, and rainwash spreads the clay rich material over the surrounding ground (Tricart, 1972, pp. 191–2). The erosion may be hastened by domestic animals and by wild mammals. Ant bears may dig into termitaria, elephants may eat the mineral rich soils in mounds, while animals like cheetah, lion, antelope and buffalo may use them as observation points. Mounds become surrounded by a wash pediment formed of soil eroded from the termitarium (Pullan, 1979). Such wash pediments are particularly well developed around the large mounds generated by the fungus cultivating termites such as *Macrotermes, Odontotermes* and *Pseudocanthotermes*. Wash pediments are less well developed around the small columnar or domed termitaria built by such genera as *Cubitermes* and *Trinervitermes*. The wash pediments of *Bellicositermes bellicosus* mounds may extend to 10–60 m in diameter (Boyer, 1973, 1975a, 1975b). On slopes the wash pediments may become deformed, creating a 'peacock feather-like' pattern. For example, on the Loita plains of Kenya, Glover *et al.* (1964) found that on interfluves the degraded mounds of *Odontotermes* were circular rings with a diameter of 0.5–10 m whereas on slopes they were ovate to elongate-ellipsoid with an average length of 14 m.

However, in addition to the potential for erosion and sediment yield caused by mound formation and abandonment it is important to remember the other major consequence of termite caused soil translocation. This is the construction of covered runways or 'sheetings' on the ground surface and on vegetation (as shown in figure 5.4). These are constructed of soil particles cemented together with salivary secretions (Bagine, 1984). Three studies in Africa give an indication of the quantities of material involved in this process: in southern Nigeria, Wood and Sands (1978) calculated a rate of 300 kg ha^{-1} y^{-1}; in Senegal, Lepage (1974) found that *Macrotermes subhyalinus* moved 675–900 kg ha^{-1} y^{-1}; while in Kenya Bagine (1984) estimated a rate of 1,059 kg ha^{-1} y^{-1}.

Although the rates of soil translocation by sheeting formation reported above are significant, they are not as important as cast by worms, which seems on average to operate at a whole order of magnitude faster rate. Watanbe and Ruaysoongnern (1984) show on a global basis that cast production by worms operates at rates between 2.5 and

$2,600 \, t \, ha^{-1} \, y^{-1}$, with most studies reporting values between 20 and $200 \, t \, ha^{-1} \, y^{-1}$.

Even when one combines the amount of soil translocation involved in the formation of both mounds and sheetings the rates still tend to be relatively low in comparison with worm cast formation. Josens (1983) reports combined rates of 1,800 and $1,200 \, kg \, ha^{-1} \, y^{-1}$ from Senegal and Upper Volta respectively.

By continuing studies of rates of termite-caused soil translocation and estimates of rates of land surface denudation it is possible to estimate the speed at which termites can build a new soil layer (table 5.8). There is a great variability in the proposed rates from different areas, and on this basis a 1 m thick soil layer might take anything between 1,000 and 40,000 years to form.

Figure 5.4 Although termite mounds are the most dramatic manifestation of the termite's geomorphological role, termites also translocate large amounts of surface material to construct their covered walkways or 'sheetings'. These examples come from the lowveld in Swaziland (10 cm penknife for scale)

Table 5.8   Estimates of speed at which erosion of termite mounds builds a new soil layer

Location	Source	Rate (mm 1000 y^{-1})
Nigeria	Nye, 1955	25
Uganda	Pomeroy, 1976	40–115
Ivory Coast	Lepage, 1984	750–1000
Australia	Lee and Wood, 1971b	80–400
Upper Volta	Roose (in Josens, 1983)	60
Australia	Williams, 1968	200–300

Earthworms

In common with termites, worms play a major role in modifying the action of surface processes. They modify soil profiles by burrowing and construction, moving material within and between soil horizons, with accompanying mixing of organic and inorganic soil constituents, disintegration and reformation of aggregates, and changes in porosity, aeration and water infiltration (Lee, 1983a). They also play a significant role in decomposition processes and nutrient cycling.

Earthworms are segmented annelid worms, belonging to several families of the Oligochaeta, and some 3,000 species are known. They are probably among the most ancient groups of terrestrial animals. Their distribution is wider than that of termites and they are only entirely lacking in zones of extreme drought or cold. Thus they are common in cool temperate (see chapter 2) and even in sub-arctic regions where termites are virtually unknown, though termites for their part are rather more common than worms, in the driest parts of the tropics.

Earthworm populations can be stratified into species associations that are associated for most of the time with specific depth zones in the soil. There are the litter species, which have no burrows, topsoil species that have burrows in the soil, but feed in the litter layer and frequently cast at the surface, and subsoil species that make extensive horizontal burrows and do not usually cast on the surface.

The dominant earthworms in the tropics are representatives of the families Almidae, Kynotidae, Glossocolecidae, Megascolecidae, Eudrilidae and Ocnerodrilidae (Lee, 1983b). Within the tropics there is a steady increase in numbers and biomass between desert and savanna as rainfall increases, and they are more abundant in tropical savannas than in nearby tropical forests (Lavelle, 1983).

This difference between temperate and tropical areas in terms of earthworm abundance and biomass is brought out in table 5.9, where the rainforest values tend to be rather low in comparison with many temperate

Table 5.9 Earthworm population densities and mean annual live biomass

Source	Location	No. m^{-2}	Biomass (g wet wt m^{-2})
(a) *Tropics*			
Krishnamoorthy, 1985	India, grassland	322	58.7
	India, woodland	273	63.9
Lavelle and Kohlmann, 1984	Mexico, rainforest	132	9.8
	Mexico, rainforest	7.9	10.7
	Sarawak, rainforest	31	0.68
	Nigeria, rainforest	34	10.2
Németh and Herrera, 1982	Venezuela, rainforest	32.7–68.4	8.7–16.6
Lavelle, 1979	Ivory Coast (Lamto), savanna	188–400	35.9–38.0
	Ivory Coast (Lamto), savanna	460–582	17.0–22.3
Dash and Patra, 1979	India, grassland	64–800	30.2
Lee, 1983a	Australia, savanna woodland	13.8	10.2
(b) *Temperate*			
Watanabe, 1975	Japan, grassland	54.2	11.3
Hoogerkamp *et al.*, 1983	Netherlands, old grassland	300–900	2500
	England, pine woodland	400	17
	Germany, beech forest	164–192	101–126
Lee, 1983b	Canada, maple forest	240–780	38–109
	Australia, sub-alpine woodland	15–106	5.7–36

grasslands and woods. However, some of the savanna values are as high as those recorded under temperate conditions.

The biomass of earthworms tends to be rather higher than that of termites, though individual abundance tends to be less (compare table 5.7 with table 5.9).

Earthworm constructions and soil characteristics

Earthworms do not make constructions of the same magnificence and prominence as termites. Nonetheless their burrows and casts are of great importance. For example, in northeastern Thailand the Megascolecid earthworm *Pheretima* makes casts 20–30 cm in height and 5 cm in diameter (Watanabe and Ruaysoongnern, 1984), while in Africa *Dichogaster jaculatrix*, makes casts 10–12 cm high and 4 cm in diameter. Such casts are larger than those for most temperate species (Edwards and Lofty, 1972). Likewise, the giant *Microchaetus microchaetus* of southern Africa (Reinecke, 1983), makes burrows down to depths of 70 cm or more. Certain micro-relief features have been attributed to such burrowing and casting activities. In southern Africa, there are elliptical cups or bowls, approximately 1 m in diameter and 30–100 cm deep, that are called 'kommetjies' (Lungström and Reinecke, 1969), but Lee (1983b) doubts whether earthworms have the social organization to create such features. This lack of social organization is in stark contrast to termites.

However, of more importance than the constructional and excavational forms themselves, is the quantity of soil translocation which they represent. An increasing amount of quantitative data are becoming available on this subject (table 5.10), though some of the data may be under-estimates, for as Watanabe and Ruaysoongnern (1984) have pointed out, some casts are undoubtedly deposited in the earthworm's burrows, and owing to heavy showers surface casts may be readily destroyed and washed away. Thus the actual amounts of soil material turned over may be greater than the results indicate.

The figures for worm cast production for tropical areas are impressive, and seem to be rather higher than for temperate areas. The amount of soil turned over appears to be of the order of 5–25 mm per year, which is considerably greater than that achieved by termites. Casting tends to take place in wet season, so that worms, may contribute a large quantity of material for erosion by surface splash and wash processes.

However, like termites worms are selective in the material which they move, and ingest: casts contain little or no gravel and usually have more clay and less sand than the soil from which they are derived. Thus soils that have been turned over by worms generally have a fine-textured surface horizon overlying sandy or stony horizons and stone-lines (Lee, 1983a).

It is likely that another important pedological and geomorphological consequence of earthworm activity is a change in the infiltration capacity

Table 5.10 Earthworm cast production

Source	Location	kg m^{-2} y^{-1}	mm y^{-1}
(a) *Tropics*			
Madge, 1969	Nigerian, grassland	17.3	15–20
Watanabe and Ruaysoongnern, 1984	Thailand, grassland	13.3–22.5	9–14
Nye, 1955	Ghana, rainforest	5.04	—
Krishnamoorthy, 1985	Indian, grassland	13.51	—
	Indian, woodland	10.04	—
Dash and Patra, 1979	Indian, grassland	7.79	—
Roose, 1980	Ivory Coast, rainforest	5	—
Lavelle, 1974	Ivory Coast, savanna	20–30	c. 25
Lee, 1983a	Sudan, grassland	26.8	c. 25
(b) *Temperate*			
Darwin, 1881	France, grassland	3.6	—
	UK, turf	1.9–4.0	—
Watanabe and Ruaysoongnern, 1984	Switzerland, meadow	3.0–7.7	—
	Switzerland, mixed wood	2.0	—
Watanabe, 1975	Japan, grassland	3.8	3.1

of soils. There are surprisingly few precise data on this phenomenon for the tropics, but it is likely that tropical soils with high earthworm activity have high infiltration capacities compared with soils where activity is low. Studies on grasslands in the Netherlands are suggestive: for soils without worms infiltration capacities are only 0.039–0.047 m per day; for soils where worms have been active for 8–10 years the infiltration capacities are 4.6–6.4 m per day (Hoogerkamp *et al.*, 1983). Likewise, various authors have shown that soils with earthworms drain two to ten times faster than soils without earthworms, and this effect seems to be especially important during heavy rain storms. The evidence overwhelmingly implies that where earthworm activity is not specially restricted by seasonal aridity, their burrows are extremely important in promoting water penetration into the soil (Lee, 1983a). Furthermore, worm casts probably contain more water-stable aggregates than the soil from which they are derived, and this may also promote a soil structure favourable to infiltration. Various mechanisms may be responsible: cementing materials such as calcium humate produced in the gums produced by bacteria in the gut, the reinforcing effect of included plant fibres, and the growth of fungal hyphae in the casts after they are deposited.

Worms tend to be less efficient than termites in terms of their digestive processes, and so play a less important role in organic matter decomposition. Nonetheless, they contribute to decomposition processes by the comminution and mixing of plant litter into the soil and the effects that this process has on further organic matter decomposition by the soil microflora.

Charles Darwin (1881) recognized the power of earthworms over a century ago (p. 310):

> The removal of worm-castings . . . leads to results which are far from insignificant. It has been shown that a layer of earth, 0.2 of an inch in thickness, is in many places brought to the surface; and if a small part of this amount flows, or rolls, or is washed, even for a short distance, down every inclined surface, or is repeatedly blown in one direction, a great effect will be produced in the course of ages. . . . Thus a considerable weight of earth is continually moving down each side of every valley, and will in time reach its bed.

and concluded (p. 316):

> The plough is one of the most ancient and valuable of man's inventions; but long before he existed the land was in fact regularly ploughed, and still continues to be thus ploughed by earthworms. It may be doubted whether there are many other animals which have played so important a part in the history of the world, as have these lowly organised creatures.

Conclusion

As a consequence of recent ecological studies in low latitudes, an increasing quantity of data is becoming available which permits a preliminary assessment of the geomorphological role of termites and earthworms. While there are still major gaps in our knowledge of the distribution of termite constructions and worm castings in different environments, in our data on certain crucial soil properties, and in long-term monitoring of rates of soil translocation, it is nonetheless apparent that the work achieved by termites and earthworms in tropical landscapes is considerable. This is especially true in the savannas. Indeed, to quote Maeterlinck (1927), 'the disproportion between the work and the worker is almost incredible'. They have a very high biomass (worms more than termites), occur in large numbers (termites more than worms), move major quantities of soil in the process of making their nests, sheetings, burrows and casts, consume very appreciable amounts of litter, create new soil horizons, cover sands and stone-lines, and in the case of termites contribute to the formation of such phenomena as calcium carbonate concretions and, more debatably, laterites. In combination with other lowly creatures such as ants, termites and earthworms probably exercise a major control on the rate of operation of such important physical processes as infiltration, soil creep, surface wash and rainsplash detachment in the world's savannas.

References

Adepegba, D. and Adegoke, E. A. 1974: A study of the compressive strength and stabilising chemicals of termite mounds in Nigeria. *Soil Science*, 117, 175–9.

Anderson, J. M. and Wood, T. G. 1984: Mound composition and soil modification by two soil-feeding termites (Termitinae, Termitidae) in a riparian Nigerian forest. *Pedobiologia* 26, 77–82.

Arshad, M. A. 1981: Physical and chemical properties of termite mounds of two species of *Macrotermes* (Isoptera, Termitidae) and the surrounding soils of the semiarid savanna of Kenya. *Soil Science*, 132, 161–74.

Bagine, R. K. N. 1984: Soil translocation by termites of the genus *Odontotermes* (Holmgren) (Isoptera: Macrotermininae) in an arid area of northern Kenya. *Oecologia* 64, 263–6.

Bodot, P. 1964: Etudes écologiques et biologiques des termites dans les savanes de Basse Côte d'Ivoire. In Bouillon, A. (ed.), *Etudes sur les termites africains*, pp. 251–62. Paris: Masson.

Boyer, P. 1973: Action de certains termites constructeurs sur l'évolution des sols tropicaux. *Annales des Sciences Naturelles, Zoologie*, 15, 329–498.

—— 1975a: Etude particulière des trois termitières de *Bellicositermes* et leur action sur les sols tropicaux, *Annales des Sciences Naturelles, Zoologie*, 17, 273–496.

—— 1975b: Les différents aspects de l'action des *Bellicositermes* sur les sols tropicaux. *Annales des Sciences Naturelles, Zoologie*, 17, 447–504.

Branner, J. C. 1896: Decomposition of rocks in Brazil. *Bulletin Geological Society of America*, 7, 255–314.

Brian, M. V. (ed.) 1978: *Production ecology of ants and termites*. Cambridge: Cambridge University Press.

Büdel, J. 1982: *Climatic geomorphology*. Princeton, N. J: Princeton University Press.

Buxton, R. D. 1981a: Changes in the composition and activities of termite communities in relation to changing rainfall. *Oecologia* 51, 371–8.

—— 1981b: Termites and the turnover of dead wood in an arid tropical environment. *Oecologia*, 51, 379–84.

Collins, N. M. 1981: The role of termites in the decomposition of wood and leaf litter in the southern Guinea savanna of Nigeria. *Oecologia*, 51, 389–99.

—— 1983: Termite populations and their role in litter removal in Malaysian rain forests. In Sutton, S. L., Whitmore, T. C. and Chadwick, A. C. (eds), *Tropical rainforest ecology and management*, pp. 311–25. Oxford: Blackwell Scientific Publications.

Cox, G. W. and Gakahu, C. G. 1983: Mima mounds in the Kenya highlands: significance for the Dalquest–Scheffer hypothesis. *Oecologia*, 57, 170–4.

Crossley, R. 1984: Fossil termite mounds associated with stone artefacts in Malawi, Central Africa. *Palaeoecology of Africa*, 16, 397–401.

Darlington, J. P. E. C. 1982: The underground passages and storage pits used in foraging by a nest of the termite *Macrotermes michaelseni* in Kajiado, Kenya. *Journal of Zoology, London*, 198, 237–47.

—— 1984: Two types of mound built by the termite *Macrotermes subhyalinus* in Kenya. *Insect Science Applications*, 5, 481–92.

—— 1985: Lenticular soil mounds in the Kenya highlands. *Oecologia*, 66, 116–21.

Darwin, C. 1881: *Vegetable mould and earthworms*. London: John Murray.

Dash, M. C. and Patra, U. C. 1979: Wormcast production and nitrogen contribution to soil by a tropical earthworm population from a grassland site in Orissa, India. *Revue d'écologie et de biologie du sol*, 16, 79–83.

de Dapper, M. 1978: Couverture limono-sableuses, stone-line, indurations ferrugineuses et action des termites sur le plateau de la Manika; Kolwezi, Shaba, Zaire. *Geo-Eco-Trop*, 2, 265–78.

de Ploey, J. 1964: Nappes de gravats et couvertures argilo-sableuses au Bas-Congo; leur génèse et l'action des termites. In Bouillon, A. (ed.), *Etudes sur les termites africains*, pp. 399–414. Paris: Masson.

Drummond, H. 1888: *Tropical Africa*. London: Hodder and Stoughton.

Edwards, C. A. and Lofty, J. R. 1972: *Biology of earthworms*. London: Chapman and Hall.

Erhart, H. 1951: Sur le rôle des cuirasses termitiques dans la géographie des regions tropicales. *Compte rendu de l'Academie des sciences, Paris*, 233, 804–6.

Faniran, A. and Jeje, L. K. 1983: *Humid tropical geomorphology*. Longman: London.

Gay, F. J. 1970: Isoptera (termites). In CSIRO, *The insects of Australia*, chapter 15. Melbourne: Melbourne University Press.

Glover, P. E., Trump, E. C. and Wateridge, L. E. D. 1964: Termitaria and vegetation patterns on the Loita plains of Kenya. *Journal of Ecology*, 52, 367–77.

Goffinet, G. 1976: Ecologie édaphique des écosystèmes naturels du Haut-Shaba (Zaire). III: Les peuplements en termites épigés au niveau des latosols. *Revue d'écologie et de biologie du sol.*, 13, 459–75.

Grasse, P.-P. and Noirot, C. 1959: Rapports des termites avec les sols tropicaux. *Revue de Géomorphologie Dynamique*, 10, 35–40.

Grigg, G. C. 1973: Some consequences of the shape and orientation of 'magnetic' termite mounds. *Australian Journal of Zoology*, 21, 231–7.

Harris, W. V. 1961: *Termites: their recognition and control.* London: Longmans.

Hesse, P. R. 1955: A chemical and physical study of the soils of termite mounds in East Africa. *Journal of Ecology*, 43, 449–61.

Holt, J. A., Coventry, R. J. and Sinclair, D. F. 1980: Some aspects of the biology and pedological significance of mound-building termites in a red and yellow earth landscape near Charters Towers, north Queensland. *Australian Journal of Soil Research*, 18, 97–109.

Hoogerkamp, M., Rogaar, H. and Eijsackers, H. J. P. 1983: Effects of earthworms on grassland on recently reclaimed polder soils in the Netherlands. In J. E. Satchell (ed.), *Earthworm ecology from Darwin to vermiculture*, pp. 85–105. London: Chapman and Hall.

Josens, G. 1983: The soil fauna of tropical savanna. III: The termites. In Bourlière, F. (ed.), *Tropical savannas*. Amsterdam: Elsevier Scientific.

Komanda, A. 1978: Le rôle des termites dans la mise en place des sols de plateau dans le Shaba méridional. *Geo-Eco-Trop*, 2, 81–93.

Krishna, K. and Weesner, F. M. (eds), 1970: *Biology of termites*. New York: Academic Press.

Krishnamoorthy, R. V. 1985: A comparative study of wormcast production by earthworm populations from grassland and woodland near Bangalore, India. *Revue d'écologie et de biologie du sol*, 22, 209–19.

Lavelle, P. 1974: Le vers de terre de la savane de Lamto. *Bulletin de liaison des chercheurs de Lamto*, 5, 133–66.

—— 1979: Relations entre types écologiques et profils demographiques chez les vers de terre de la savane de Lamto (Cote d'Ivoire). *Revue d'écologie et de biologie du sol*, 16, 85–101.

—— 1983: The soil fauna of tropical savannas. II: The earthworms. In Bourlière, F. (ed.), *Tropical Savannas*, pp. 485–504. Amsterdam: Elsevier Scientific.

—— and Kohlmann, B. 1984: Etude quantitative de la macrofaune du sol dans une forêt tropicale humide du Mexique (Bonampak, Chiapas). *Pedobiologia*, 27, 377–93.

Lee, K. E. 1983a: Soil animals and pedological processes. In CSIRO Division of Soils, *Soils: an Australian viewpoint*, pp. 629–44. Melbourne: CSIRO; London: Academic Press.

—— 1983b: Earthworms of tropical regions – some aspects of their ecology and relationships with soils. In Satchell, J. E. (ed.), *Earthworm ecology from Darwin to vermiculture*, pp. 179–93. London: Chapman and Hall.

—— and Wood, T. G. 1971a: *Termites and soils*. London and New York: Academic Press.

—— and Wood, T. G. 1971b: Physical and chemical effects on soils of some Australian termites and their pedological significance. *Pedobiologia*, 11, 376–409.

Lepage, M. 1974: Les termites d'une savanne sahélienne (Ferlo Septentrional, Sénégal): peuplement, populations, consommation, rôle dans l'écosystème. D.Sc. Thesis, University of Dijon.

—— 1984: Distribution, density and evolution of *Macrotermes bellicosus* nests (Isoptera: Macrotermitinae) in the north-east of the Ivory Coast. *Journal of Animal Ecology*, 53, 107–17.

Ljungström, P. O. and Reinecke, A. J. 1969: Ecology and natural history of the microchaetid earthworms of South Africa. 4: Studies of the influence of earthworms upon the soil and the parasitological question. *Pedobiologia*, 9, 152–7.

López-Hernández, D. and Febres, A. 1984: Changements chimiques et granulométriques dans les sols de Côte d'Ivoire par la présence de trois espèces de termites. *Revue d'écologie et de biologie du sol*, 21, 477–89.

Madge, D. S. 1969: Field and laboratory studies on the activities of two species of tropical earthworms. *Pedobiologia*, 9, 188–214.

Maeterlinck, M. 1927: *The life of the white ant*. London: Allen and Unwin.

Malaisse, F. 1978: High termitaria. *Monographiae biologicae* 31, 1281–1300.

Matsumoto, T. 1976: The role of termites in an equatorial forest ecosystem of West Malaysia. 1: Population density, biomass, carbon, nitrogen and calorific content and respiration rate. *Oecologia* 22, 153–78.

Miedema, R. and Van Vuure, W. 1977: The morphological, physical and chemical properties of two mounds of *Macrotermes bellicosus* (Smeathman) compared with surrounding soils in Sierra Leone. *Journal of Soil Science*, 28, 112–24.

Moeyersons, J. 1978: The behaviour of stones and stone implements buried in consolidating and creeping Kalahari Sands. *Earth Surface Processes*, 3, 115–28.

Mohindra, P. and Mukerji, K. G. 1982: Fungal ecology of termite mounds. *Revue d'écologie et de biologie du sol*, 19, 351–61.

Németh, A. and Herrera, R. 1982: Earthworm populations in a Venezuelan tropical rain forest. *Pedobiologia*, 23, 437–43.

Nye, P. H. 1955: Some soil forming processes in the humid tropics. IV: The action of the soil fauna. *Journal of Soil Science*, 6, 73–83.

Ojani, F. F. 1968: The mound topography of the Thika and Athi plains of Kenya: a problem of origin. *Erdkunde*, 22, 269–75.

Omo Malaka, S. L. 1977: A study of the chemistry and hydromorphic conductivity of mound materials and soils from different habitats of some Nigerian termites. *Australian Journal of Soil Research*, 15, 87–91.

Pendleton, R. L. 1942: Importance of termites in modifying certain Thailand soils. *Journal of the American Society of Agronomy*, 34, 340–4.

Pomeroy, D. E. 1976: Some effects of mound-building termites on soils in Uganda. *Journal of Soil Science*, 27, 377–94.

—— 1977: The distribution and abundance of large termite mounds in Uganda. *Journal of Applied Ecology*, 14, 465–75.

—— 1978: The abundance of large termite mounds in Uganda in relation to their environment. *Journal of Applied Ecology*, 15, 51–63.

Prestwich, G. D. 1984: Defense mechanisms of termites. *Annual review of Entomology*, 29, 201–30.

Pullan, R. A. 1979: Termite hills in Africa: their characteristics and evolution. *Catena* 6, 267–91.

Reinecke, A. J. 1983: The ecology of earthworms in southern Africa. In Satchell, J. E. (ed.), *Earthworm ecology from Darwin to vermiculture*, pp. 195–207. London: Chapman and Hall.

Roose, E. J. 1980: Dynamique actuelle de quelques types de sols en Afrique de l'ouest. *Zeitschrift für Geomorphologie*, 35, 32–9.

Sands, W. A. 1965: Termite distribution in man-modified habitats in West Africa, with special reference to species segregation in the genus *Trinervitermes*. *Journal of Animal Ecology*, 34, 557–71.

Skaife, S. H. 1955: *Dwellers in darkness*. London: Longmans.

Smeathman, H. 1781: Some account of the termites which are found in Africa and other hot climates. *Philosophical Transactions of the Royal Society of London*, 71, 139–92.

Snyder, T. E. 1956: Annotated subject-heading bibliography of termites: 1350 BC to AD 1954. *Smithsonian Miscellaneous Collection* 130.

Spain, A. V., Okello-Oloya, T. and Brown, A. J. 1983: Abundances, above-ground masses and basal areas of termite mounds at six locations in tropical north-east Australia. *Revue d'écologie et de biologie du sol*, 20, 547–66.

Taltasse, P. 1957: Les cabeças de jacaré et le rôle des termites. *Revue de Géomorphologie dynamique*, 6, 166–70.

Thomas, M. F. 1974: *Tropical geomorphology*. London: Macmillan.

Trapnell, C. G., Friend, M. T., Chamberlain, G. T. and Birch, H. F. 1976: The effect of fire and termites on a Zambian woodland soil. *Journal of Ecology*, 64, 577–88.

Tricart, J. 1957: Observations sur le rôle ameublisseur des termites. *Revue de Géomorphologie dynamique*, 6, 170–2, and 179.

—— 1972: *The landforms of the humid tropics, forests and savannas*. London: Longman.

UNESCO/UNEP/FAO 1979: *Tropical grazing land ecosystems*. Paris: UNESCO.

Vallachmedov, B. V. 1981: Termites *Anacanthotermes ahngerianus* (Isoptera, Hodotermitidae) and their influence on takyr formation in south-western Tadjikstan (central Asia). *Pedobiologia*, 21, 242–56.

Watanabe, H. 1975: On the amount of cast production by the Megascolecid earthworm *Pheretima hupiensis*. *Pedobiologia*, 15, 20–8.

—— and Ruaysoongnern, S. 1984: Cast production by the megacolecid earthworm *Pheretima* sp. in northeastern Thailand. *Pedobiologia*, 26, 37–44.

Watson, J. P. 1974: Calcium carbonate in termite mounds. *Nature*, 247, 74.

—— 1975: The composition of termite (*Macrotermes* spp.) mounds in soil derived from basic rock in three rainfall zones of Rhodesia. *Geoderma*, 14, 147–58.

Williams, M. A. J. 1968: Termites and soil development near Brocks Creek, Northern Territory. *Australian Journal of Science*, 31, 153–4.

—— 1978: Termites, soils and landscape equilibrium in the Northern Territory of Australia. In Davies, J. L. and Williams, M. A. J. (eds), *Landform evolution in Australasia*, pp. 128–41. Canberra: Australian National University Press.

Wood, T. G. and Sands, W. A. 1978: The role of termites in ecosystems. In Brian, M. V. (ed.), *Production ecology of ants and termites*, pp. 245–92. Cambridge: Cambridge University Press.

6 The biogeomorphology of arid and semi-arid environments

David S. G. Thomas

Introduction

Scientific definitions and criteria used to delimit arid and semi-arid environments have varied according to the nature of the investigation (Heathcote, 1983), but that of Meigs (1953) is probably most widely used. This is based upon a moisture index relating precipitation and potential evapotranspiration, but excludes dry areas in high latitudes, which are too cold for crop growth. Three subdivisions of the arid zone are recognized: extremely arid, or hyper-arid, where precipitation often does not fall during 12 consecutive months or more and mean annual amounts are less than 25 mm; arid, with 25–200 mm p.a.; and semi-arid, with 200–500 mm p.a. (precipitation figures from Grove, 1977). These environments represent 4.3 per cent, 16.2 per cent and 15.8 per cent of the global land surface respectively, predominantly in the tropics and subtropics, but also cover significant areas of the continental interiors of Asia and North America.

The arid zone has been both more extensive and more contracted in the recent geological past than it is today. Widespread dryness was a characteristic of the Devensian (most recent) glaciation (e.g. Goudie, 1983a), resulting in the expansion of the realm of geomorphological processes such as dune building (perhaps up to 50 per cent of the land area between 30°N. and 30°S.: Sarnthein, 1978), and colluvium deposition (Watson et al., 1984). In contrast, periods of wetter climates during the Quaternary are indicated by evidence of high lake levels in today's arid zones (e.g. Street-Perrott and Roberts, 1983), whilst in the short term the unreliability and variability of rainfall events have a significant bearing on the ecology and geomorphology of arid environments.

Natural arid zone vegetation is mainly composed of mesophyte and xerophyte plant types, with all species being adapted to either drought avoidance or drought endurance through special survival strategies (e.g. Gupta, 1979). Plant cover ranges from almost none in hyper-arid areas, through sparse desert scrub, to savanna grassland and thorn savanna. Geomorphologically the vegetation type and its density, both above and below the ground surface (e.g. Chew and Chew, 1965), has an important influence upon sediment movement by water and wind, whilst biocrusts, formed by algae and lichens, are also important determinants of sediment transport rates in some arid environments (e.g. West and Skujins, 1978).

The role of microorganisms is being increasingly recognized in the formation of varnishes and crusts in desert areas where conditions are generally too inhospitable for the extensive existence of other life forms. As the range of life forms increases along the hydrological gradient from arid to semi-arid environments, so the interaction of biological factors with geomorphological processes increases. Whilst lichens can protect rock surfaces from weathering in humid environments, they are a most effective agent of rock breakdown in arid environments (Krumbein, 1969). Termites are an important and effective agent of erosion in savanna grasslands (see chapter 5 in this volume). The relatively greater biomass of the semi-arid areas supports a diverse population of herbivores, some of which contribute to the formation of landforms.

Botanical influences on sediment transport

Whilst rates of erosion and sediment transport are influenced by a range of variables, there is certainly a broad inverse relationship between the amount of vegetation cover and sediment movement. Various aspects of the biomass affect erosion through the protection afforded to the ground surface by canopy cover, plant spacing, litter fall, and the binding effect which roots and organic matter exert in a sediment. Many of the variables which Cooke and Doornkamp (1974) identified as affecting the erodibility of a surface by wind are related to vegetation cover (table 6.1), and also apply to sediment movement by water. The variability of water availability through time and space in arid environments affects sediment movement

Table 6.1   Ground surface variables affecting erodibility by wind and water

Sediment variables		Surface variables	
Particle size	±	Vegetation: residue	+
Cohesive properties	+	height	±
Abradability	−	orientation	+
Transportability	−	density	+
Organic content	+	fineness	+
		total cover	+
		Surface roughness	+
		Surface length from protective cover	−
		Slope angle	±

− erosion reduced if variable value reduced
+ erosion reduced if variable value increased
modified after Cooke *et al.*, 1982; Cooke and Doornkamp, 1974

directly, but especially through the intermediary of control over vegetation development.

Vegetation and sediment movement by water

Sediment movement by water in arid environments is episodic because of the unreliable nature of precipitation events. Whilst rainfall events in semi-arid areas are also variable in space and time, they do display a seasonality which increases the effective erosion and movement of material.

Amongst the studies of generalized sediment yields in rivers in different global environments, those by Fournier (1949) and Langbein and Schumm (1958) identified peak yields near the semi-arid/arid boundary, or 300 mm of effective animal precipitation according to Langbein and Schumm (1958). Whilst Fournier (1949) also identified a second peak in the monsoon tropics, pointing to the significance of rainfall seasonality for erosion rates, and Wilson (1973) suggested that the semi-arid peak only applied in continental areas, these studies demonstrated the general relationship between sediment yield (or erosion), effective precipitation and vegetation cover.

Different vegetation communities afford different degrees of protection to the ground surface against erosion by raindrop impact and overland flow. Arid and semi-arid scrub, along with tundra vegetation, affords the least natural surface coverage (Young, 1972), due to the strategies adopted to survive in environments deficient in moisture. Of especial importance are plants with episodic growth strategies, linked to immediate water availability, the density of the vegetation cover, and the small production of litter.

The arid environment plant classification of Shantz (1956) is a useful basis for assessing their geomorphological significance. Based on the ecological means of coping with drought conditions, drought-escaping, drought-evading, drought-resistant, and drought-enduring species were recognized. Drought evaders form the bulk of arid-land flora (Gupta, 1979), and are ephemerals and annuals which only retain resistant 'reserve organs' such as seeds and rhizomes during moisture deficient seasons. The trigger for growth is an external moisture source, and therefore they respond to rainfall events. Thus the vegetation cover is very sensitive to local moisture variations (Dunne et al., 1978), and although the growth response is triggered by moisture availability in the upper 5–30 cm of the soil there is normally a lag in response time (Gupta, 1979) with ephemerals appearing in 2–3 days and annuals a few days later (Heathcote, 1983).

This is very significant geomorphologically. In the semi-arid Mtera area of Tanzania, the dry season vegetation covers only 3 per cent of the ground surface, but the coverage rises to 87.5 per cent in the wet season. The rainfall events which trigger plant growth at the end of the dry season are therefore also going to be erosionally very effective, especially in areas of high intensity tropical thunderstorms (Hudson and Jackson, 1959). This effect is probably further enhanced by the need for a single rainfall event in excess of 25 mm to trigger the germination of many desert plants (Went, 1959; Gupta, 1975).

Drought-evading plant species make economical use of limited moisture by maintaining a wide spacing between individuals (Shantz, 1956), and drought-resistant succulents are also often widely spaced. Thus these arid land plants do not provide a great deal of ground protection through interception. Even semi-arid annual grasses are usually clumped, resulting in less protection against rainsplash than in humid grasslands (Young, 1972). Nevertheless, in Tanzania the protective role of a grass cover in semi-arid environments is mainly one of interception (Hudson and Jackson, 1959). Grass is a relatively more effective interceptor than scrub vegetation, which allows 76–96 per cent of rainfall to pass through to the ground surface during high intensity events, and tree canopies, which permit the passage of 98 per cent of rainfall.

Although the wide spacing of plants may provide little protection against raindrop impact, the root systems in arid environment shrub communities may be very extensive, even when the canopy cover extends over no more than 5 per cent of the area (Gupta, 1979). Roots therefore exhibit a binding effect on the soil (e.g. Chew and Chew, 1965). In addition, the plants can affect the overland transport of sediment by water, and Van Rensburg (1955) and Temple (1972) note that both grasses and shrubs are effective in controlling sheet and rill erosion in arid areas. On experimental plots in Tanzania, grass planted in strips across slopes effectively reduced erosion, and enhanced infiltration (Temple, 1972). Some arid zone vegetation communities, for example *Acacia aneura*, naturally adopt a banded growth pattern, which has been termed 'brousse tigre' in some regions (e.g. Goudie and Wilkinson, 1977). This is thought to be a strategy to reduce competition for moisture, as it catches runoff and therefore may inhibit downslope sediment movement.

The general lack of leaf litter in arid environments compared with humid areas is another way in which surface protection against erosion is reduced (e.g. Temple, 1972). Whilst this is often attributed to the low productivity of arid biomass (e.g. Jansson, 1982) it is also due to warm desert plants tending not to shed leaves which die from moisture deficiency until new growths are triggered (e.g. Went, 1979). The occurrence of sclerophyllic plants, and the rapid breakdown of litter (Cooke and Warren, 1973) also reduce ground cover, although the litter production in deserts may in fact be greater than in coniferous forests (Rodin and Brazilevich, 1965).

Vegetation and sediment movement by wind

The formation of coastal dune systems in many environments indicates how aeolian processes are effective when there is a deflatable and vegetation-free source of sediment. In such cases the inter-tidal zone is the vegetation-free area, marine processes supply sand sized sediment to that zone, and onshore winds created by land/sea temperature differences move the sediment to the landward side of the beach zone.

In arid environments the precipitation conditions which militate against extensive vegetation growth, and the vegetation–surface conditions discussed above, can favour sediment transportation by wind, and the formation of dunefields or sand sheets. This is so provided a source of deflatable material is available, which may be the bed of an ephemeral or dry river, as in Australia (Twidale, 1972; Bowler, 1976) and south western Botswana (Thomas, 1986a), the floor of a dry lake or pan, for example in Tunisia (Coque, 1979), a weathered bedrock source, or the reworking of sand deposited in a previous arid episode.

Active aeolian deposits cover about 20 per cent of the world's arid zones (Ahlbrandt and Fryberger, 1982). Active and fossil dunes are often delimited on the basis of mean annual precipitation values (table 6.2)

Table 6.2   Examples of rainfall limits for the formation of desert dunes, and rainfall amounts in areas of fossil desert dunes

Area	Dune type	Annual rainfall limit for dune activity (mm)	Annual rainfall, fossilized dunes (mm)	Source
Western USA				
Arizona	Parabolic, transverse linear	238–254	305–380	Hack, 1941
N. E. Colorado USA	Parabolic	—	282–464	Muhs, 1985
Africa				
Mauritania	Various, mainly linear	25–50	—	Sarnthein and Diester-Hass, 1977
Southern Sahara	Various	150	750–1000	Grove, 1958
Namib, S. W. Kalahari	Mainly linear	150	—	Lancaster, 1979
N. Kalahari, Zimbabwe, Angola, Zambia	Linear	—	up to 972	Thomas, 1984
Australia	Mainly linear	100	—	Mabbutt, 1971
W. Australia	Linear	200	1000	Glasford and Killigrew, 1976
India				
North west	Parabolic and linear	200–275	800	Goudie *et al.*, 1973

Source: Adapted from Goudie 1983b with additional information

Figure 6.1 Dust haze around a pan, western Zimbabwe: high game animal concentrations have destroyed the vegetation cover and increased the potential for aeolian processes to take effect

because a well developed vegetation cover inhibits aeolian processes by raising the wind velocity profile above the ground surface (Cooke *et al.*, 1982). As Ash and Wasson (1983) note, though, it is important to recognize that the transformation from aeolian activity to inactivity in arid and semi-arid areas is gradual, with no precise boundary being present. This gradation may influence the spatial distribution of dune forms, as noted by Wasson *et al.* (1983) in the Thar desert, India. The sensitivity of the vegetation–deflation relationship in semi-arid areas is highlighted by the renewed sand movement which can occur due to pressures of overgrazing, both by domestic livestock (e.g. Grainger, 1982) and wild animals (figure 6.1; Thomas 1986b).

Feedback also occurs between sand mobility and the plant cover, for just as vegetation encourages aeolian stability, so the less active dune forms, or the more stable flanks of dunes, may encourage vegetation growth. It was observed by Jutson (1934) that the most active downwind end of longitudinal dunes in western Australia were devoid of vegetation, whereas the more stable upwind parts were colonized. In the Simpson Desert,

Australia, the cane grass *Zygochloa paradoxa* occurs on mobile dune crests, whereas the spinifex *Triodia basedowii* is found on more stable flanks and interdune areas (Purdie, 1984). In general, species with shallow roots are less successful on mobile sand than those with anchoring tap roots, which give better support and are able to withstand shifting sand.

On stable sand however, shallow rooted species dominate because they are able to intercept rainfall in the top layers of sand before it percolates to the depths required by species with tap roots. Thus, in the southwestern USA Hack (1941) observed that the deep rooting sage brush *Artemisia filifolia* and Rabbit brush, *Chrysothamus* spp. colonized mobile sand, whereas grasses were able to survive more effectively on more stable dune forms.

The relationship between dune forms and vegetation in Australia is shown in figure 6.2. In areas with up to 250 mm mean annual rainfall ground cover is up to 10 per cent: dune crests are colonized by drought-escaping ephemeral grasses and deep rooting, drought-evading plants are present on dune flanks. Where precipitation rises up to 350 mm p.a.

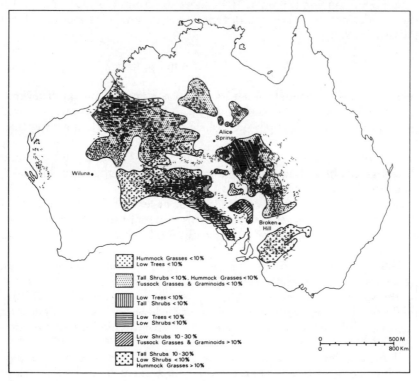

Figure 6.2   Vegetation types and densities in the dunefields of the Australian arid and semi-arid zone
(Utilizing information from Jennings, 1968, and Ash and Wasson, 1983)

two-storey 'mallee' vegetation grows, the under storey being grasses and the upper storey species of *Eucalyptus* and *Acacia* (Ash and Wasson, 1983); ground cover correspondingly increasing to 30 per cent. Marshall (1970) indicated that 32–60 per cent cover of vegetation would inhibit aeolian movement on evenly vegetated surfaces, and the same author later noted that erosion rates would rise rapidly as vegetation cover fell below 15 per cent (Marshall, 1973). Whilst these figures might apply to sand sheets, Ash and Wasson's (1983) investigation of sand movement in the Australian dunefield found that a cover density as high as 30 per cent was unable to inhibit aeolian entrainment, largely because of the effects of the unevenness of dune topography on both wind flow and vegetation distribution. As vegetation cover is a function of moisture availability, the relationship may be expressed by the mobility index M (Ash and Wasson, 1983):

$$M = 3.8 \times 10^{-4}(U)^4(Ea/Ep)$$

where U = mean wind speed at 10 m above the ground surface, and Ea/Ep the ratio of actual to potential evapotranspiration; or in modified form (Wasson, 1984):

$$M = 0.21(0.13W + \ln Ep/P)$$

where P = mean annual precipitation, Ep = Annual potential evapotranspiration, and W = per cent of days with sand shifting winds, measured as the percentage of days per year which have wind speeds at greater than $8 \, \text{ms}^{-1}$ at 3 pm.

Sand mobility occurs when M is greater than 1.0, and the total sand discharge can be estimated using a modified form of the Bagnold (1941) equation (Wasson and Nanninga, 1986).

Ash and Wasson (1983) concluded that in semi-arid and arid parts of Australia today, the vegetation cover is insufficient to stop sand movement, but significant mobilization is currently inhibited by insufficient wind velocities.

The type of vegetation affects the ability of a cover to inhibit aeolian processes. As 90 per cent of aeolian sediment is carried in the 0.5 m closest to the ground (Heathcote, 1983), surface grasses are more effective inhibitors than trees without branches on the lower stems even if densities are less, whilst leafless sclerophytes must also provide little obstruction to wind blown sand.

Plants also influence aeolian processes by modifying the aerodynamic roughness of a surface, a characteristic which is enhanced by the lack of a total ground cover in arid areas. In this respect larger plants such as perennials (Ash and Wasson, 1983) are more effective than smaller annual grasses. Whilst the latter are more effective stabilizers, the survival strategy of shrubs and bushes, involving a wider spacing in order to make efficient use of the available moisture, contributes to their greater modification of airflow characteristics. As well as creating a mobilization shadow on the lee side, which Marshall (1970) recognized was a function of the height and depth of a plant, velocities are increased as air movement is deflected around the side of the obstacle (figure 6.3; Ash and Wasson, 1983). This process may result in the excavation of a moat around the sides of the vegetation clump, and the deposition of sediment due to the drop in wind speed in the lee side cavity of the plant.

The root-exposure which may result from the scouring activity around the flanks of the plant explains why species with anchoring tap roots are more able to survive in areas of mobile sand. Some plants, such as the woody shrub *Zygophyllum album* found in North Africa, are able to adjust to the relative shifting of sand about their bases by extending a branching tap root system in the direction of sand accumulation. Gimingham (1955) observed that *Z. album* was present on the north and west sides of the Jalo Oasis in Libya, and with sand-moving winds blowing from the north

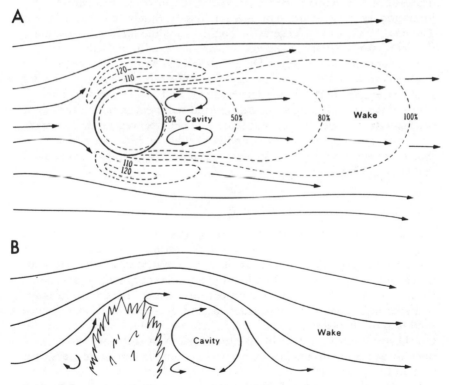

Figure 6.3 Percentage wind speed and stream lines around a bush, (A) in plan and (B) in section
(after Ash and Wasson, 1983)

west, the stabilizing effect of this plant resulted in the formation of hummocky dunes. Shrubs of the *Haloxylon* species, found for example, in the Kara Kum desert, USSR, have a similar effect on mobile sand.

Hack (1941) expressed the view that whilst aerodynamic forces are responsible for the form of a dune, the actual type was determined by physiographic conditions, including vegetation. Whilst it is now widely recognized that windflow regimes play a very significant role in determining dune type (see, for example, Fryberger, 1979), vegetation does contribute to the form of some dunes. Rebdon and Nebkha, respectively large and small coppice dunes, are vegetation obstacle forms (e.g. Melton, 1940; Bourcart, 1928) with the actual form dependent on the species and shape of the vegetation (Cooke and Warren, 1973).

Parabolic dunes are associated with blow-outs on vegetated semi-arid surfaces (Hack, 1941; Verstappen, 1968), the arms of the dune being anchored by vegetation, limiting aeolian erosion to the central corridor where the cover has been disrupted. The high sodium content of some South African soils discourages vegetation growth on the surface of pan depressions (Le Roux, 1978), providing a source of sediment for the formation of downward pan margin lunette dunes (e.g. Goudie and Thomas, 1985, 1986). Vegetation on the less saline surfaces surrounding the depressions assists in the accumulation of the dune sediments.

A recent study in the Negev Desert, Israel, has shown how the presence of a vegetation cover can influence linear dune morphology by impeding sand movement on the dune surface (Tsoar and Møller, 1986). The role of vegetation in trapping wind-transported sediment in the Negev is also demonstrated by the absence of loess deposits in arid, vegetation-free areas, except near springs with large tamarix shrubs, and its accumulation on the better vegetated slopes of the less arid central Negev highlands (Yaalon and Dan, 1974).

Biocrusts and sediment movement

Sediment movement in arid environments is not only affected by the vegetation cover, but also by biocrusts formed from lichens and algae (e.g. Bond and Harris, 1964). These are distinct from duricrusts, the formation of which may be enhanced by plant life and microorganisms, and desert varnish, which forms on rock surfaces, and which are discussed later.

Algae and lichens are important in arid and semi-arid soils for the provision of nitrogen (e.g. Fletcher and Martin, 1948; Fuller *et al.*, 1960) but they may also create a filamentous crust that stabilizes sandy soils, not only in dry environments but on coastal dunes in temperate areas (Van Den Anker *et al.*, 1985).

In the White Sands of New Mexico, the unicellular green algae *Palmogloea protuberans* dominates the 2–5 mm thick crust of cemented gypsum particles (Shields *et al.*, 1957), and may in fact cause the surface concentration of CaSO₄ (Martin and Fletcher, 1943). Such crusts, which may undergo lichenization, may be closely related to the clay content of desert soils. As biocrusts tend to develop away from vegetation (Shields *et al.*, 1957), they may be important for stabilizing dune surfaces and reducing erosion by wind and water. In the North American arid zone, aeolian erosion is considerably less where 80 per cent of the surface is covered by a biocrust, as in the Great Basin desert of Utah and Nevada, especially when compared with the Sonoran Desert of Arizona and Mexico, where there is only 4 per cent crustal protection (West and Skujins, 1978). It has been suggested, however, that the biocrust on fixed sand dunes in Niger *enhances* erosion by overland flow and gully formation because of the associated reduction in infiltration capacity, especially during heavy tropical storms (Talbot and Williams, 1978).

The geomorphological effects of mammals

Mammals, especially large herbivores, affect their environments in a number of ways which can have geomorphological consequences (figure 6.4). In the savannas and semi-arid lands of Africa, these effects are probably greater today than ever before. Although large indigenous herbivores are now confined to much smaller areas than say a hundred years ago – for

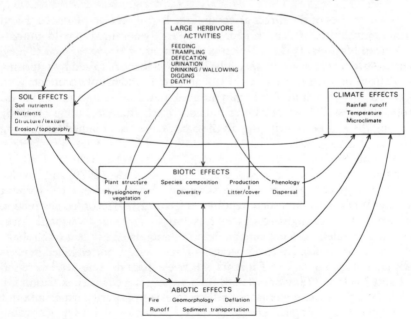

Figure 6.4 The activities of large herbivores and their influences on the environment
(adapted, after Cumming, 1982)

example, to less than 12 per cent of Zimbabwe (Cumming 1981) – animal densities are much higher. In the great game reserves and National Parks in Botswana, Namibia, and western Zimbabwe, the geomorphological effects are being exacerbated. In Hwange National Park, Zimbabwe, for example, which has an area of 13,561 km^2, elephant numbers have risen from 1,000 in the 1920s, to 5,000 in the 1960s and by the early 1980s to 24,000 (Thomas, 1986b). Beyond the game reserves, numbers of domestic livestock have been increasing rapidly in semi-arid lands during the twentieth century, with management techniques tending to concentrate upon increasing absolute numbers rather than quality (e.g. Cooke, 1985). The result is environmental damage, and overstocking has frequently been pointed to as one of the probable causes of desertification (e.g. Mabbutt, 1977; Grainger, 1982).

The geomorphological impact of large mammals in semi-arid areas is both direct and indirect. Indirect effects are biotic, through the destruction of vegetation and litter which may reduce ground cover or alter community structure, thereby enhancing erosion rates. Direct abiotic effects include digging, wallowing, trampling and drinking activities which can both enhance erosion rates and create distinct landforms.

Biotic effects

Detailed studies of the direct affect of large herbivores on plant cover and plant communities in semi-arid areas are largely confined to wild animals in African National Parks. Elephants (*Loxodonta africana*) appear to have particularly severe effects, especially when densities are raised by imposing restrictions on their movements. Elephants are found in environments with between 300 mm and 2,000 mm rainfall p.a., or semi-arid to humid tropical forest. Their browsing habits, which include killing trees by pushing them over or stripping off the bark, have been blamed for turning woodland into grassland (e.g. Laws, 1970), although Cumming (1982) points out that such vegetation changes may well be cyclic.

In semi-arid ecosystems in Kenya and Tanzania, Laws (1970) and Savidge (1968) report changes from bush savanna to grassland due to elephant activity. Watson (unpublished, reported in Laws, 1970) recorded destruction rates of 2,011 drought-resistant trees km^{-2} in Tsavo National Park, Kenya, equivalent to 6 per cent per annum, and also 2 per cent of baobabs (*Adansonia digitata*). Up to 55 per cent of trees were reported killed in parts of Ruaha National Park in Tanzania, where elephant densities were between 1.2 and 3.0 km^{-2} (Savidge, 1968). In a variety of ecosystems, Cumming (1982) suggests the general tree felling rate is 4–5 per cent per annum.

The geomorphological effects of such destruction will vary according to the original density of trees and whether an under storey vegetation exists. Rainfall interception rates will be altered, and the impact of raindrops on the ground surface will increase due to the concomitant reduction of litterfall. An average adult elephant, weighing 1,700 kg, consumes 102 kg of biomass per day (Laws, 1970). With the average natural size of elephant communities being about six, the geomorphological impact will be small if elephant densities are low. The unnatural confinement of elephants to relatively small areas today and the high numbers which may ensue in the National Parks, can have a potentially devastating effect on vegetation cover. Under near natural conditions too, elephants may group together in clans of up to 500, and even on a temporary basis the ensuing consumption of 50,000 kg of biomass per day can have a very severe impact on ground cover and soil erosion.

Whilst light grazing may improve plant growth, the destruction of grasses and other ground cover plants by heavy grazing may be more significant in terms of erosion rates, because of their interception effects (e.g. Hudson and Jackson, 1959) and limitation of rill and sheet wash. Although elephants are preferentially browsers, up to 90 per cent of the content of elephant dung may be grass (reported in Laws, 1970), suggesting that once woody species have been destroyed, they are forced to become grazers.

Hippopotamus (*Hippopotamus amphibius* L.) are grazers that have horny lips which may tug grasses by the roots from soft sediments (Lock, 1972), or crop them down to the ground surface (Field, 1970). High concentrations of domestic livestock have similar effects. In the early 1960s the effects of overgrazing in the Tribal Trust Lands of Zimbabwe led to 77 per cent of grasslands being laid bare or having a severely reduced ground cover (Cleghorn, 1966, reported in Cumming, 1982). The effects, whether induced by domestic or wild animals, become concentrated in semi-arid lands because of the restriction of the range of animals in the dry season: for example, elephants cannot range more than 24 kilometres from waterholes (Laws, 1970) because of their daily drinking requirements, and serious vegetation destruction can ensue, especially during periods of enhanced drought conditions (Thomas, 1986b).

Abiotic effects

As well as destroying the vegetation cover, ungulates can enhance erosion rates by the trampling effect. Infiltration rates are reduced by soil compaction, and in Australia, Heathcote (1983) reports that this is significant because sheep and cattle exert pressures equivalent to 0.65 to 1.7 kg cm^{-2}, whereas the heaviest natural grazer, the kangaroo, is comparatively lightfooted, exerting only 0.1 kg cm^{-2} pressure. Trampling can also break up soil aggregates and increase the susceptibility to deflation (Goudie, 1986), especially when the vegetation cover has been destroyed.

The activities of large mammals contribute to the formation of pan depressions, which are an important component of many arid and semi-arid landscapes (Goudie and Thomas, 1985). Although a range of factors

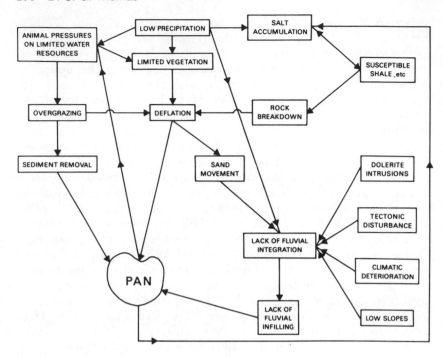

Figure 6.5   A model of pan development
(after Goudie and Thomas, 1985)

contribute to the formation of these annular depressions (figure 6.5), notably deflation, herbivores can play a role both by enhancing the susceptibility of a surface to aeolian processes by vegetation destruction and by the loosening of surface materials, and by the direct removal of sediment on and within their bodies.

The scouring action of animals has been commented on by Alison (1899), Passarge (1904) and Jaeger (1939) in southern Africa, and Chamberlain (in Gilbert, 1895), Green (1951) and Reeves (1966) in western Texas, whilst the processes involved have been described from both western Zimbabwe (Weir, 1960) and Kenya (Ayeni, 1977; see figure 6.6).

Elephants and other animals extract dietary requirements of sodium and other minerals from localities where they are found in above average concentrations in the soil. These are often in conjunction with termite mounds because their large surface areas enhance evaporation rates and therefore salt accumulation. Elephants dig out material with their foretoes and tusks and then eat it, creating small depressions which may be up to 1.5 m deep and 25 m in diameter (figure 6.7; Weir, 1969). Rubbing activity by other animals, for example rhinocerus (*Diocenus bicornis* L.) and hartebeest (*Alcelaphus buselaphus cokei* Gunther), contributes to the

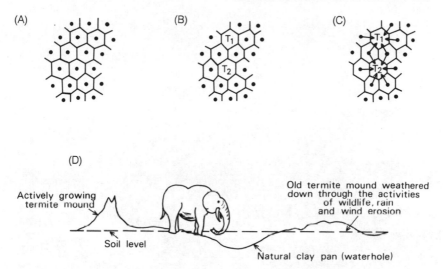

Figure 6.6 The evolution of pans through animal activity (A) Termitaria equally spaced; (B) abandoned termitaria T_1 and T_2 weathered by rainwash, wind and animals which creates a depression that accumulates water (C) and attracts animals which enlarge the depression (D)
(adapted, after Ayeni, 1977)

diminution of abandoned mounds and depression formation (Ayeni, 1977). These activities are concentrated in the dry season, with water accumulating in the depressions in the wet season. This is then utilized for drinking by a range of animals, and the puddling effect of the large number of hooves creates an impermeable clay seal, which increases water retention. Buffaloes (*Syricerus caffer* Sparrman) and elephants use the depression for wallowing in the wet season, deepening the pan by removing from 0.3 to $1.0 \, m^3$ of mud per animal per wallow (Flint and Bond, 1968). In addition to these means of enlarging the pan, 'an elephant carries away mud from a special sort of mudbath. He sprays himself with water from a pan and then blows dust from the pan margin over his wet hide. The resulting mud dries, and as the animal moves off, starts to flake away, removing sweat, scurf and insects' (Flint and Bond, 1968), thereby laterally extending the extent of the depression (figure 6.8).

In the wet season, overland flow washes material into the depression (Weir, 1969) but the quantity of sediment removed by animals is sufficient to maintain the presence of a pan even without the contribution of deflation (Weir, 1960). The Kenyan pans formed by these zoogenous processes range in size from 0.45 to $0.59 \, km^2$ (Ayeni, 1977), and those in western Zimbabwe are rarely larger than 200 m across. Of these, the largest sometimes display a depressed, churned clay rim in the dry season (see figure 5 in Goudie and Thomas, 1985) which probably results

Figure 6.7 The creation of small depressions by elephant excavations, Hwange National Park, Zimbabwe

Figure 6.8 A small pan depression created by animal activity. Note the reduced vegetation cover surrounding the pan, and the trees (middle left) killed by elephant bark-stripping

from the wet season wallowing activities being concentrated around the rims of pans.

Biological influences on the formation of crusts

Desert varnish and microorganisms

Desert varnish is a glossy, dark coloured, hydrated manganese- and iron-rich coating on rock surfaces, which is normally found in moisture-deficient environments. As these may include high latitude arctic and antarctic locations (Dorn and Oberlander, 1982), the term 'rock varnish' is sometimes preferred. Possible processes of desert varnish formation have been reviewed by Dorn and Oberlander (1982), and include hypotheses pointing to the source of the manganese being from within the rock mass itself (e.g. Krumbein and Jens, 1981), from atmospheric sources (e.g. White, 1924), and from ground water rise (e.g. Hume, 1925).

Importantly, endolithic microorganisms are significant in the formation of the varnish, as they are the only life forms which are able to colonize bare rock surfaces in harsh, arid environments. Laudermilk (1931) believed that lichens drew iron and manganese out of the rock body. Today, it is widely accepted that atmospheric dust is the source of the constituent minerals (Perry and Adams, 1978; Bauman, 1976) and of the original colonizing microorganisms (Dorn and Oberlander 1981). This source was convincingly demonstrated by Allen (1978) who showed that Fe and Mg concentrations do not decrease outwards from the rock surface, which could be expected if an internal source occurred.

Table 6.3   The chemical composition of desert varnish

| Compound | Percentage oxide weight, range of values | | |
	USA (Lakin et al., 1963)	USA (Engel and Sharp, 1958)	New South Wales (Dragovich, 1984)
MgO	1.5–7.0	0.00–5.47	1.27–1.64
Fe_2O_3	11.0–35.0	4.29–18.56	9.71–17.84
SiO_2		28.19–61.95	31.92–37.36
Na_2O		1.27–3.43	0.05–1.13
Al_2O_3		15.28–21.26	17.81–25.92
K_2O		1.31–1.64	0.65–23.60
CaO	0.0–2.4	0.31–4.46	0.45–2.02
TiO_2	0.15–1.0	0.33–2.48	0.72–1.30
P_2O_5		0.00–2.68	0.78–2.56
H_2O and other compounds		8.85–11.66	7.45–10.03

Source: Adapted from Dragovich, 1984

The varnished surface may be up to 5 mm thick (Dorn and Oberlander, 1982), but is commonly of the range 10–30 μm. Despite the difference in form, Bauman (1976) demonstrated that the chemistry of desert varnish was almost identical to that of hydrated marine ferromanganese nodules, which are themselves the product of biological fixation, and Zapffe (1931) recognized that some bacteria selectively promote Mg and Fe deposition. Mineral concentrations from desert varnishes are shown in table 6.3.

The biogeochemical process of desert varnish formation is summarized in figure 6.9. Following the colonization of a rock surface by microorganisms, clay minerals, which form up to 70 per cent of desert varnish (Potter and Rossman, 1977), are adsorbed. This has the effect of protecting the microorganisms against both desiccation and temperature extremes (Burns, 1979): this process is the mechanism which affords survival in an otherwise inhospitable environment. Mixotrophic species are responsible for desert varnish formation, being able to survive in environmental niches otherwise

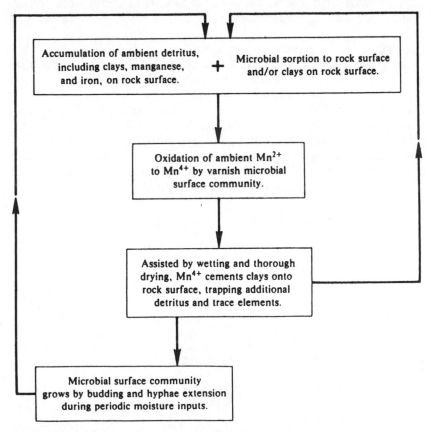

Figure 6.9 Simplified biogeochemical model of rock varnish formation (after Dorn and Oberlander, 1982)

unsuitable for heterotrophs (Dorn and Oberlander, 1981). The widespread cyanophytes of arid environments (Bauman, 1976), such as *Mettallogenium* and *Pedomicrobium* are probably dominant in the process of varnish formation, as they scavenge, concentrate and fix divalent manganese from atmospheric fallout, which is oxidized and precipitated in the tetravalent form Mg_4 (Ehrlich, 1978) within the lattices of the adsorbed clays.

The process is enhanced by microhydrological change on the rock surface, caused by large diurnal temperature fluctuations and importantly, by the evaporation of early morning dewfall, which enhances microorganism activity and the reduction of minerals. Dorn and Oberlander (1981) believe that desert varnish can only result from biochemical processes, whereas Bauman (1976) suggests it is the result of geochemical processes which are continued by bacterial action. Elvidge and Moore (1980) were able to simulate the reactions purely geochemically, suggesting therefore that desert varnish may be able to form without any biological input.

The varnish naturally found in arid environments may range from dark, almost black forms to much lighter coloured varieties. Although such variations may reflect age differences, they are more likely a function of differences in rock surface pH. According to Dorn and Oberlander (1981), black, manganese-rich varnish forms in acidic conditions which favour the reduction of manganese in preference to iron. More alkaline conditions inhibit bacterial growth, and may also allow heterotrophs, poor fixers of dust-borne manganese, to dominate. The result is paler, orange coloured varnish. Brown desert varnish indicates intermediate conditions, whilst the laminated varnishes observed by Perry and Adams (1978) result from cyclic fluctuations in rock surface conditions.

Desert varnish formation may be exceedingly slow. Although Engel and Sharp (1958) reported a varnish forming in less than 50 years in the Mojave Desert, USA, others (e.g. Carter, 1980; Dorn and Oberlander, 1982) regard periods in excess of 5,000 years as necessary for the formation of a complete surface varnish.

Non-algal, rock inhabiting microorganisms which have many of the characteristics of those associated with desert varnish have been described from Australia, the USA and the Gobi Desert by Staley *et al.* (1982). These black-brown micro colonies of fungi attain densities of 200 cm^{-2} and fix atmospheric nutrients to the rock surface. Preferred microtopographical sites for colonization vary from rock crevices in the Mojave Desert to flat exposed surfaces in the Simpson Desert, with no lithological preferences being expressed. These findings suggest that algae may not be necessary for the biogeochemical formation of rock varnish, thereby increasing the number of ecological niches in which this feature of arid landscapes may form.

The role of microorganisms and plants
in the formation of calcrete

Calcrete (caliche) is an important component of arid and semi-arid landscapes, and is widely reported from many of the world's drylands. Whilst it is frequently extremely difficult to draw a distinction between actively forming and relict forms, there is a general delimitation of the occurrence of calcrete in many regions by the 500 mm annual isohyet (Goudie, 1973). A range of models have been proposed for calcrete formation (Goudie, 1973), and undoubtedly each is appropriate under different topographical and hydrological conditions and $CaCO_3$ sources. It is now becoming clear though that organic processes can contribute in a number of ways to its formation with perhaps the most significant being in the relationship between microorganisms and plant roots, and the role of lichen colonies on subaerially exposed calcrete surfaces.

The remains of blue-green algae, bacteria and fungi have been identified within calcretes (e.g. Krumbein, 1969; James, 1972; Estaban, 1976) with over 40 per cent of some calcretes from the western Mediterranean region being composed of fossil *Microcodium* grains (Johnson, 1953). A series of papers by Klappa (1978, 1979, 1980) has elucidated upon the role of these *microcodia*, or *mycorrhizal* organisms, in what has been termed the biolithogenesis (Klappa, 1978) of calcrete. Lichens, which are symbiotic associations of algae and fungi, are seen to play a major part in the formation of the laminated calcretes studied by Klappa, which occurred within soil horizons above bedrock that is being actively weathered.

The term 'lichen stromatolite' has been applied to these laminar calcretes (Klappa, 1979) because of the similarity of the processes involved in their formation with those which form algal stromatolites in marine and lacustrine environments (e.g. Walter, 1976). Using a modification of Walter's (1976) definition, lichen stromatolites may be defined as 'organosedimentary structures produced by sediment trapping, binding and/or precipitation as a result of the growth and metabolic activity of microorganisms, principally lichens' (Klappa, 1979).

The processes involved are summarized in figure 6.10. The laminae are induced by the cyclic telogenesis–pedogenesis–diagenesis sequence caused by lichen growth and replacement under fluctuating hydrological conditions in arid and semi-arid conditions.

Rhizoconcretionary calcretes form in association with plant roots, and according to Klappa (1978) result from a symbiotic relationship between calcific mycorrhizae and plant root cells. The capillary attraction of moisture in arid zone soils to roots leads to calcium carbonate precipitation in the zone around the root. Initially, this may form a sheath with a diameter of 5–25 mm around the living root (Semeniuk and Meagher, 1981). This may eventually thicken to such an extent that the reduction in nutrient and moisture supply to the root contributes to its death.

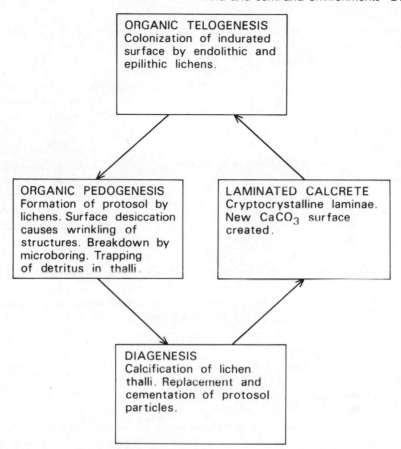

Figure 6.10 A model of lichen stromatolite (Laminated Calcrete) formation (based on information from Klappa, 1979)

Ultimately the dead root can be replaced by a complete calcium carbonate cast. The calcrete rhizoconcretion therefore preserves the former root position and orientation, is often in excess of 1 metre long, and can be a hollow tube or a solid rod.

The processes of calcrete formation which V. Semeniuk and co-workers studied in southwestern Australia were found to be 'accelerated, if not largely dependent on vegetation' (Semeniuk and Meagher, 1981, p. 61). Consequently, the widely held relationship between calcrete occurrence and aridity was not found to hold, because of better vegetation cover development in the more humid parts of the study area. Two general calcrete types were identified in Holocene coastal sands: capillary rise zone calcrete and vadose zone calcrete, with the former being most developed because of the higher water turnover. Mottled to massive thinly laminated

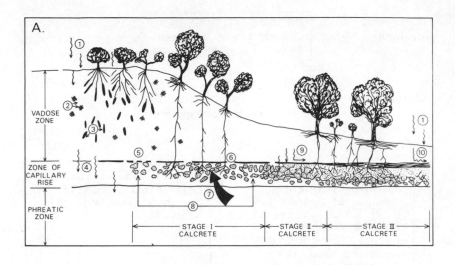

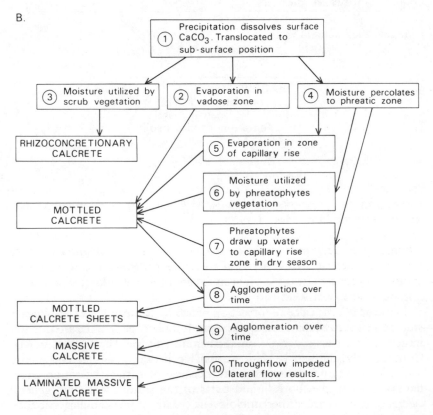

Figure 6.11 (A) The formation of calcrete in association with vegetation (after Semeniuk and Searle, 1981), explained in (B)

calcrete forms in the zone of capillary rise, in association with roots of phreatophytic woodland trees, whilst rhizoconcretions are precipitated around the root systems of scrub vegetation that derives water from the deep vadose zone (figure 6.11).

Different plant species draw upon water and transpire at different times and from different sources. Consequently, Semeniuk and Meagher (1981) report that the two major plant assemblages produce different calcrete developments. The large volume of moisture transpired from a mature woodland draws upon more groundwater, and precipitates more calcrete, than that produced by the more limited moisture demands of the scrub assemblage, obtained from pellicular and gravitational sources (Semeniuk and Searle, 1985). The calcrete which forms in the zone of capillary rise may pass through a series of development stages (figure 6.11) to a massive laminated form (Semeniuk and Meagher, 1981).

In western Australia this effect may be translated into a vegetation–climate relationship. Tropical subhumid conditions (mean precipitation nearly 900 mm p.a., potential evaporation 1,200 mm p.a.) support sufficient vegetation to cause the development of both vadose and capillary rise-zone calcretes (Semeniuk and Searle, 1985). Semi-arid and arid conditions (less than 630 mm p.a., less than 200 mm p.a. respectively) support a reduced vegetation cover without phreatophytic species, and consequently calcrete is both rare, and limited to the vadose zone. The trend towards more evaporation and less precipitation is therefore not conducive to calcrete development where it is controlled by vegetation.

Conclusions

Biological interactions with geomorphological processes are wide-ranging in arid and semi-arid environments, in terms of extent, scale of influence and the resultant impact upon landforms. The effects may be process-delimiting, process-enhancing or process-catalytic. The major process-delimiting influence is the role played by vegetation cover in inhibiting sediment movement by water and wind. Empirical findings are relatively limited concerning the erosivity–vegetation cover – moisture availability relationships present in desert environments though, and the complex range of adaptation strategies adopted by different plant communities suggest that these relationships are both complex and variable from area to area. The Langbein and Schumm (1958) model should not, therefore, be viewed as anything other than a theoretical generalization. Similarly, Ash and Wasson (1983) have demonstrated that the interaction between aeolian processes and vegetation cover is more complex than it is often perceived to be. Modifications to the aerodynamic roughness of a surface by an individual vegetation clump can, at the metre^2 scale, create an aeolian process-enhancement effect, and not the overall surface stability effect which is produced by a vegetation cover at a larger scale.

Large herbivores can contribute to the sediment movement–vegetation cover relationship by reducing plant cover density levels to below the maximum that can be supported by the available moisture. Erosion rates may therefore be enhanced. Other activities of these animals can be process-enhancing or process-catalytic in the formation of pan depressions. Microorganisms are catalysts in the geomorphological processes resulting in calcrete formation, whereas they are seen by some authorities as the only agents through which desert varnish can form (Dorn and Oberlander, 1981). Whilst this view may remain controversial in some quarters, it is now widely apparent that in the drylands, where moisture and temperature conditions can be critical for the survival of life forms, biological and geomorphological interactions have a significant, if imperfectly understood, influence upon the nature of processes and landforms.

References

Ahlbrandt, T. S. and Fryberger, S. G. 1982: Introduction to eolian deposits. In Scholle, P. A. and Spearing, D. (eds), pp. 11–47. *Sandstone depositional environments*. Tulsa: American Association of Petroleum Geologists.

Alison, M. S. 1899: On the origin and formation of pans. *Transactions of the Geological Society of South Africa* IV, 159–61.

Allen, C. C. 1978: Desert varnish of the Sonoran Desert: optical and electron probe micro analysis. *Journal of Geology* 86, 743–52.

Ash, J. E. and Wasson, R. J. 1983: Vegetation and sand mobility in the Australian desert dunefield. *Zeitschrift für Geomorphologie Supplementband* 45, 7–25.

Ayeni, J. S. O. 1977: Waterholes in Tsavo National Park, Kenya. *Journal of Applied Ecology* 14, 369–78.

Bagnold, R. A. 1941: *The Physics of Blown Sand and Desert Dunes*. London: Methuen.

Bauman, A. J. 1976: Desert varnish and marine ferromanganese oxide nodules: congeneric phenomena. *Nature* 259, 387–8.

Bond, R. D. and Harris, J. R. 1964: The influence of the microflora on the physical properties of soils. 1. Effects associated with filamentous algae and fungi. *Australian Journal of Soil Research* 2, 111–22.

Bourcourt, J. 1928: L'action du vent a la surface de la terre. *Revue Géographie Physique et Geologie Dynamique* 1, 26–54, 194–265.

Bowler, J. M. 1976: Aridity in Australia: age, origin and expression in aeolian landforms and sediments. *Earth Science Reviews* 12, 279–310.

Burns, R. G. 1979: Interaction of microorganisms, their substrates and their products with soil surfaces In Ellwood, D. C., Melling, T. and Rutter, P. (eds), pp. 109–38. *Adhesion of microorganisms to surfaces*. New York: Academic Press.

Carter, G. F. 1980: *Earlier than you think*. College Station: University Press.

Chew, R. M. and Chew, A. E. 1965: The primary productivity of a desert scrub (*Larrea tridentata*) community. *Ecological Monographs* 35, 355–75.

Cleghorn, W. B. 1966: Report on the conditions of grazing in the tribal trust land. *Rhodesia Agricultural Journal* 63, 57–67.

Cooke, H. J. 1985: The Kalahari today: a case of conflict over resource use. *Geographical Journal* 151, 75–85.

Cooke, R. U. and Doornkamp, J. C. 1974: *Geomorphology in Environmental Management*. Oxford: Oxford University Press.

Cooke, R. U. and Warren, A. 1973: *Geomorphology in deserts*. London: Batsford.

Cooke, R. U., Brunsden, D., Doornkamp, J. C. and Jones, D. K. C. 1982: *Urban geomorphology in drylands*. Oxford: Oxford University Press.

Coque, R. 1979: Sur la place du vent dans l'érosion en milieu aride. L'example des lunettes (bourrelets éoliens) de la Tunisie. *Mediterranée* 1 & 2, 15–21.

Cumming, D. H. M. 1981: The management of elephant and other large mammals in Zimbabwe. In Jewell, P. A., Holt, S. and Hart, D. (eds), *Problems on Management of locally abundant wild mammals*, pp. 91–118. New York: Academic Press.

—— 1982: The influence of large herbivores on savanna structure in Africa. In Huntley, B. J. and Walker, B. H. (eds), *Ecology of Tropical Savannas*, pp. 217–45. Ecological Studies 42. Berlin: Springer-Verlag.

Dorn, R. J., and Oberlander, T. M. 1981: Rock varnish origin, characteristics and usage. *Zeitschrift für Geomorphologie N.F.* 25, 420–36.

—— 1982: Rock varnish. *Progress in Physical Geography* 6, 317–67.

Dragovich, D. 1984: The survival of desert varnish in subsurface positions, western New South Wales, Australia. *Earth Surface Processes and Landforms* 9, 425–34.

Dunne, T., Dietrich, W. E. and Brunengo, M. J. 1978: Recent and past erosion rates in semi-arid Kenya. *Zeitschrift für Geomorphologie Supplementband* 29, 130–40.

Ehrlich, H. L. 1978: Inorganic energy sources for chemolithotrophic and mixotrophic bacteria. *Geomicrobiology Journal* 1, 65–83.

Elvidge, C. D. and Moore, C. B. 1980: Restoration of petroglyphs with artificial desert varnish. *Studies in Conservation* 25, 108–17.

Engel, C. E. and Sharp, R. P. 1958: Chemical data on desert varnish. *Bulletin of the Geological Society of America* 69, 487–518.

Estaban, C. M. 1976: Vadose pisolite and caliche. *Bulletin of the American Association of Petroleum Geologists* 60, 2048–57.

Field, C. R. 1970: A study of the hippopotamus (*Hippopotamus amphibius* L.) in the Queen Elizabeth National Park, with some management implications. *Zoologica Africana* 5, 71–86.

Fletcher, J. E. and Martin, N. P. 1948: Some effects of algae and moulds in the rain crust of desert soils. *Ecology* 29, 95–100.

Flint, R. F. and Bond, G. 1968: Pleistocene sand ridges and pans in western Rhodesia. *Bulletin of the Geological Society of America* 789, 299–314.

Fournier, M. F. 1949: Les facteurs climatiques de l'érosion du sol. *Bulletin de l'Association Geographique Français* 203, 97–103.

Fryberger, S. G. 1979: Duneform and wind regime. In McKee, E. D. (ed.), *A study of global sand seas*, pp. 137–69. U.S. Geological Survey Professional Paper 1052.

Fuller, W. H., Cameron, R. E. and Raioa, N. Jr. 1960: Fixation of nitrogen in desert soils by algae. *Transactions, 7th International Congress of Soil Science, Madison* 2, 617–24.

Gilbert, G. K. 1895: Lake basins created by wind erosion. *Journal of Geology* 3, 47–9.

Gimingham, C. H. 1955: A note on water table, sand-movement and plant distribution in a North African oasis. *Journal of Ecology* 43, 22–5.

Glasford, D. K. and Killigrew, L. P. 1976: Evidence for Quaternary westward extension of the Australian Desert into southwestern Australia. *Search* 7, 394–6.

Goudie, A. S. 1973: *Duricrusts of tropical and subtropical landscapes*. Oxford: Clarendon Press.

—— 1983a: The arid earth. In Gardner, R. A. M. and Scoging, H. (eds), pp. 152–71. *Mega Geomorphology*. Oxford: Oxford University Press.

—— 1983b: *Environmental Change*, 2nd edition. Oxford: Oxford University Press.

—— 1986: *The Human Impact*, 2nd edition. Oxford and New York: Basil Blackwell.

—— and Thomas, D. S. G. 1985: Pans in southern Africa with particular reference to South Africa and Zimbabwe. *Zeitschrift für Geomorphologie* NF 29, 1–19.

—— and Thomas, D. S. G. 1986: Lunette dunes in southern Africa. *Journal of Arid Environments* 10, 1–12.

—— and Wilkinson, J. C. 1977: *The warm desert environment*. Cambridge: Cambridge University Press.

——, Allchin, B. and Hegde, K. T. M. 1973: The former extensions of the Great Indian Sand Desert. *Geographical Journal* 139, 243–57.

Grainger, A. 1982: *Desertification*. London: Earthscan.

Green, F. E. 1951: Geology of sand dunes, Lamb and Hale counties, Texas. Unpublished M.Sc. thesis, Texas Technical College.

Grove, A. T. 1958: The ancient ergs of Hausaland and similar formations on the south side of the Sahara. *Geographical Journal* 124, 528–33.

Grove, A. T. 1977: The geography of semi-arid lands. *Philosophical Transactions of the Royal Society of London*, B178, 457–75.

Gupta, R. K. 1975: Plant life in the Thar desert. In Gupta, R. K. and Prakash, I. (eds), pp. 202–36. *Environmental analysis of the Thar desert*. Dehro Dune English Book Depot, India.

—— 1979. Integration. In Goodall, D. W. and Perry, R. A. (eds), *Arid land ecosystems: structure, functioning and management*, volume 1, pp. 661–75. Cambridge: Cambridge University Press.

Hack, J. T. 1941: Dunes of the western Navajo country. *Geographical Review* 31, 240–63.

Heathcote, R. L. 1983: *The arid lands: their use and abuse*. London: Longman.

Hudson, N. W. and Jackson, D. C. 1959: Results achieved in the measurement of erosion and runoff in Southern Rhodesia. *Third Inter-African Soils Conference. Dalaba CCTA*, 575–83.

Hume, W. F. 1925: *Geology of Egypt*, volume 1. Cairo: Government Press.

Jaeger, F. 1939: Die Trockenseen der Erde. *Petermanns Geographische Mitteilungen Erg* 236, 1–160.

James, N. P. 1972: Holocene and Pleistocene calcareous crust (caliche) profiles, criteria for subaerial exposure. *Journal of Sedimentary Petrology* 42, 817–36.

Jansson, M. B. 1982: *Land erosion by water in different climates*. Uppsala Universitet Naturgeografiska Institutionen rapport 57.

Jennings, J. N. 1968: A revised map of the desert dunes of Australia. *Australian Geographer* 10, 408–9.

Johnson, J. H. 1953: *Microcodium* Gluck est-il un organisme fossile? *Compte rendu hebdomadaire des séances de l'Academie des Sciences, Paris* 237, 84–6.

Jutson, J. T. 1934: The physiography (geomorphology) of Western Australia. *Geological Survey of Western Australia Bulletin* 95.

Klappa, C. F. 1978: Biolithogenesis of Microdium: elucidation. *Sedimentology* 25, 489–522.

—— 1979: Lichen stromatolites: criterion for subaerial exposure and a mechanism for the formation of laminar calcretes (caliche). *Journal of Sedimentary Petrology* 49, 387–400.

—— 1980: Rhizoliths in terrestrial carbonates: classification, recognition, genesis and significance. *Sedimentology* 27, 613–29.

Krumbein, W. E. 1969: Uber den Einfluss der Mikroflora auf die Exogene Dynamik (Verwitterung und Krustenbildung). *Geologische Rundschau* 58, 333–63.

—— and Jens, K. 1981: Biogenic rock varnishes in the Negev Desert (Israel): an ecological study of iron and manganese transformation by cyanobactcria and fungi. *Oecologia* 50, 25–8.

Lakin, H. W., Hunt, C. B., Davidson, D. F. and Oda, U. 1963: Variation in minor-element content of desert varnish. *US Geological Survey Professional Paper* 4424-B, B28–B31.

Lancaster, I. N. 1979: Quaternary environments in the arid zone of southern Africa. *Department of Geography and environmental studies, Occasional Paper* 22. *University of Witwatersrand.* Johannesburg.

Langbein, W. B. and Schumm, S. A. 1958: Yield of sediment in relation to mean annual precipitation. *Transactions of the American Geophysical Union* 39, 1076–84.

Laudermilk, J. D. 1931: On the origin of desert varnish. *American Journal of Science* 21, 51–66.

Laws, R. M. 1970: Elephants as agents of habitat and landscape change in East Africa. *Oikos* 21, 1–15.

Le Roux, J. S. 1978: The origin and distribution of pans in the Orange Free State. *South African Geographer* 6, 167–76.

Lock, J. M. 1972: The effects of hippopotamus grazing on grasslands. *Journal of Ecology* 60, 445–67.

Mabbutt, J. A. 1971: The Australian arid zone as a prehistoric environment. In Mulvaney, D. J. and Golson, T. (eds), *Aboriginal man and environment*, pp. 66–79. Canberra, Australian National University Press.

—— 1977: Climatic and ecological aspects of desertification. *Nature and Resources* 13, 3–9.

Marshall, J. K. 1970: Assessing the protective role of shrub-dominated rangeland vegetation against soil erosion by wind. *Proceedings of the XI International Grassland Congress*, 19–23.

—— 1973: Drought, land use and soil erosion. In Lovett, J. V. (ed.), pp. 55–80. *Drought.* Sydney: Angus & Robertson.

Martin, W. P. and Fletcher, J. E. 1943: Vertical zonation of great soil groups on Mount Graham, Arizona, as correlated with climate, vegetation and profile characteristics. *University of Arizona Agricultural Experimental Station Technical Bulletin* 99, 89–153.

Meigs, P. 1953: World distribution of homoclimates. In *Reviews of research on arid zone hydrology*, pp. 203–9. Paris: UNESCO.

Melton, F. A. 1940: A tentative classification of sand dunes. *Journal of Geology* 48, 113–73.

Muhs, D. R. 1985: Age and palaeoclimatic significance of Holocene dune sands in northeastern Colorado. *Annals of the Association of American Geographers* 75, 566–82.

Passarge, S. 1904: *Die Kalahari*. Berlin: Reimer.

Perry, R. S. and Adams, J. 1978: Desert varnish: evidence of cyclic deposition of Manganese. *Nature* 276, 489–91.

Potter, R. M. and Rossman, G. R. 1977: Desert varnish: the importance of clay minerals. *Science* 196, 1446–8.

Purdie, R. 1984: *Land systems of the Simpson Desert region*. Natural Resources Series No. 2, Division of Water and Land Resources. Australia, CSIRO.

Reeves, C. C. Jr. 1966: Pluvial lake basins in west Texas. *Journal of Geology* 74, 269–91.

Rodin, L. E. and Brazilevich, N. I. 1965: In Fogg, G. E. (ed.), *Production and mineral cycling in terrestrial vegetation*, pp. 184–207. Edinburgh: Oliver and Boyd.

Sarnthein, M. 1978: Sand deserts during glacial maximum and climatic optimum. *Nature* 272, 43–6.

—— and Diester-Hass, L. 1977: Eolian sand turbidities. *Journal of Sedimentary Petrology* 47, 868–90.

Savidge, J. M. 1968: Elephants in the Ruaha National Park, Tanzania: management problems. *East African Agriculture Journal* 33, 191–6.

Semeniuk, V. and Meagher, T. D. 1981: Calcrete in Quaternary coastal dunes in southwestern Australia: a capillary-rise phenomenon associated with plants. *Journal of Sedimentary Petrology* 51, 217–68.

Semeniuk, V. and Searle, D. J. 1985: Distribution of calcrete in Holocene coastal sands in relationship to climate, southwestern Australia. *Journal of Sedimentary Petrology* 55, 86–95.

Shantz, H. L. 1956: History and problems of arid lands development. In White, G. F. (ed.), *The future of arid lands*. American Association for the Advancement of Science Publication 43, Washington DC.

Shields, L. M., Mitchell, C. and Drouet, F. 1957: Alga- and lichen-stabilized surface crusts as soil nitrogen sources. *American Journal of Botany* 44, 489–98.

Staley, J. M., Palmer, F. and Adams, J. B. 1982: Microcolonial fungi–common inhabitants of desert rocks? *Science* 215, 1093–5.

Street-Perrott, F. A. and Roberts, N. 1983: Fluctuations in closed-basin lakes as an indicator of past atmospheric circulation patterns. In Street-Perrott, F. A., Beran, M. and Ratcliffe, R. (eds), *Variations in the Global Water Budget* pp. 331–45. Dordrecht: Reidel.

Talbot, M. R. and Williams, M. A. T. 1978: Erosion of fixed sand dunes in the Sahel, Central Niger. *Earth Surface Processes* 3, 107–13.

Temple, P. H. 1972: Measurements of runoff and soil erosion at an erosion plot scale with particular reference to Tanzania. *Geografiska Annaler* 54A, 203–20.

Thomas, D. S. G. 1984: Ancient ergs of the former arid zones of Zimbabwe, Zambia and Angola. *Institute of British Geographers, Transactions* New Series 9, 75–88.

—— 1986a: Dune pattern statistics applied to the Kalahari Dune Desert, Southern Africa. *Zeitschrift für Geomorphologie NF* 30, 231–42.

—— 1986b: Ancient deserts revealed. *Geographical Magazine LVIII*, 11–15.

—— and Goudie, A. S. 1984: Ancient ergs of the southern hemisphere. In Vogel,

J. C. (ed.), *Late Cainozoic palaeoclimates of the southern hemisphere*, pp. 407–18. Rotterdam: Balkema.

Tsoar, H. & Møller, J. T. 1986: The role of vegetation in the formation of linear sand dunes. In: Nickling, W. G. (ed.) *Aeolian Geomorphology*, pp. 75–95. Boston: Unwin Hyman.

Twidale, C. R. 1972: Evolution of the sand dunes in the Simpson Desert, central Australia. *Institute of British Geographers, Transactions* 56, 77–109.

Van Den Ancker, J. A. M., Jungerius, P. D. and Mur, L. R. 1985: The role of algae in the stabilization of coastal dune blowouts. *Earth Surface Processes and Landforms* 10, 189–92.

Van Rensburg, H. J. 1955: Runoff and soil erosion tests, Mpwapwa, central Tanganyika. *East African Agricultural Journal* 20, 228–31.

Verstappen, H. Th. 1968: On the origin of longitudinal (seif) dunes. *Zeitschrift für Geomorphologie* NF 12, 200–20.

Walter, M. R. (ed.), 1976: *Stromatolites*. Developments in sedimentology 20. Amsterdam: Elsevier.

Wasson, R. J. 1984: Late Quaternary palaeoenvironments in the desert dunefields of Australia. In Vogel, J. C. (ed.), *Late Cainozoic Palaeoclimates of the Southern Hemisphere*, pp. 419–32. Rotterdam: Balkema.

—— and Nanninga, P. H. 1986: Estimating wind transport of sand on vegetated surfaces. *Earth Surface Processes and Landforms* 11, 505–11.

——, Rajaguru, S. N., Misra, V. N., Agrawal, D. P., Dhir, R. P., Singhvi, A. K. and Kameswara Rao, K. J. 1983: Geomorphology, late Quaternary stratigraphy and palaeoclimatology of the Thar dunefield. *Zeitschrift für Geomorphologie Supplementband* 45, 117–51.

Watson, A., Price Williams, D. and Goudie, A. 1984: The palaeoenvironmental interpretation of colluvial sediments and palaeosols of the late Pleistocene hypothermal in southern Africa. *Palaeogeography, Palaeoclimatology, Palaeoecology* 45, 225–49.

Weir, J. S. 1960: A possible course of evolution of animal drinking holes (pans) and reflected changes in their biology. *First federal science conference*. Salishury, Rhodesia: Mardon 301–6.

—— 1969: Chemical properties and occurrences on Kalahari sands of salt licks created by elephants. *Journal of Zoology* 158, 293–310.

Went, F. K. 1959: Ecology of desert plants. II: The effect of rain and temperature on germination and growth. *Ecology* 30, 1–13.

—— 1979: Germination and seedling behavior. In Goodall, D. W. and Perry, R. A. (eds), *Arid land ecosystems: structure, functioning and management*, volume 1, pp. 477–90. Cambridge: Cambridge University Press.

West, N. E. and Skujins, J. 1978: *Nitrogen in desert ecosystems*. US/IBP Synthesis Series 9. Strandsburg: Dowden, Hutchinson and Ross.

White, C. H. 1924: Desert varnish. *American Journal of Science* 9, 413–20.

Wilson, L. 1973: Variations in mean annual sediment yield as a function of mean annual precipitation. *American Journal of Science* 273, 335–49.

Yaalon, D. H. and Dan, J. 1974: Accumulation and distribution of loess derived deposits in the semi-desert and desert fringe areas of Israel. *Zeitschrift für Geomorphologie Supplementband* 20, 91–100.

Young, A. 1972: *Slopes*. Edinburgh: Oliver & Boyd.

Zapffe, C. 1931: Deposition of Manganese. *Economic Geology* 26, 799–832.

7 The biogeomorphology of periglacial environments

R. B. G. Williams

Introduction

Periglacial environments provide a formidable challenge to plants and animals. The coldest temperatures in the Northern Hemisphere (circa −70 °C) are recorded in the periglacial zone, and not, as might be expected, on the Greenland ice-cap or the frozen seas around the North Pole. Comparatively few plant species (less than 3 per cent of the world's total) have managed to evolve a sufficient degree of frost hardiness to survive in the periglacial zone (Billings, 1974). These plants have also had to evolve a rapid metabolism to cope with the often very short growing seasons. In many places, buffeting winds expose the plants to severe wind-chill, and carry sand or silt, abrading unprotected leaves and stems. Snow, when it falls, may be a blessing, covering the plants and shielding them from wind blast. Because precipitation is generally low, and the soil is frequently frozen, the plants risk desiccation. Many species have evolved dwarf, hairy or scaly leaves to conserve moisture, and prostrate or cushion forms that offer minimum wind resistance (Bliss, 1962; Savile, 1972). In addition to the climatic hazards, the plants have to cope with soils that are often infertile and liable to frost-stirring (Rieger, 1974). When thawed, the soils may become badly waterlogged and unstable. In many areas, soil movements occur so repeatedly that they prevent the development of true climax vegetation.

Because of the harshness of periglacial environments, plant communities are functionally simpler and structurally less diverse than those in more hospitable regions. There are fewer ecological niches available for animals, and food chains possess relatively small numbers of linkages (Irving, 1972; Hoffman, 1974; Remmert, 1980). The simplicity of community structure may be the reason why some animal populations are subject to sudden fluctuations in numbers. The cyclical changes in lemming populations in tundra areas are particularly violent and have profound effects on the plant cover.

Permafrost is characteristic of many, but not all, periglacial environments. The most extensive permafrost region in the world borders the Arctic Ocean in Northern Canada, Alaska, and Siberia (figure 7.1).

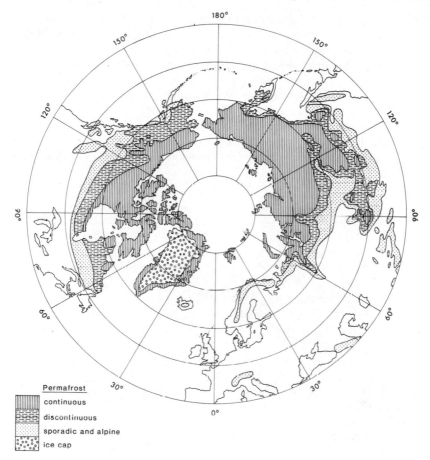

Figure 7.1 Distribution of permafrost in the Northern Hemisphere
(adapted from Harris, 1986)

Small, isolated patches exist in some high mountain ranges in lower
latitudes (for example, the Alps and the Rockies), and much of the Tibetan
Plateau is underlain by permafrost. Permafrost also occurs in the
unglaciated parts of Antarctica and on some of the adjacent islands.

It is customary to subdivide the permafrost region in the Northern
Hemisphere into two or three zones. Nearest the pole (and further south
on the highest mountains) is a zone of 'continuous permafrost'. Permafrost
is found everywhere, except in recently deposited sediments that have not
had time to freeze, and beneath rivers and lakes that do not freeze to the
bottom in winter. Further from the pole (and at somewhat lower elevations
in the mountains) the permafrost becomes thinner and interspersed with
areas that lack permafrost and where the ground freezes only in winter.

This is the zone of 'discontinuous permafrost'. At first the areas of seasonally frozen ground are restricted in size and form isolated bodies or 'islands' surrounded by a 'sea' of permafrost. In lower latitudes (or further down the mountains) the permafrost becomes increasingly broken up, existing as islands within a sea of seasonally frozen terrain. Many writers (e.g. Harris, 1986) distinguish this outer fringe of the discontinuous zone as a separate zone of 'sporadic permafrost'.

The continuous permafrost zone coincides roughly with the arctic tundra and polar deserts, but includes some small areas of Boreal forest (figure 7.2). The discontinuous and sporadic zones are associated in the lowlands

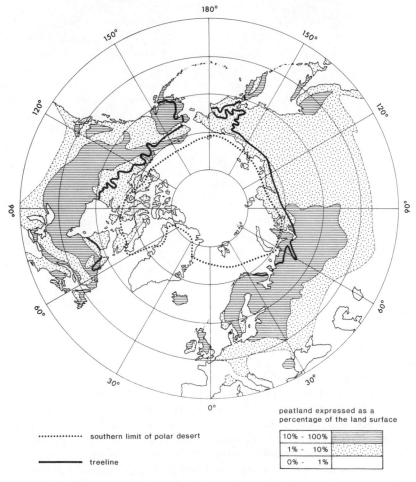

peatland expressed as a
percentage of the land surface

.... southern limit of polar desert

10% - 100%	
1% - 10%	
0% - 1%	

———— treeline

Figure 7.2 Extent of peatlands, tundra and polar desert in the Northern Hemisphere
(modified from Sjörs, 1961; Rieger, 1974; and Gore, 1983)

with Boreal forest, meadow and bogland (muskeg). Alpine tundra vegetation and rockfields are found in the mountains.

Beyond the limits of the permafrost region is a peripheral zone where the temperatures are not quite low enough to cause permafrost, but solifluction and other cold climate geomorphological processes are widespread. In high latitudes this peripheral zone is found near sea level, and the dominant vegetation is Boreal forest and bogland. With increasing distance from the pole, the zone rises steadily in elevation, being found above about 2,000 m in the Central Alps and 4,000 m at the Equator (Troll, 1958). The lower boundary of the zone in mid- and low latitudes coincides broadly with the tree-line and the vegetation is of alpine tundra type, except in the tropics where subarborescent life forms dominate at the lowest levels. In mid- and high latitudes, the zone is characterized by prolonged winter frosts and night-time frosts in summer, but within the tropics the climate displays little seasonality and night-time frosts occur all year round.

Although arctic and alpine tundra vegetation is floristically very similar, often including identical species, the physical environments show important differences (Ives and Barry, 1974). In high mountain areas such as the Alps, the thin, dust-free atmosphere ensures considerable day-time insolation and heating of the ground surface. Night-time frosts are frequent, but the ground tends to freeze only at the surface because the frosts are short-lived. Needle ice (pipkrakes) often develops on the surface, drying out the soil beneath. Wind speeds are often high, and snow avalanching is important on steep slopes. Arctic areas, by contrast, generally experience more intense and prolonged frosts, which may last the whole winter. The number of freeze–thaw cycles is restricted. A period of twenty-four-hour darkness in winter is followed by a period of continuous sunlight in summer, but the sun has little heating power because of its low angle of elevation. Winds tend to be less severe than in alpine areas, and precipitation is usually much lower.

Vegetational modifications to the environment

In periglacial regions the natural vegetation has been little disturbed by man, and a close relationship exists between vegetation types and environmental factors. Climate is the over-riding factor responsible for major regional differences in vegetation, but relief and rock type often dominate at the local scale, modifying or even reversing the influence of regional climate.

Although the vegetation is profoundly affected by environmental factors, the environment in turn is partly controlled by vegetation. The characteristics of the vegetation directly influence the micro-climates, soil properties, and geomorphological processes.

The influence of vegetation on periglacial micro-climates is complex and not well understood. Everyone agrees, however, that vegetation increases surface roughness and helps to reduce the force of the wind. This is particularly important in mid and high latitudes in winter when low growing scrub and herbaceous vegetation may increase the depth of snow covering the ground. Trees, especially if closely spaced, may have the opposite effect, trapping snow in their crowns and reducing the snow cover on the ground. During the summer growing season, transpiration by vegetation may lower the temperature of the air and increase its humidity (Tyrtikov, 1964). Wet moss is apparently a particularly effective cooling agent.

Even more important than the influence of vegetation on micro-climates is its effect on the moisture and temperature regimes in the soil. Vegetation has an important insulating role, helping to shield the soil from solar radiation. Even a thin layer of lichen or dry moss can greatly alter the soil temperatures (Benninghoff, 1952). In summer, the soil is kept cool and thawing of ground-ice is delayed or even prevented. Evaporation rates are reduced, and the soil after thawing tends to be maintained in a saturated or near saturated condition, despite transpiration. The wetness of the soil increases its thermal capacity which makes it slow to warm up. In autumn, refreezing is delayed, sometimes by over a month, because of the insulation provided by the vegetation and the generally high moisture content of the soil (Tyrtikov, 1964). During winter, the soil may cool down quite slowly if the vegetation traps enough snow, because snow is an excellent insulator.

The accumulation of organic matter in soils greatly increases their ability to retain moisture. Peat can retain a quantity of water equal to 60 per cent of its dry weight, and forest leaf litter also has a high moisture capacity. A soil rich in organic matter may have a moisture content 5 to 10 times that of a mineral soil from which it is derived.

Because vegetation affects micro-climate and soil properties, it indirectly affects a great many geomorphological processes. It also has an important direct effect, due to the binding action of plant roots. Both mass movement and frost heaving, for example, are strongly inhibited by deep and dense networks of roots.

In the tundra zone, vegetation is believed to exert less influence on permafrost and geomorphological processes than in the forest zone since its biomass is much smaller (Tyrtikov, 1964; Brown and Péwé, 1973).

Peat formation

Plant remains decompose slowly under periglacial conditions because of the low temperatures and generally poor drainage. As a result surface peats tend to accumulate, often over large areas. Within the tundra region, peat formation is relatively slow, and the most extensive peats occur in the Low Arctic where there is most precipitation (see figure 7.2). In the High Arctic the polar desert climate ensures that peats are scarce or non-existent. The most extensive and deepest peatlands in the world are found within the Boreal forest zone. Sjörs (1961, p. 220) records that 'south of Hudson Bay there is a continuous lowland of about the same area as Great Britain. This lowland, with almost the only exception of lakes and river-beds, is completely covered by wet peat'. Other vast bogs exist in Scandinavia, Western Siberia and Alaska. About one-third of Finland is covered by peat, the highest proportion for any country, but the largest continuous peat area in the world is in Western Siberia, between the Ob and Irtysh rivers (Gore, 1983).

Many bogs develop as result of the invasion of lakes by vegetation (Drury, 1956). Plants grow out from the shores, or form floating communities, and their remains accumulate on the lake bottom. Gradually, the lake shrinks and a quaking bog is formed, then a more solid bog. This infilling of lakes by plants is one of the most important biogeomorphological processes in periglacial regions.

Bogs also develop directly from forest. Mosses colonize the forest floor, especially in depressions or in permafrost areas where drainage is impeded. The mosses hold water, and the underlying soil becomes steadily more saturated, until the trees are killed off and a bog is formed.

The surface of many bogs becomes gently domed as a result of greater accumulation of peat towards the centres of the bogs. Such 'raised bogs', as they are called, often spread laterally into adjacent forest. Water drains to the margin of each bog where it tends to collect, forming a kind of moat (the 'lagg'), up to 2 or 3 m deep. The trees along the edge of the bog become swamped and die, allowing the bog to expand outwards. If the forest is underlain by permafrost, the expansion process is accelerated because the frozen ground tends to thaw and subside, aiding the swamping. Bogs that are actively expanding in permafrost terrain tend to have inward-facing banks that are broken by numerous slumps.

Drury (1956) provides a detailed account of bog formation and forest destruction on river floodplains in Central Alaska. He points out that, as bog surfaces become raised and better drained, forest can colonize the bogs and replace them. He suggests that forests and open bogs alternately displace each other forming a cyclo-climax. The phases of creation and destruction of forest and bog move laterally across the floodplains in repeated cycles.

Vegetation and permafrost distribution

At its southern limits in the Boreal lowlands, permafrost is mainly confined to the drier peat bogs. The surface layers of peat become saturated during autumn rains and then rapidly freeze as winter progresses (Brown and Johnston, 1964; Brown, 1969). Saturated frozen peat has a high thermal conductivity (33 times that of dry peat and 5 times that of unfrozen saturated peat), and heat is easily lost from the bogs during the winter. In the spring and summer, the surface layers thaw and start to dry out. Because the dry peat is a poor conductor, little solar heat enters the bogs, and thus heat has to be used to melt the ice and evaporate the water. The bogs thus function as a kind of greenhouse in reverse, losing relatively large amounts of heat in winter and gaining only small amounts in summer. Patches of permafrost start to form in the driest parts of the bogs, and, as the water is converted to ice, it expands and tends to push up the bog surface, often allowing colonization by trees. The trees shade the ground and intercept snow in winter, reducing the insulating blanket of snow on the ground, and furthering the development of ground-ice.

In the sporadic permafrost zone, the wettest bogs are unaffected by permafrost because the peat remains sodden during the summer and readily transmits heat. Permafrost develops only as peat accumulates and the bog surfaces rise and begin to dry out. Permafrost first appears under cushions or hummocks of plants such as *Sphagnum* which become further raised above the general level of the bogs. The permafrost may subsequently grow downwards through the peat and enter the underlying sediments.

Palsas are the most distinctive relief features produced by permafrost in peatlands. They begin life as low flat-topped hummocks and develop into conical mounds up to 10 m high. They have cores consisting of alternating layers of frozen peat and segregation ice (ice created by freezing of pore water and water transferred by capillarity). It is generally assumed that palsas develop where snow cover in winter is thinner than normal (Seppälä, 1982), or the vegetation is patchy, and the peat freezes more deeply than in surrounding parts of the bog. As a palsa grows upwards, its projecting form helps to ensure that it is blown free of winter snow, and the freezing of the peat is thus accelerated. The upper part of the palsa begins to dry out and the original bog vegetation may be killed off. The weight of the ice causes the palsa to subside slightly within the peat bog, and a moat of water may develop round the mound, especially on its south side (Lundqvist, 1962). As the palsa continues to grow higher, its sides become steeper, and slumping or cracking may expose the ice core, allowing melting to take place. The palsa may then collapse and be replaced by a shallow pool.

Palsas generally provide the first indications of permafrost that one encounters travelling northwards from temperate areas into the Boreal zone (figure 7.3). Other small bodies of permafrost exist in peatlands at the southernmost limits of permafrost but they have no surface expression and can only be discovered by drilling (Brown, 1969). At first, palsas are quite rare, but further north, towards the boundary with the discontinuous zone, they become much more common, often occurring in large groups. Alongside the palsas, more extensive bodies of permafrost begin to develop in the peatlands, raising the surface and forming 'peat plateaus' or 'permafrost islands'. They are often covered by trees, in contrast to the

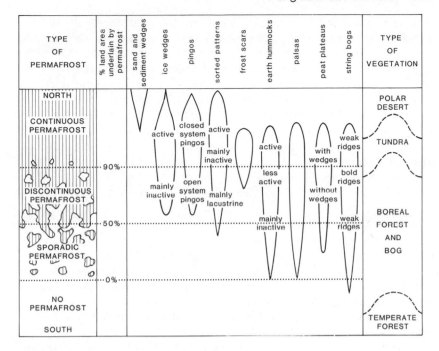

Figure 7.3 North–south zonal distribution of permafrost in North America and relationships with selected geomorphological and biological phenomena

surrounding bog. Most are rounded in outline, or have a tear-drop shape with the tail directed downslope (Sjörs, 1959). Usually, they are flat-topped or gently domed, but some are atoll-shaped with an outer raised rim colonized by trees and a wet treeless centre.

Also in the northern part of the sporadic zone, one starts to find permafrost developing as isolated patches in silts and clays. Favoured locations are steep north-facing slopes which receive only low amounts of insolation, and frost hollows into which cold air tends to drain. As the discontinuous zone is crossed, permafrost becomes more and more common, starting to appear in sands and gravels, and on south facing slopes.

In the sporadic and discontinuous zones, mean ground temperatures are generally close to 0 °C, and the surface vegetation plays a decisive role in determining whether permafrost is present or absent. Even a single tree may create an individual lens of permafrost if it intercepts winter snow in its crown leaving the ground unprotected, or if it shades the soil in summer promoting the growth of a insulating moss mat (Viereck, 1965). The lens forms under the tree roots, lifting the tree and pushing up the ground surface into a mound as much as a metre high.

In areas where ground temperatures are marginal, clearance of vegetation, or removal of surface layers of leaf litter, may cause rapid

Table 7.1   Thickness of the active layer in metres, Central Alaska

Type of vegetation	Peat	Type of soil Silt/clay	Sand/gravel
Paper birch (*Betula papy-rifera*) and aspen (*Populus tremuloides*)	no sites	1.0–3.0	1.8–6.0
White spruce (*Picea glauca*)	no sites	0.8–1.5	1.0–3.5
Black spruce (*Picea mariana*)	0.2–0.7	0.3–1.0	0.5–1.5
Grassland, sedge meadows (cleared forest land)	no sites	2.0–10.0	2.0–15.0
Sphagnum bog	0.3–0.8	no sites	no sites

Sources: Rieger *et al.*, 1963; Péwé, 1965 and 1982; Kreig and Reger, 1982; Brown and Kreig, 1983; and personal fieldwork

thawing of existing permafrost, and sometimes its complete disappearance. In Central Alaska, for example, destruction of forest for agriculture or construction purposes has caused the permafrost to down-melt by as much as 13 m in places (Péwé, 1982).

In permafrost terrain, the surface layer of soil that thaws in summer and refreezes in winter is called the 'active layer'. Its thickness depends on summer temperatures and decreases polewards. Also important are the nature of the vegetation and type of soil. The effect of vegetation is much more marked in the forest zone than in the tundra because the plants have a more variable mass and density. Table 7.1 records some typical thaw depths in the forest zone in Central Alaska.

Although vegetation partly controls thaw depths, the depth of thawing also influences the vegetation. Black spruce, for example, shades the soil and encourages the growth of a thick insulating mat of moss, which greatly restricts amounts of thawing. The spruce is able to root successfully within the shallow active layer, but deeper-rooting species, such as paper birch and aspen, are unable to compete. The birch and aspen grow in relatively open stands and shield the soil much less effectively than the black spruce. Their fallen leaves suppress the growth of mosses that would promote permafrost development. Permafrost, if present, is found only at depth, and, as there is a deep rooting zone, the birch and aspen can continue to flourish.

A further complication is that vegetation type and thaw depths are closely related to other environmental factors such as aspect. In Central Alaska, for example, paper birch tends to favour south facing slopes, which are relatively warm and have a deep active layer, whereas north facing slopes, which are cold and have a shallow active layer, are usually covered by black spruce. Cold, wet soils on river floodplains encourage the spruce, but the birch favours well drained hillsides and upland sites.

Forest succession also needs to be taken into account (Viereck, 1970; Brown and Kreig, 1983). Some of the youngest sediments on the river

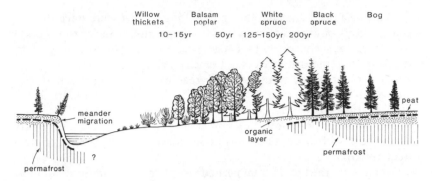

Figure 7.4 Diagrammatic section across a river meander in Central Alaska showing zonation of vegetation and distribution of permafrost (modified from Brown and Kreig, 1983; and Viereck, 1970)

floodplains in Central Alaska are found on the inside of meander bends (figure 7.4). As the rivers migrate across the floodplains, sand and silt are deposited as a series of point bars on the slip-off slopes inside the meanders. These bars soon become colonized by willows and alders, which begin to be replaced by balsam poplars (*Populus balsamifera*) within 10 to 15 years. Seedlings of white spruce, which are shade tolerant, become established under the poplars about 50 years after the initial colonization of the bars by the willows and alders. After 75 to 100 years of growth the spruce overtops the poplars and begins to replace them. During these early successional stages permafrost is usually absent. However, as the spruce matures, it begins to cast a deep shade promoting the development of a thick insulating moss mat and surface organic layer. The soil gradually becomes colder and wetter, and permafrost develops 200 to 300 years after the initial colonization of the bars. The white spruce is slowly replaced by black spruce which tolerates a shallower root zone and wetter soil. Eventually, the soil may become so sodden that even the black spruce begins to die and open bog vegetation develops.

Towards the northern part of the discontinuous zone, even the wettest peat bogs are usually underlain by permafrost. Peat plateaus become increasingly common, and palsas, it has often been said, begin to become scarce. Recent work suggests, however, that palsas are not uncommon but often coalesce to form ridges and peat plateaus (French, 1976, Brown, 1970). Small mounds that resemble palsas occur quite widely even in the continuous permafrost zone, but their origin is somewhat controversial (Jahn, 1986).

Much better studied than palsas are another class of frozen mounds known as 'pingos' or 'bulganyakhs'. These develop in mineral soil as well as in peat, and range from a few metres to over 60 metres in height. They have a core of ground ice that is usually much more massive than the ice core of a palsa, and considerably purer. Pingo cores were once thought to consist solely of injection ice (water that froze after being intruded under pressure), but it seems likely that segregation ice is also present, perhaps exclusively in some instances (see French, 1976).

Often present in the same general areas as pingos are much smaller 'winter frost mounds' that develop within the active layer during cold winter weather (Brown *et al.*, 1983; Pollard and French, 1984). They disappear during the following summer or persist only for a few years. Their mode of formation is not fully understood and is the subject of much current research.

Ice wedges are the best known permafrost phenomena in the continuous and discontinuous zones (see French, 1976 and Washburn, 1979, for general details). They develop when the permafrost attempts to contract in cold winters and becomes fissured. A thick vegetation cover, or a deep layer of snow, inhibits ice wedge formation. The most favourable conditions are found in tundra areas where the vegetation is low growing and there is little snow accumulation in winter. Many wedges in the discontinuous zone exist under forest and are largely inactive. Péwé (1966) suggests that they are mostly relic features that developed under colder conditions in the past.

Thermokarst

Thawing of ice-rich permafrost leads to subsidence and collapse of the ground surface. An uneven topography ('thermokarst') results, with cave-in lakes, steep walled pits, underground passageways, gullies and mounds (French, 1976; Péwé, 1982). Man often unintentionally creates thermokarst by clearing natural vegetation for agricultural or constructional purposes. Other thermokarst is natural in origin, resulting from climatic change, overgrazing by wild animals, changes in surface drainage and other environmental disturbances. Fires started by lightning are thought to be a major cause of natural thermokarst. Even seemingly trivial environmental disturbances may be sufficient to initiate thawing of permafrost. Mackay (1970) records how an Eskimo dog in Northern Canada, tethered to a stake for 10 days, so trampled and destroyed the herbaceous vegetation that in two years the active layer deepened by at least 10 cm and the ground surface subsided about 20 cm. Even hibernating bears have sometimes been blamed for causing thermokarst!

Mass wasting

Mass wasting processes in periglacial regions are greatly influenced by the nature of the vegetation cover. Talus creep, for example, is limited to talus or scree that vegetation has failed to stabilize. Large-scale solifluction is confined to areas where trees are rare or absent (Harris, 1981). It is widespread in the High Arctic where the sparsity of vegetation allows the formation of continuous 'solifluction sheets'. In the Low Arctic and in alpine areas, the cover of tundra vegetation is more complete and solifluction is more localized, forming 'lobes' and 'terraces'. The vegetation is generally thinnest on the treads and thickest on the fronts of the lobes and terraces (although unvegetated, stone-banked features are also found). Measurements suggest that rates of surface movement are greater on the treads than on the fronts, reflecting the different amounts of vegetation. The vegetation mat on the fronts holds the soil together and provides a covering that gradually 'stretches' to allow frontal advance to take place.

In theory, mass wasting ought to be more rapid in permafrost areas than in areas of purely seasonal frost because the permafrost provides a slip surface and impedes drainage, thus increasing the saturation of the soil. However, in practice, many non-permafrost areas seem to have just as rapid mass wasting because they tend to experience greater frost heaving and often have more snow, which helps maintain soil saturation as it thaws. Shallow permafrost reduces amounts of surface heave (French, 1976), which tends to restrict mass wasting.

Patterned ground

Ice-wedge networks

In periglacial regions the surface of the ground is often strikingly patterned. In some localities, the patterns are due to micro-relief features which are arranged geometrically. Elsewhere, the patterns are caused by differences in soil or vegetation which recur in a systematic manner. The collective name for all the different types of surface patterning is 'patterned ground' (Washburn, 1979).

The most common patterns are the polygonal networks of surface troughs that develop above ice-wedges in permafrost terrain. The troughs, which are often quite shallow, tend to be bordered by low ridges created by lateral expansion of the wedges. The ridges enclose the low-lying polygon centres which often become flooded in summer by thaw water. Polygons with this type of surface relief are described as 'low-centred'. They become 'high-centred' if thaw water collects in the troughs and causes down-melting of the wedges, thus lowering the troughs and leaving the polygon centres as mounds or hummocks. This conversion process often begins when peat accumulates within the low-centred polygons and raises their surface. Continued thawing of high-centred polygons produces conical thermokarst mounds (also known as 'cemetery mounds' or 'baydzherakhi') separated by a network of trenches 2 or 3 m deep. Some of the best examples are found in the discontinuous permafrost zone where ground temperatures are close to 0 °C and clearance of forest underlain by ice-wedge systems has caused deep thawing and settlement of the ground surface. In several areas, cultivated fields have been abandoned, and allowed to revert to forest, because mounds have become so pronounced that the fields can no longer be worked by farm machinery (Péwé, 1982; French and Heginbottom, 1983).

Sorted patterns

Very different patterns result from the size-sorting of stones and fines within the active layer. On flat ground, the stones are usually arranged in a polygonal network with the fines delimiting the centres of the polygons. Patterns of this type are known as 'stone polygons' or 'sorted nets'. More rarely, 'stone circles' are found, consisting of roughly circular patches of fines ringed with stones. The circles may touch one another or be separated by strips of unsorted ground (Hallet and Prestrud, 1986). On slight slopes, stone polygons and stripes become elongated, and with increasing gradient pass into 'stone stripes' – alternating bands of fines and stones directed downslope.

Although frost action seems to be responsible for the majority of sorted patterns, some examples develop as a result of other processes, such as wetting and drying. Vegetation has a strong inhibiting effect that is often overlooked in the literature, and deserves greater emphasis. Sorted patterns develop only where there is little or no plant cover to shield and bind the soil. In forested areas actively-forming examples are almost never found, except on lake shores and at the bottom of shallow lakes. Although the lake patterns are often supposed to form underwater in response to special processes (see, for example, Ray et al., 1983), they may in fact develop subaerially at times when the lakes are dry or water levels are low (Walters, 1983). Repeated desiccation and inundation prevents vegetation from colonizing the ground, and the absence of vegetation allows frost action to proceed unchecked.

Even in tundra areas, the vegetation mat is often dense enough to stop sorted patterns forming. Many of the patterns that can be found seem to be mostly fossil or inactive. The stones are frequently covered by slow-growing lichens and appear not to have moved for many years. Sedges, grasses and other plants have frequently encroached on the patterns. The sorting may have developed when the climate was slightly wetter and colder than now and the vegetation cover was less (Ballantyne and Matthews, 1982). Or perhaps it developed following a tundra fire.

Frost scars

Sorted patterns, although often studied by researchers, are relatively uncommon. Much more widespread, but curiously little studied, are

Figure 7.5 Frost scars in tundra vegetation from Dempster Highway, Northern Yukon

patterns in which vegetation is an essential component. In tundra areas, the low growing, herbaceous vegetation is often very patchy. On flat ground roughly circular spreads of mineral soil are found, ranging from 0.5 to 5 m in diameter. They are known variously as 'frost scars', 'frost boils', 'mud spots', or 'nonsorted circles'. Sometimes the scars are widely scattered, but usually they are closely spaced and form the centres of polygons (see figure 7.5). The vegetation is confined to the borders of the polygons, where the dead plant-remains tend to accumulate forming a shallow layer of peat. On gentle slopes, frost scars become somewhat elongated and are often arranged in lines; on steeper slopes they form continuous bare stripes, separated by vegetation stripes.

Frost scars are restricted to sediments that are susceptible to frost heaving, in particular silts and clays. Stones may migrate upwards through the sediments to form a surface layer, but there is no radial sorting as with stone polygons. Excavations show that the sediments under the scars are frequently arched up, and buried stones tend to be oriented vertically. The scars are commonest in windy locations where little or no snow accumulates in winter.

Frost scars are thought to develop when continuous spreads of low-growing vegetation thin or start to die back in irregular patches (figure 7.6). Wind

A : Initial stage. Thawing in summer is exceptionally deep where the vegetation cover is thin.

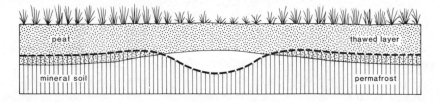

B : Early autumn. Expansion of the mineral soil on freezing pushes aside the surface peat.

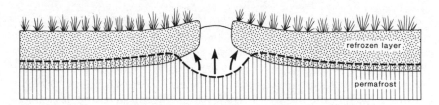

C : Many years later. The frost scar continues to widen in subsequent winters as the mineral soil freezes and expands. A peat ring begins to form around the scar.

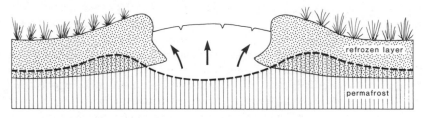

D : Mature stage. Continued lateral thrusting has forced wedges of mineral soil into the peat ring causing some peat to move radially inwards beneath the scar.

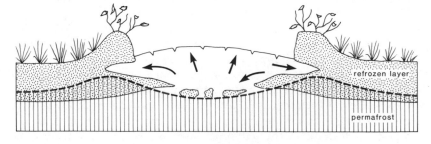

Figure 7.6 Development of a frost scar by frost heaving and lateral thrusting (modified from Hopkins and Sigafoos, 1951)

erosion and over-grazing may speed the death of the plants. Hopkins and Sigafoos (1951) note that closely grazed reindeer corrals in Alaska possess numerous frost scars which are scarce or absent on the surrounding ungrazed tundra.

Hopkins and Sigafoos believe that frost scars result from vertical frost heaving and lateral thrusting. The bare soil of the scar thaws more deeply in summer than the soil beneath the adjacent vegetation, which is better insulated. In autumn, the bare soil freezes more swiftly, and water migrates radially inwards towards the scar centre, freezing to form layers or lenses of segregation ice. The ice raises the scar surface, which becomes convex. Because of its extra relief, the scar tends to be blown free of snow and hence it continues to freeze more deeply than its surroundings. The doming of the surface is accompanied by dilation cracking and lateral thrusting. The surrounding vegetation mat or peat cover may be pushed back to form a turf of peat ring (Rawlinson, 1983, figure 85). In summer, the ice layers melt and the surface of the scar subsides and flattens. Needle ice formed by overnight frost helps to destroy the vegetation at the edge of the scar.

An alternative suggestion is that frost scars originate in summer, not winter, and are caused by diapirism (Shilts, 1978; Washburn, 1979). During thawing, the subsoil becomes highly saturated and unstable. Because of density differences, pore water pressures or for other reasons, it may migrate upwards as dome- or plug-like intrusions, erupting at the surface to form the scars. French (1976, pp. 43–4) and other writers have observed localized eruptions of thawed subsoil material on tundra surfaces, but there is as yet no clear evidence that all frost scars result from diapirism.

Frost scars survive only as long as they can prevent the surrounding vegetation encroaching. Biological and geomorphological processes are engaged in a continuing contest. Sometimes the plants gain the upper hand and the scars are colonized by seedlings or runners from the marginal plants. At other times, frost heaving or diapirism dominates and the plants are driven off. Some frost scars persist with little change for 20 years or more (Johnson and Neiland, 1983), but others are likely to be more ephemeral.

Earth hummocks

In periglacial regions, small, dome-shaped hummocks or mounds often develop, especially on flat ground. The hummocks, which are separated by narrow furrows, range up to about a metre in height and 1.5 or 2 m in diameter.

In peat bogs, many of the hummocks are of purely biological origin. They result from the growth of tussock plants *Sphagnum* moss cushions. Dead leaves and rhizomes accumulate at the base of the tussock plants, forming a tightly packed mass, which turns gradually into peat. The plants rise slowly on their own debris, and drainage water becomes channelled between the growing hummocks, causing erosion at some sites (Hopkins and Sigafoos, 1951).

Frost heaving plays a part in raising hummocks in some peat bogs. The 'pounus' or 'pounikko' of Finnish bogs, for instance, are *Sphagnum* hummocks heaved up above the general surface of the bogs by winter frosts (Seppälä, 1983; Ruuhijärvi, 1983).

Hummocks also develop in mineral soil, usually under a vegetation cover. They are commonly called 'earth hummocks', 'turf hummocks' or 'thurfurs'. The term 'earth hummocks' is a reference to the fact that the interior of the hummocks is formed of earth or mineral soil, rather than peat. The earth core, however, is often concealed by a thick layer of surface peat or humus, and its existence can often be confirmed only by digging. Lundqvist (1962) suggests that there exists a complete gradation of forms between earth and peat hummocks.

Earth hummocks are widely distributed in both permafrost and non-permafrost areas. They occur under forest (unlike frost scars) as well as on the tundra. In the Lower Mackenzie Valley in Canada, about 80 per cent of the mineral soil terrain south of the tree-line is affected by earth hummocks (Tarnocai and Zoltai, 1978). North of the tree-line, on the tundra, about 95 per cent of the mineral soil terrain is hummocky.

In tundra areas, the tops of the hummocks are sometimes bare and plants grow only on the sides and in the inter-hummock furrows. These bare-centred forms, which have been called 'mud hummocks' (Mackay, 1980), appear to be caused by disruption of the vegetation cover either by wind-blasting in winter and subsequent die-back, or by the eruption of mineral soil from below. They greatly resemble frost scars and may actually be transitional forms. In the High Arctic other hummocks occur that are virtually devoid of vegetation, even in the furrows.

On flat ground, earth hummocks are generally circular, but they gradually become elliptical with increasing slope (Tarnocai and Zoltai, 1978). They lose their circularity more slowly than frost scars, which become strikingly distorted on quite gentle slopes.

Earth hummocks commonly develop on ill-drained silts and clays, especially where there is a shallow surface covering of peat. The soil layers under the hummocks are often domed up, paralleling the surface (figure 7.7). Other hummocks have a vertical core or central lens of coarser soil. In many areas a subsurface layer of peat or humus extends under the hummocks, at the base of the active layer (Zoltai and Tarnocai, 1974). This layer may link up with the surface peat in the inter-hummock depressions, thus totally enclosing the mineral soil cores of the hummocks and forming isolated cells. In other examples, the subsurface organic layer is disrupted under the hummocks, appearing as vertical tongue-like intrusions. Similar disturbances may affect the soil parent material.

The base of the active layer in late summer is generally lowest under the hummock tops and highest under the inter-hummock depressions. The

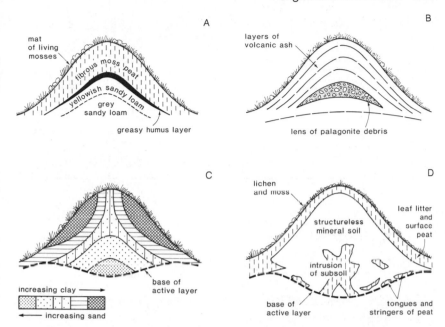

Figure 7.7 Generalized cross-sections of earth hummocks (A) with uparched layers, cf. Raup, 1965 (Northeast Greenland); (B) with lens-like core, cf. Schunke, 1975 (Iceland); (C) with central vertical core, cf. Mackay, 1980 (tundra, Canada); (D) with discontinuous, subsurface organic layer, cf. Tarnocai and Zoltai, 1978 (Boreal forest, Canada)

permafrost surface, in other words, is a reverse image of the ground surface. Thaw water accumulates in summer in the hollows in the permafrost under the tops of the hummocks. After winter freezing, the base of the hummocks has a high ice content, and layers of segregation ice are often present.

It is generally agreed that hummocks result from downward movement of surface soil and peat in the depressions and a corresponding upward displacement of mineral soil in the centres. The details of the process remain controversial, however. Mackay (1979, 1980) has studied tundra hummocks and suggests that they develop slowly year by year as a result of alternate contraction and dilation (figure 7.8). In autumn, the hummocks start to refreeze from the top downwards and also, at some sites, from the base upwards as the permafrost begins to freeze back towards the surface. The mineral soil, sandwiched between, experiences desiccation and contraction. In summer, conditions reverse. The ground surface and base of the permafrost subside creating dilation in the interior of the hummocks, which become highly saturated with thaw water and unstable. Because the hummocks have a convex surface and the active layer a convex base, a cell type circulation is set up with sediment rising under the hummock centres and spreading sideways at the surface.

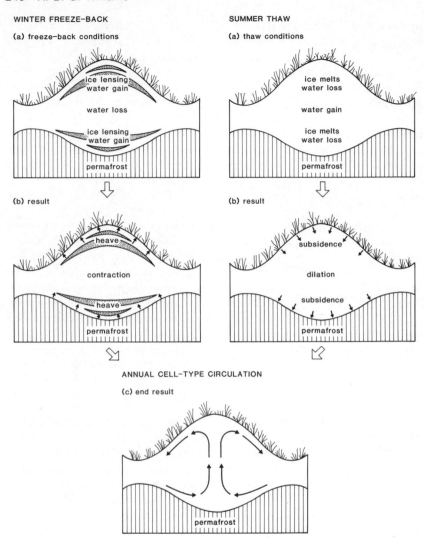

WINTER FREEZE–BACK

(a) freeze–back conditions

SUMMER THAW

(a) thaw conditions

ice lensing water gain

water loss

ice lensing water gain

permafrost

ice melts water loss

water gain

ice melts water loss

permafrost

(b) result

heave

contraction

heave

permafrost

(b) result

subsidence

dilation

subsidence

permafrost

ANNUAL CELL–TYPE CIRCULATION

(c) end result

permafrost

Figure 7.8. Mackay's equilibrium model for hummocks. Movement at the top of the hummock is down and radially outwards, while movement at the bottom of the active layer is up and radially inward

Tarnocai and Zoltai (1978) believe that hummocks in the forest zone develop very intermittently. Ring dating of tilted trees on hummocks suggests that major movements occur only every 20 or 25 years. Uplift appears to be highly localized. One side of a hummock may remain unmoved while the other side is raised; then years later, the pattern may reverse, with the uplifted side remaining stationary and the previously unmoved side undergoing movement. The irregular and intermittent uplift is associated with the intrusion of tongues of mineral soil and peat into the base of the hummocks.

The role of vegetation in the creation of earth hummocks has been little studied but is likely to be very important (Raup, 1965). The fact that the hummocks are generally vegetated suggests that plant cover aids rather than hinders their formation, at least in the initial stages. The hummock tops usually have less plant and peat cover than the intervening depressions, and as a result they freeze and thaw more deeply. The greater wetness of the depressions aids peat formation, and further slows their rate of thawing. Near the southern boundary of the permafrost region, hummocks may be caused by the growth of trees, which shelter the soil and allow individual lenses of permafrost to develop (Viereck, 1965).

Earth hummocks have been widely reported from temperate areas where frost action is limited. They occur, for example, on high ground in the Pennines and on Dartmoor (Tufnall, 1975; Waters, 1962). They are unlikely to have survived from the Pleistocene and may result from frost heaving in historic times, or even from contemporary frosts. Pemberton (1980) describes hummocks in pastures on low ground in Cumbria, which he ascribes to recent frost action. Some of the hummocks have developed within the last 150 years.

An alternative explanation of at least some temperate zone hummocks is that they are biological in origin. Most examples are too large for mole hills, but some might be ancient ant hills or sites of former clumps of vegetation. In the New Forest many of the stream-side 'lawns' are strikingly hummocky, and Tubbs (1986) believes that the hummocks develop from *Molinia* tussocks as a result of intensive grazing by horses and other animals. The *Molinia* is slowly destroyed by the grazing and is replaced by other, non-tussock forming grasses and herbs, but its remains can often be found inside the hummocks. For these and other temperate zone occurrences biological explanations seem no less plausible than those invoking frost action.

Tussock-birch-heath polygons

Tussock–birch–heath polygons were first described from tundra on the Seward Peninsula, Alaska (Hopkins and Sigafoos, 1951). Since then they have been identified in many other unforested areas in Alaska (see, for example, Hopkins *et al.*, 1955; Brown and Kreig, 1983, p. 111), but not apparently elsewhere.

In cross-section the polygons consist of low mounds or centres of silty mineral soil separated by peat filled troughs (figure 7.9). There is very little surface relief: the tops of the mounds are almost level with the peat in the troughs. The layering of the mineral soil beneath the mounds is often highly disturbed, and stringers of peat commonly extend under the

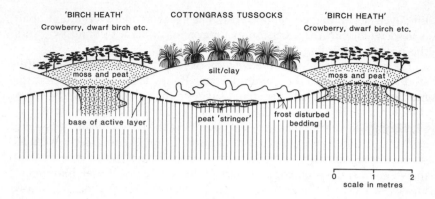

Figure 7.9. Diagrammatic cross-section through a tussock–birch-heath polygon (after Hopkins and Sigafoos, 1951)

mounds at the base of the active layer. Cottongrass (*Eriophorum vaginatum*) grows as tussocks on the mounds, while 'birch-heath' (*Betula nana, Empetrum nigrum, Ledum palustre*, etc.) grows on the peat in the troughs.

Tussock–birch-heath polygons much resemble earth hummocks in underground structure, but are altogether larger features, ranging from 2 to 4.5 m in diameter. On slopes of 4 ° to 5 ° they become noticeably elongate, and on still steeper slopes pass into alternating stripes of peat and mineral soil running downslope.

Hopkins and Sigafoos suggest that tussock–birch-heath polygons develop from coalescing frost scars, but the details of the process remain vague. It is generally agreed, however, that vegetation plays an important, and possibly essential, role in the formation of the patterns.

Patterned bogs

Many bogs in periglacial regions have patterned surfaces that are particularly conspicuous when seen from the air. Ridges of peat, called 'strings' or 'kermis', enclose shallow depressions, known as 'flarks' or 'rimpis', which tend to be very wet and often contain pools of water. The ridges are usually about 1 or 2 m wide, and have flat or undulating crests and characteristically steep sides. They accommodate plants that are intolerant of repeated flooding. Any shrubs or trees invading the bogs are confined to the ridge crests. Bog mosses and other plants that tolerate swampy conditions grow in the depressions, except where there are permanent pools, which are largely devoid of vegetation.

Steeply sloping bogs generally have the best developed patterns. The strings and flarks lie across the slope at right angles to the drainage, creating a staircase effect. The flarks are narrow (typically only 2 to 10 m wide) and

Figure 7.10 Silver Flowe Bog, Galloway, Scotland: a valley floor bog with a strikingly patterned surface. The hummocks and water-filled hollows are aligned across the main drainage direction
(University of Cambridge Collection)

are often flooded because the strings act as a series of dams impounding drainage water. On bogs that are only slightly sloping, the strings and flarks are again aligned along the contours, but the flarks are much wider and are often drier. Where the flarks are flooded they are curiously reminiscent of a system of terraced paddy fields (Sjörs, 1961). Bogs that are almost level are frequently unpatterned or have an irregular hummocky surface. Where distinct strings occur, they can be seen to wind in a rather random fashion, joining up to create an irregular network. The flarks are roughly circular or polygonal in plan, and often very large. Dome-shaped raised bogs have strings and flarks arranged in concentric patterns, suggesting tree-rings.

Patterned bogs, or 'string bogs' as they are often called, are commonest in the Boreal forest zone in areas of sporadic or discontinuous permafrost. The ridges tend to be highest and best developed in the northern parts of the zone where winter frosts are most intense. However, the patterns become rarer and less distinct further north on the tundra. According to Foster and King (1984, p.116), the reason is that 'permafrost disrupts pattern development and the shorter growing season restricts the depth of peat accumulation'. A few string bogs occur well south of the Boreal forest, for example in Central Germany (Troll, 1958) and Southern

Scotland (the Silver Flowe Bog, Galloway–see figure 7.10, and Taylor, 1983).

The cause of the patterning has not been firmly established, despite extensive research. Useful recent reviews of the problem include Moore and Bellamy (1974), Washburn (1979), Moore (1982), and Seppälä and Koutaniemi (1985). Ecologists have often assumed that geomorphological processes are paramount, but many geomorphologists have ignored string bogs, believing the patterning to be purely biotic in origin. The biotic theory is backed up by evidence that vegetation growth on the ridges is often greater than in the flarks, allowing more rapid peat accumulation. Many of the plants on the ridges have a tendency to form hummocks or tussocks, though why they should grow in lines and not clusters remains unexplained.

A second possibility is that the patterning is a hydrological phenomenon (Ivanov, 1981). On many bogs it is noticeable that the patterning is best developed on the parts that carry most drainage water. Drury (1956) has argued that spring meltwater flowing across a smoothly sloping bog would tend to form a series of shallow benches transverse to the slope. At the outer margin of each bench, plants would grow and peat accumulate, thus forming a string. Once the pattern of strings was initiated, it would tend to be self-perpetuating. The strings would function as dams, holding back the drainage water, and thus furthering the growth of the bog plants and enhancing peat formation. The bog pools would enlarge during rain only to shrink during dry spells, and this would help to buffer the bog against moisture changes.

A less convincing suggestion is that the patterns result from the slow downslope movement of semi-liquid peat which causes wrinkling of the bog surfaces. Measurements confirm that some steeply sloping bogs experience creep or slow slippage, but, as Sjörs (1961) has pointed out, strings and flarks are frequently found on slopes that are too slight to permit any peat movement. Moreover, excavations in gently sloping patterned bogs have not revealed any disruption of the peat layers that could be interpreted as evidence of movement (Seppälä and Koutaniemi, 1985).

The two most popular theories of pattern formation invoke seasonal frost action. In winter the bog pools freeze more rapidly than the peat ridges because the water has a higher thermal conductivity than the peat. Several writers (e.g. Sjörs, 1961) have suggested that the ice on the pools pushes horizontally against the still unfrozen ridges, squeezing the peat and elevating the ridge tops. The sandy or gravelly shores of many arctic lakes are raised by ice action into ridges, but as yet few field observations have been made to show that similar pushing occurs on bog pools. Further difficulties with the theory are that it fails to explain how the pools are initiated, and it assumes that all flarks have at one time or another contained pools, which is probably not true. An alternative theory is that the ridges are raised by frost heaving from beneath. Excavations have shown that ice lenses often develop in the ridges in winter, causing heaving. The fact that the patterns are better developed in forest areas than on the tundra may be because this is where the number of frost cycles or amount of frost heaving is greatest.

A separate issue, now largely resolved, concerns the stability of the patterns. Von Post and Sernander (1910) and Osvald (1923) considered that wet areas of bog have faster-growing vegetation than drier areas, and hence have more rapid peat accumulation. Depressions, it was argued, tend to become infilled with peat and develop into raised areas, whereas raised areas become replaced by depressions. This idea of repeated reversal of the micro-relief on bog surfaces became widely accepted by early writers. Lenticular structures within the underlying peat were claimed as evidence of a constant alternation of hummocks and hollows (see Tallis, 1983, for a recent discussion). It is now generally agreed, however, that the geometry of string bogs is largely irreversible, because the strings normally occupy a much smaller area than the flarks and are always linear features whereas the flarks are linear only on slopes. Also, as has already been mentioned, peat accumulation on many ridges is faster than in adjacent flarks. Moreover, pollen analysis of string bogs suggests that the patterns can persist for up to 5,000 years (Moore, 1977).

Most modern writers have concluded that there is no single explanation for the patterning of string bogs. Ruuhijärvi (1983), for example, proposes that the patterns are caused by periodic flooding of the bog surfaces combined with frost action. Seppälä and Koutaniemi (1985) agree that flooding is of prime importance, but they suggest that the role of frost can easily be exaggerated, and consider that biotic processes also play a significant part.

Fire

Plant communities in periglacial areas are much less stable and long-lasting than might be supposed. As has already been explained, many lowland tracts of Boreal forest are gradually turning into open bogs and much open bogland is slowly becoming forested. In addition to the bog–forest cycle, more rapid changes occur that are often of great biogeomorphological importance. Frost heaving and thermokarst processes greatly modify the vegetation, but the most dramatic and abrupt changes are brought about by fire. Every year, large areas of forest and tundra are destroyed by fires that are usually started by lightning but are also increasingly caused by man. In interior Alaska, for example, about 400,000 ha of forest used to burn annually although fire suppression measures have now reduced the total to about 240,000 ha per year (Viereck, 1983). In the Mackenzie Valley, the average annual loss of forest is about 3.7 million ha, which approaches 0.5 per cent of the total forested area (Heginbottom, 1978). The frequency of fires suggests that much of the Boreal forest is a fire climax with destruction intervals as short as 300 years (Zoltai and Pettapiece, 1974). It is thought that forest fires help to determine the location of the tree line in many districts (Greene, 1983).

Although most fires occur in forested areas, some affect tundra, especially where there is a surface cover of dry peat or inflammable tussock vegetation. More than 380,000 ha of tundra burned in northwest Alaska in the exceptionally dry summer of 1977 (Racine *et al.*, 1983). Areas of wet tundra or sparse vegetation are largely immune from damage.

The geomorphological effects of tundra and forest fires vary according to the intensity of the burn. Fires that sweep rapidly across an area, fanned by strong winds, may burn only part of the vegetation. Surface plant litter or peat may escape with slight charring and the permafrost beneath may be unaffected. Other fires that burn slowly and intensely may remove all the vegetation and surface organic material, initiating deep thawing of the permafrost that may continue for 5 to 10 years. Thawing is likely to be most severe if ground temperatures prior to burning are near 0 °C. The thawing of ice bodies in the permafrost may cause irregular surface subsidence and the formation of thermokarst pits and mounds. Slope materials with a high ice-content may liquify following a fierce fire and become highly unstable (McVee, 1973). Sudden releases of water may trigger mud flows, slumping or gulley erosion. Large amounts of colluvium may be swept on to valley floors, and streams may become highly charged with sediment. Fire breaks constructed to combat the fire may cause more damage than the fire itself.

Many fires occur in remote wilderness areas and are difficult to study. Perhaps the best documented fire is the August 1968 burn which destroyed an area of forest and tundra near Inuvik in the Northwest Territories. The preceding months were unusually dry and the fire burned intensely, causing serious slumping, gulleying, and surface subsidence. Streams that had previously run clear became highly turbid. During the five years following the fire, the active layer nearly doubled in thickness, and it continued to deepen, although more slowly, for at least a further three years (Mackay, 1977). Similar delayed increases in thaw depths have been reported at other burn sites in Alaska and Canada (Viereck, 1982; Johnson and Viereck, 1983; Racine *et al.*, 1983).

Faunal modifications to the environment

In the Boreal forest zone, beavers are unsurpassed among mammals as modifiers of their environment (Wilsson, 1971; Blanchet, 1977; Ryden, 1986). There are two closely related species: the Old World *Castor fiber* and the North American *Castor canadensis* (now introduced into Finland).

Both species are expert swimmers and spend much of their time in water. In remote areas and where they are seldom persecuted by man, they often dam small streams with sticks, stones and mud, forming ponds that may reach 3 m deep and extend over several hectares (see figure 7.11). They also build stick and mud houses, called 'lodges', on the edges of the ponds, or out in the water. Where there is a deep river or lake, or heavy persecution by man, beavers live in burrows in earth banks close to the water, and do not build dams or lodges.

In summer, beavers mainly eat aquatic and marsh plants, but they also fell trees and chew the bark. Branches and trunks are often cut into portable lengths and dragged into the water near the burrows or lodges to provide a winter food store. During mild spells in winter, or if the food store becomes depleted, they continue with tree felling. One particularly industrious pair in Canada succeeded in demolishing 266 trees in just 15 months!

Sometimes, beavers dig shallow canals, up to 100 m or more in length, in order to float branches and tree trunks down to their pond from a

Figure 7.11   A small beaver dam in the Lake Superior Provincial Park, Ontario, Canada

logging area further up valley. If the timber supply gives out, they may move a short distance along the stream and build a new dam and lodge. In some valleys, repeated damming operations have created a staircase or chain of lakes, some in active use and others abandoned and undergoing infilling by vegetation.

Beaver dams help considerably to regulate stream flow, lowering flood peaks and raising minimum discharges. The animals also improve water quality because some of the suspended sediment carried by the streams settles out on the floors of the ponds. Abandoned ponds act as foci for bog development (Sjörs, 1983).

In tundra regions, several mammals have important geomorphological effects. Grizzly bears excavate shallow dens or holes, up to 3 m in diameter, often in large numbers in sandy or gravelly terrain (Walker, 1973). In the Rocky Mountains, pocket gophers (*Thomomys talpoides*) construct earth-filled burrows under snow banks during winter, and in the following spring and summer the loose earth is quickly blown or washed away (Johnson and Billings, 1962). Gopher disturbed sites sometimes become badly eroded and take years to revegetate. On the arctic tundra, the 3 to 4 year lemming cycle causes major changes in the plant biomass (Pitelka, 1973; Remmert, 1980) and doubtless also affects thaw rates and geomorphological processes, although this has still to be investigated in detail.

Concluding remarks

Research into the biogeomorphology of periglacial environments is still very much at the pioneer stage. Already it is clear, however, that plants and animals greatly influence a variety of geomorphological processes, including ground freezing, mass movement and stream flow. Periglacial environments appear to be particularly fragile, and even minor modifications to the vegetation can have far-reaching consequences. Much greater attention needs to be given to biogeomorphological factors in future periglacial research.

References

Ballantyne, C. K. and Matthews, J. A. 1982: The development of sorted circles on recently de-glaciated terrain, Jotunheimen, Norway. *Arctic and Alpine Research*, 14, 341–54.
Benninghoff, W. S. 1952: Interaction of vegetation and soil frost phenomena. *Arctic*, 5, 34–44.

Billings, W. D. 1974: Arctic and alpine vegetation: plant adaptations to cold summer climates. In Ives, J. D. and Barry, R. G. (eds), *Arctic and alpine environments*, pp. 403–44. London: Methuen.

Blanchet, M. 1977: *Le castor et son royaume*. Basle: Ligue Suisse pour la Protection de la Nature.

Bliss, L. C. 1962: Adaptations of arctic and alpine plants to environmental conditions. *Arctic*, 15, 117–44.

Brown, J. and Kreig, R. A. (eds) 1983: Elliott and Dalton Highways, Fox to Prudhoe Bay, Alaska: Guidebook to permafrost and related features. *Fourth International Conference on Permafrost, Fairbanks, Alaska, Guidebook 4.*

Brown, J., Nelson, F., Brockett, B., Outcalt, S. I. and Everett, K. R. 1983: Observations on ice-cored mounds at Sukakpak Mountain, South Central Brooks Range, Alaska. *Proceedings of the Fourth International Permafrost Conference, Fairbanks, Alaska*, 91–6.

Brown, R. J. E. 1969: Factors influencing discontinuous permafrost in Canada. In Péwé, T. L. (ed.), *The periglacial environment*, pp. 11–54. Montreal: McGill-Queen's University Press.

—— 1970: *Permafrost in Canada: its influence on northern development*. Toronto: University of Toronto Press.

—— and Johnston, G. H. 1964: Permafrost and related engineering problems. *Endeavour*, 23, 66–72.

—— and Péwé, T. L. 1973: Distribution of permafrost in North America and its relationship to the environment: a review, 1963–1973. In *The North American contribution to the Second International Permafrost Conference, Yakutsk, USSR*, 71–100.

Drury, W. H. Jr. 1956: Bog flats and physiographic processes in the Upper Kuskokwim River Region, Alaska. *Contributions from the Gray Herbarium*, 178.

Foster, D. R. and King, G. A. 1984: Landscape features, vegetation and developmental history of a patterned fen in south-eastern Labrador, Canada. *Journal of Ecology*, 72, 115–44.

French, H. M. 1976: *The periglacial environment*. London: Longman.

—— and Heginbottom, J. A. (eds) 1983: North Yukon Territory and Mackenzie Delta, Canada: Guidebook to permafrost and related features. *Fourth International Conference on Permafrost, Fairbanks, Alaska, Guidebook 3.*

Gore, A. J. P. (ed.) 1983: *Mires: Swamp, bog, fen and moor* (Ecosystems of the World 4A and B). Amsterdam: Elsevier.

Greene, D. F. 1983: Permafrost, fire, and the regeneration of white spruce of treeline near Inuvik, Northwest Territories, Canada. *Proceedings of the Fourth International Permafrost Conference, Fairbanks, Alaska*, 374–9.

Hallet, B. and Prestrud, S. 1986: Dynamics of periglacial sorted circles in Western Spitzbergen. *Quaternary Research*, 26, 81–99.

Harris, C. 1981: *Periglacial mass-wasting: a review of research*. Norwich: Geobooks.

Harris, S. A. 1986: *The permafrost environment*. London: Croom-Helm.

Heginbottom, J. A. 1978: *Lower Mackenzie River Valley*. Field Trip No. 3. Third International Conference on Permafrost, Edmonton, Alberta.

Hoffman, R. S. 1974: Terrestrial vertebrates. In Ives, J. D. and Barry, R. G. (eds), *Arctic and alpine environments*, pp. 475–570. London: Methuen.

Hopkins, D. M. and Sigafoos, R. S. 1951: Frost action and vegetation patterns on Seward Peninsula, Alaska. *US Geological Survey Bulletin*, 974-C, 51–100.

Hopkins, D. M., Karlstrom, T. N. V. and others 1955: Permafrost and groundwater in Alaska. *US Geological Survey Professional Paper*, 264-F.

Irving, L. 1972: *Arctic life of birds and mammals, including man.* Berlin: Springer-Verlag.

Ivanov, K. E. 1981: *Water movement in mirelands.* London: Academic Press.

Ives, J. D. and Barry, R. G. (eds) 1974: *Arctic and alpine environments.* London: Methuen.

Jahn, A. 1986: Remarks on the origin of palsa frost mounds. *Biuletyn Peryglacjalny*, 31, 123–30.

Johnson, A. W. and Neiland, B. J. 1983: An analysis of plant succession on frost scars 1961–1980. *Proceedings of the Fourth International Permafrost Conference, Fairbanks, Alaska*, 537–42.

Johnson, L. and Viereck, L. 1983: Recovery and active layer changes following a tundra fire in northwestern Alaska. *Proceedings of the Fourth International Permafrost Conference, Fairbanks, Alaska*, 543–7.

Johnson, P. L. and Billings, W. D. 1962: The alpine vegetation of the Beartooth Plateau in relation to cryopedogenic processes and patterns. *Ecological Monographs*, 32, 105–35.

Kreig, R. A. and Reger, R. D. 1982: Air-photo analysis and summary of landform soil properties along the route of the Trans-Alaska pipeline system. *Alaska Geological and Geophysical Surveys, Geologic Report 66.*

Lundqvist, J. 1962: Patterned ground and related frost phenomena in Sweden. *Sveriges Geologiska Undersokning Årsbok 55*, 7.

Mackay, J. R. 1970: Disturbances to the tundra and forest tundra environment of the Western Arctic. *Canadian Geotechnical Journal*, 7, 420–32.

—— 1977: Changes in the active layer from 1968 to 1976 as a result of the Inuvik fire. *Report of Activities, Part B; Geological Survey of Canada*, Paper 77-1B, 273–5.

—— 1979: An equilibrium model for hummocks (nonsorted circles), Garry Island, Northwest Territories. *Current Research, Part A, Geological Survey of Canada*, Paper 79-1A, 165–7.

—— 1980: The origins of hummocks, western Arctic coast, Canada. *Canadian Journal of Earth Science*, 17, 996–1006.

McVee, C. V. 1973: Permafrost considerations in land use planning. In *The North American contribution to the Second International Permafrost Conference, Yakutsk, USSR*, 146–50.

Moore, P. D. 1977: Stratigraphy and pollen analysis of Claish Moss, North-West Scotland: significance for the origin of surface pools and forest history. *Journal of Ecology*, 65, 375–97.

Moore, P. D. 1982: Pool and ridge patterns in peat mires. *Nature*, 300, 5888, 110.

—— and Bellamy, D. J. 1974: *Peatlands.* London: Elek.

Osvald, H. 1923: Die Vegetation des Hochmoores Komosse. *Svenska Växtsociologiska Sallskapets Handlingar*, 1, 1–436.

Pemberton, M. 1980: Earth hummocks at low elevation in the Vale of Eden, Cumbria. *Transactions of the Institute of British Geographers*, NS 5, 487–501.

Péwé, T. L. 1965: *Central and South Central Alaska.* Guidebook for Field Conference F. INQUA VIIth Congress, Lincoln, Nebraska: Nebraska Academy of Sciences.

Péwé, T. L. 1966: Ice-wedges in Alaska: Classification, distribution, and climatic significance. *Proceedings of the International Permafrost Conference.* National Academy of Sciences–National Research Council Publication 1287.
—— 1982: Geologic hazards of the Fairbanks area, Alaska. *Alaska Geological and Geophysical Surveys,* Special Report 15.
Pitelka, F. A. 1973: Cyclic pattern in lemming populations near Barrow, Alaska. In Britton, M. E. (ed.), *Alaskan arctic tundra.* Arctic Institute of North America, Technical Paper 22, 199–215.
Pollard, W. H. and French, H. M. 1984: The groundwater hydraulics of seasonal frost mounds, North Fork Pass, Yukon Territory. *Canadian Journal of Earth Sciences,* 21, 1073–81.
Racine, C. H., Patterson, W. A. and Dennis, J. G. 1983: Permafrost thaw associated with tundra fires in northwest Alaska. *Proceedings of the Fourth International Permafrost Conference, Fairbanks, Alaska,* 1024–9.
Raup, H. M. 1965: The structure and development of turf hummocks in the Mesters Vig district, N. E. Greenland. *Meddelelser om Grønland,* 166, 3, 112p.
Rawlinson, S. E. 1983: Prudhoe Bay, Alaska: Guidebook to permafrost and related features. *Fourth International Conference on Permafrost, Fairbanks, Alaska, Guidebook 5.*
Ray, R. J., Krantz, W. B., Caine, T. N. and Gunn, R. D. 1983: A mathematical model for patterned ground: Sorted polygons and stripes, and underwater polygons. *Proceedings of the Fourth International Permafrost Conference, Fairbanks, Alaska,* 1036–41.
Remmert, H. 1980: *Arctic animal ecology.* Berlin: Springer-Verlag.
Rieger, S. 1974: Arctic soils. In Ives, J. D. and Barry, R. G. (eds), *Arctic and alpine environments,* pp. 749–70. London: Methuen.
——, Dement, J. A. and Sanders, D. 1963: Soil survey of Fairbanks area, Alaska. *US Department of Agriculture Series 1959,* 25.
Ruuhijärvi, R. 1983: The Finnish mire types and their regional distribution. In Gore, A. J. P. (ed.), *Mires: Swamp, bog, fen and moor* (Ecosystems of the World, 4B), pp. 47–67. Amsterdam: Elsevier.
Ryden, H. 1986: *The beaver.* New York: Putnam.
Savile, D. B. O. 1972: *Arctic adaptations in plants.* Canadian Department of Agriculture, Monograph 6.
Seppälä, M. 1982: An experimental study of the formation of palsas. In French, H. M. (ed.), *The Roger Brown Memorial Volume,* 36–42. *Proceedings of the Fourth Canadian Permafrost Conference, Calgary, Alberta.*
—— 1983: Seasonal thawing of palsas in Finnish Lapland. *Proceedings of the Fourth International Permafrost Conference, Fairbanks, Alaska,* 1127–32.
—— and Koutaniemi, L. 1985: Formation of a string and pool topography as expressed by morphology, stratigraphy and current processes on a mire in Kuusamo, Finland. *Boreas,* 14, 287–310.
Shilts, W. W. 1978: Nature and genesis of mudboils, Central Keewatin, Canada. *Canadian Journal of Earth Sciences,* 15, 1053–68.
Sjörs, H. 1959: Bogs and fens in the Hudson Bay lowlands. *Arctic,* 12, 3–19.
—— 1961: Surface patterns in Boreal peatland. *Endeavour,* 20, 217–24.
—— 1983: Mires of Sweden. In Gore, A. J. P. (ed.), *Mires: Swamp, bog, fen and moor* (Ecosystems of the World, 4B). Amsterdam: Elsevier.

Tallis, J. H. 1983: Changes in wetland communities. In Gore, A. J. P. (ed.), *Mires: Swamp, bog, fen and moor* (Ecosystems of the World, 4B). Amsterdam: Elsevier, 311–47.

Tarnocai, C. and Zoltai, S. C. 1978: Earth hummocks of the Canadian Arctic and Subarctic. *Arctic and Alpine Research*, 10, 581–94.

Taylor, J. A. 1983: The peatlands of Great Britain and Ireland. In Gore, A. J. P. (ed.), *Mires: Swamp, bog, fen and moor* (Ecosystems of the World, 4B). Amsterdam: Elsevier.

Troll, C. 1958: Structure soils, solifluction and frost climates of the Earth. *Translation 43, US Army Snow, Ice and Permafrost Research Establishment, Illinois*.

Tubbs, C. R. 1986: *The New Forest*. London: Collins.

Tufnall, L. 1975: Hummocky microrelief in the Moor House area of the northern Pennines, England. *Biuletyn Peryglacjalny*, 24, 353–68.

Tyrtikov, A. P. 1964: The effect of vegetation on perennially frozen soil. *National Research Council of Canada, Technical Translation 1088*. Ottawa.

Viereck, L. A. 1965: Relationship of white spruce to lenses of perennially frozen ground, Mount McKinley National Park, Alaska. *Arctic*, 18, 262–7.

—— 1970: Forest succession and soil development adjacent to the Chena River in interior Alaska. *Arctic and Alpine Research*, 2, 1–26.

—— 1982: Effects of fires and firelines on active layer thickness and soil temperatures in interior Alaska. In French, H. M. (ed.), *The Roger Brown Memorial Volume*, 123–35. *Proceedings of the Fourth Canadian Permafrost Conference, Calgary, Alberta*.

—— 1983: Vegetation. In Brown, J. and Kreig, R. A. (eds), Elliott and Dalton Highways, Fox to Prudhoe Bay, Alaska: Guidebook to permafrost and related features, 26–31. *Fourth International Conference on Permafrost, Fairbanks, Alaska, Guidebook 4*.

Von Post, L. and Sernander, R. 1910: *Pflanzenphysiognomische Studien auf Torfmooren in Närke*. Livretguide des excursions en Suède du XI Congress Géologique Internationale 14, Stockholm.

Walker, H. J. 1973: Morphology of the North Slope. In Britton, M. E. (ed.), *Alaskan arctic tundra*. Arctic Institute of North America, Technical paper 22, 49–92.

Walters, J. C. 1983: Sorted patterned ground in ponds and lakes of the High Valley/Tangle Lakes Region, Central Alaska. *Proceedings of the Fourth International Permafrost Conference, Fairbanks, Alaska*, 1350–5.

Washburn, A. L. 1979: *Geocryology: A survey of periglacial processes and environments*. London: Methuen (New York: Wiley, 1980).

—— 1983: Palsas and continuous permafrost. *Proceedings of the Fourth International Permafrost Conference, Fairbanks, Alaska*, 1372–7.

Waters, R. S. 1962: Altiplanation terraces and slope development in West Spitzbergen and South-West England. *Biuletyn Peryglacjalny*, 11, 89–101.

Wilsson, L. 1971: Observations and experiments on the ethology of the European Beaver (*Castor fiber* L.). *Vitrevy: Swedish Wildlife*, 8, 115–266.

Zoltai, S. C. and Pettapiece, W. W. 1974: Tree distribution on perennially frozen earth hummocks, *Arctic and Alpine Research*, 6, 403–11.

—— and Tarnocai, C. 1974: Soils and vegetation of hummocky terrain. *Environmental-Social Committee Northern Pipelines, Task Force on Northern Development*, Report 74–5.

Part III
Coastal and Karst Environments

8 Coastal biogeomorphology

T. Spencer

Introduction and approach

Coastal environments splendidly demonstrate the ubiquity and diversity of biological influences on geomorphology. Shallow marine waters include some of the world's most productive plant systems and support a rich diversity of meiobenthic and macrobenthic[1] organisms trophically linked to microfloral and microbial processes in the nearshore zone.

There are several ways in which such environments can be viewed. From geomorphology, coastal sedimentary environments have been seen (e.g. by Steers, 1957) as 'physiographic' units which reflect the interactions between physical, biological and chemical processes and which are capable of creating, maintaining and transforming their own topography. Thus, for example, sand dune development is clearly linked to plant succession and the evolution of mangrove swamps can be intimately related to changing patterns of deltaic sedimentation (e.g. Thom, 1967; Thom *et al.*, 1975). No causality is implied in these statements; it is the product of the interactions which create the landform whole. Important interrelationships between biological activity and geomorphology have been reviewed for saltmarshes and tidal flats (e.g. Chapman, 1977; Frey and Basan, 1978; Reise, 1985), for mangrove swamps (e.g. Thom, 1982, 1984), and for coral reefs (e.g. Hopley, 1982). Such interactions probably also characterize seaweed-based systems (although the biogeomorphology of kelp has yet to be written) and they are certainly pertinent in the deep sea (e.g. Rowe, 1974) and in subaerial beach crest environments (e.g. Psuty, 1965).

Biology has taken a rather different view. Here, considerable insights into the functioning of geomorphology's physiographic units have been achieved through the principles and techniques of ecosystem research. There has been a concern both with ecosystem structure, and the search for recognizable hierarchical patterns of interaction and organization, and with ecological processes, requiring the determination of the fluxes of energy and matter through such structures (e.g. Mann, 1982). An ecosystem approach, in the broadest sense, does offer the possibility of

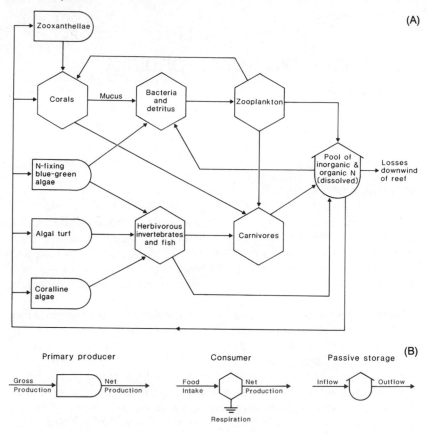

Figure 8.1   (A) Probable relationships of functional groups in a coral reef ecosystem; (B) symbols from energy circuit language of Odum (1971) (from Mann, 1982)

defining the precise nature and rate of operation of physical–biological–chemical linkages in the littoral environment; this is perhaps best done at the small scale and through experimentation and ecosystem manipulation. In fact, a disparate literature using such methods already exists and can be focused towards questions of geomorphological interest.

As space precludes a complete and comprehensive review of coastal biogeomorphology, this chapter concentrates upon those intertidal and shallow subtidal (less than 20 m) environments characterized by rocky shores; algal mats and seagrass meadows; and bioturbated sediments. Particular attention is paid to limestone shores and carbonate sediments for they provide much of our quantitative information in these areas (e.g. Bathurst, 1975). There are good reasons for choosing these examples which, in ecosystem terms, are not as different as might be supposed from

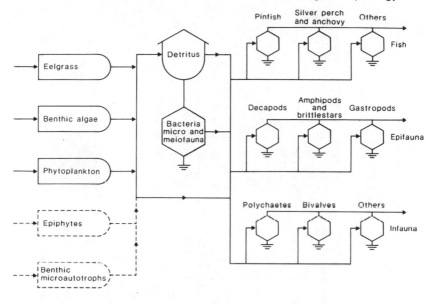

Figure 8.2 Energy flow in a seagrass community in California: for symbols see figure 8.1
(from Mann, 1982)

a purely geomorphological viewpoint. In all three cases, either a microfloral or a microbial/meiofaunal community plays a key role, through macrophytes or macroinvertebrates, in mediating erosional, depositional or transport processes.

It is now becoming clear that on limestone coasts, and on substrates of live and dead coral and coralline algae, the cyanobacteria (terminology of Carr and Whitton, 1982; also known as cyanophytes, blue-green algae) greatly influence the style and rate of 'bioerosion' (e.g. Schneider, 1976; Föcke, 1978; Spencer, 1985a; Trudgill, 1985, 1987). This control is achieved by trophic flows which are largely channelled through grazing pathways (figure 8.1). Estimates from the Great Barrier Reef suggest that 20–90 per cent of the net daily production of epilithic algae, and up to 6 per cent of the standing crop per day, may be removed in this way (Hatcher, 1983). A number of experimental studies have shown that the exclusion of grazers, primarily fish and echinoids, results in an increase in the standing crop of the epilithic algal community and that algal mats transplanted into areas where grazers are common are rapidly consumed (e.g., Wanders, 1977; Van den Hoek et al., 1978; Brock, 1979).

Micro-algal communities are similarly important on shallow subtidal sediments where, with macrophytes, they trap and bind sediments and through 'baffle' effects encourage sedimentation (Ginsburg and Lowenstam, 1958; Neumann et al., 1970; Scoffin, 1970a). Such processes are

Table 8.1 Quantitative estimates of subtidal and intertidal erosion rates

Location	Substrate	Environment	Erosion rate mm yr^{-1}	Author	Comments
(a) *Tropical environments/reef substrates*					
Aldabra Atoll	Reef limestones	Intertidal	0.6–4.0	Trudgill (1976)	Micro-erosion meter technique
			1.5–2.7	Viles and Trudgill (1984)	Re-measurements of Trudgill's (1976) sites
Grand Cayman	Reef limestones	Subtidal	1.12–1.79	Spencer (1985a, 1985b)	Micro-erosion meter technique
			0.46	Spencer (1983)	Weight-loss tablet method
		Intertidal	0.88–1.23	Spencer (1985a, 1985b)	Micro-erosion meter technique
			0.56	Spencer (1983)	Weight-loss tablet method
Grand Bahama	Reef limestones	Intertidal	6.40	Warthin (1959)	Soluble/insoluble blocks
Bermuda	Reef limestones	Subtidal	2.50	Warthin (1959)	Soluble blocks
Puerto Rico	Reef limestones	Intertidal	1.30	Bromley (1978)	Rock pedestals
Norfolk Is. and W. Australia	Eolianite and beachrock	Intertidal	1.00	Kaye (1959)	Dated surfaces
			0.6–1.00	Hodgkin (1964)	Erosion pegs/dated surfaces
S. W. Australia	Reef limestones	Intertidal	270–670 cm^3 100 cm^{-2} yr^{-1}	Revelle and Fairbridge (1957)	
Heron Is.	Beachrock	Intertidal	0.5	Stephenson (1961)	
Lizard Is., Great Barrier Reef	Coral	Subtidal	0.63–1.26	Kiene (1985)	Slices of coral *Porites*, upper surface lowering
	Coral	Intertidal reef flat	0.006–0.02	Kiene (1985)	
	Coral	Lagoon	0.22–0.93	Kiene (1985)	
Bikini Atoll	Beachrock	Intertidal	0.3	Revelle and Emery (1957)	

(b) *Temperate environments*

Location	Rock type	Zone	Rate	Reference	Method
N. Adriatic	Limestone	Intertidal	0.63	Torunski (1979)	Micro-erosion meter technique
		Intertidal	0.25	Schneider (1976)	Historical evidence
Co. Clare, Ireland	Limestone	Intertidal	0.20	Trudgill et al. (1981)	Micro-erosion meter technique
South Island, New Zealand	Limestone	Intertidal	0.38–1.35	Kirk (1977)	Micro-erosion meter technique
N. Yorks., UK	Mudstone	Intertidal	0.64–2.50	Robinson (1977)	Micro-erosion meter technique
	Shales	Intertidal	1.0		
Victoria, Australia	Greywackes and siltstone	Intertidal	0.37	Gill and Lang (1983)	Micro-erosion meter technique
S. Devon, UK	Greenschist	Supratidal	0.61	Mottershead (1982)	Micro-erosion meter technique
Plymouth breakwater, UK	Devonian limestone	Intertidal	0.5	Southward (1964)	Limestone v. granite blocks

intimately related to the fact that these ecosystems are structured around detritus-based food chains (e.g. seagrasses: figure 8.2); thus they provide a strong contrast with the biogeomorphology of rocky shores.

Detritus, in various stages of colonization by microbes, is processed in a thin benthic boundary layer (Rhoads and Boyer, 1982) a few centimetres above the sediment surface to a few decimetres (in general) below this level. Although superficial, this zone has a major effect on sedimentation and sediment transport. Animal burrows can increase the surface area of the sediment–water interface by three orders of magnitude or more, and thus have an important impact on geochemical fluxes and the early stages of sediment diagenesis. These physical and chemical effects may in turn have ecological consequences.

Studies of coastal biogeomorphology are not just of academic interest. At the broad scale, all the marine flowering plants, for example, have the tendency to form dense, sediment-trapping stands. The large amounts of sediment and biomass associated with these plants act as efficient buffers between the marine and terrestrial environments. Three examples can be used to demonstrate this ability. First, mangroves cushion storm waves (e.g. Fosberg, 1971) and have been used in shoreline restoration schemes (e.g. Teas, 1977). Secondly, natural sand dune systems which allow for the dissipation of wave energy and the short-term redistribution of sediment during storm events are more successful in absorbing wave attack than many artificial sea defences on barrier islands (Dolan, 1972). Thirdly, naturally vegetated coral islands have been shown to withstand and recover from severe hurricane forces while neighbouring islands cleared for coconut plantations have been destroyed (Stoddart, 1964). This chapter shows that the properties of algal mats are potentially of significance to low cost, 'bioengineering' (Schiechtl, 1980) solutions to problems of sediment stability in estuaries and harbours. Furthermore, the way in which the benthos processes seafloor sediments can determine whether potentially harmful radionuclides, heavy metals or other contaminants are mixed and trapped within near-surface sediments or, alternatively, transported and dispersed (e.g. Phelps, 1966; Lee and Schwartz, 1980; Schwartz and Lee, 1980).

Erosional biogeomorphology of rocky shores

Measured erosion rates in subtidal and intertidal environments on tropical and temperate coasts generally average between 0.5 and 2.0 mm yr^{-1} (table 8.1; see also Trudgill, 1985). Our knowledge of this subject is largely restricted (but ought not to be) to limestone substrates, although this umbrella covers a wide range of lithologies from contemporary beachrock to massive, recrystallized limestones. Heterogeneous, relatively unconsolidated tropical limestones tend to be characterized by mean

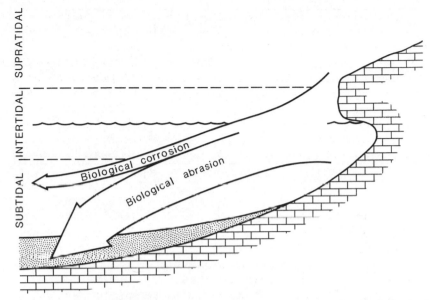

Figure 8.3 Bioerosion on limestone coasts: biogenic destruction by biological corrosion (dissolved erosion product) and biological abrasion (granular product) and contribution to nearshore sedimentation. On Adriatic coast (Torunski, 1979), 10-30 per cent biological corrosion, 70-90 per cent biological abrasion of total erosion

erosion rates in excess of 1.0 mm yr^{-1} while older, temperate limestones retreat at less than 0.5 mm yr^{-1}. Nevertheless, within these two groups the similarity of erosion rates from different localities and determined by different methods has been used as evidence for an ecological balance between a rock-dwelling, micro-organic community acting as a food source on the one hand and an exploitative community of grazing macro-organisms on the other (Spencer, in press).

Although limestone erosion is clearly the product of a combination of physical and chemical processes the balance described above is also represented in an erosional equilibrium between 'biological corrosion', a solutional process undertaken by micro-organisms and some macro-borers which modifies the substrate but provides no erosion product, and 'biological abrasion', a series of physical processes carried out by grazing, burrowing and boring organisms which result in particulate debris production (figure 8.3). The following three subsections review present knowledge on biocorrosion and then the boring and grazing processes which contribute to biological abrasion.

The micro-organic community: characteristics and distribution

The diverse micro-organic community on rocky shores is often dominated by cyanobacteria (Golubic and Schneider, 1979; Schneider and Torunski,

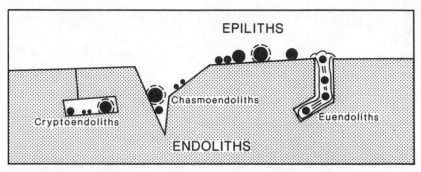

Figure 8.4 Niche types, illustrated by cyanobacteria (after Golubic *et al.*, 1981)

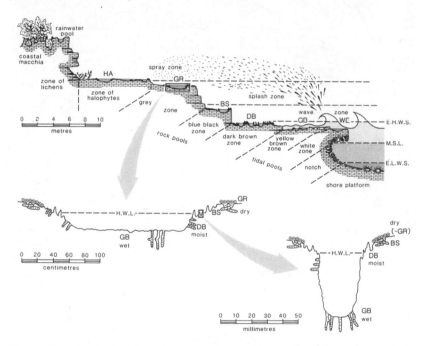

Figure 8.5 Nested scales of topography and cyanobacterial colour zonation (after Schneider, 1976) epiliths on rock surfaces, endoliths within substrate. Colour zones WE-HA; E.H.W.S. = extreme high water springs, M.S.L. = mean sea level, E.L.W.S. = extreme low water springs; H.W.L. = high water line. Maximum micro-relief in areas subject to wetting and drying, above H.W.L

1983) but also includes chlorophytes, rhodophytes, heterotrophic fungi (Kohlmeyer, 1969; Rooney and Perkins, 1972) and also lichens and bacteria (Risk and MacGeachy, 1978; Trudgill, 1987, figures 4–6). Several niche types are present: epiliths dwell on the rock surface whereas endoliths live within the limestone substrate (Golubic *et al.*, 1981). This latter group

Figure 8.6 Intertidal colour zonation, Grand Cayman Island, West Indies (photograph by T. Spencer)

consists of euendoliths which actively bore into the rock, chasmoendoliths which inhabit existing cracks and cavities and cryptoendoliths (Purdy and Kornicker, 1958; Ginsburg, 1953; Highsmith, 1981) which occur as discrete layers (e.g. the '*Ostreobium* layer') within porous lithologies (figure 8.4).

Marked zonation amongst lithophytic communities has been described by many authors (for review: Whitton and Potts, 1982). For Adriatic limestones, Schneider (1976) has recognized 7 zones, the lower 5 colour zones being defined by micro-organisms (figure 8.5). Comparisons with the French Mediterranean, Bermuda and the Caribbean (e.g. figure 8.6) have led Schneider to suggest that this 'phenomenological classification' (Schneider, 1976, 13) is generally applicable (and see also Van den Hoek *et al.*, 1972; Potts and Whitton, 1980). This colour zonation is, however, only partly due to the presence or absence of indicator species. Boring by cyanobacteria is positively hydrotropic, towards within-substrate moisture, and, to some degree, negatively phototropic. Thus average cell size and depth of carbonate penetration tend to decrease along the gradient of falling submergence (Golubic *et al.*, 1975; Le Campion-Alsumard, 1970). Furthermore, colouration and niche type are strongly interrelated. The supratidal and upper intertidal zones (figure 8.5, DB–GR) are characterized by strongly-pigmented epiliths, responding to insolation, desiccation and illumination, and shallow-boring (100–200 μm) endoliths which pit the surface ('caries'; Fremy, 1945) rather than forming true

Table 8.2 Bioerosion by the cryptofaunal coral reef community: sponges and polychaetes

Carbonate removal (kg m^{-2} yr^{-1})	Locality	Method of estimation and substrate type	Author
Clionid sponges			
20.0–25.0 (1.0–1.4 cm^3 yr^{-1})	Bermuda	Small block transplant; 100 days	Neumann (1966)
0.25–0.30	Bermuda	Small block transplant; 12 months	Rützler (1975)
1.35 (0.21–3.50)	St Croix, Virgin Is.	Thin-section of algal ridge cores	Moore and Shedd (1977)
0.96 (0.21–1.81)	Jamaica	Deep-reef sediment traps	Moore and Shedd (1977)
2.6–3.3	Netherlands Antilles	Experimental blocks	Bak (1976)
8.0	Grand Cayman	Changes in specific gravity between bored, unbored blocks	Acker and Risk (1985)
1.35 (1.12–1.60)	Barbados	Coral head; X-ray section (1980)	Scoffin *et al.* (1980)
3.0–13.4	Florida Keys	Coral head; X-ray section	Hudson (1977) (calculation from Davies 1983)
0.17	Florida Keys	Coral head; X-ray section	Hein and Risk (1975) (calculation from Davies 1983)
0.01–0.11 (reef slope)	Lizard Is., Great Barrier Reef	Small block transplants; 2–3 years	Kiene (1985)
0.009–0.02 (reef flat)			
0.03–0.43 (lagoon)			
0.35 (lagoon)	Davies Reef, Great Barrier Reef	Manufactured carbonate sand grains	Tudhope and Risk (1985)
Polychaetes			
0.18	Florida Keys	Coral head; polydorid polychaetes	Hein and Risk (1975)
0.694 (reef front)	Lizard Is., Great Barrier Reef	Coral block transplants; cirratulids, polydorids and sabellids	Davies and Hutchings (1983)
0.843 (reef flat)			
1.788 (patch reef)			
4.08–4.72 (reef flat)	Lizard Is., Great Barrier Reef	Coral block transplants; polydoids and fabricinids	Hutchings and Bamber (1985)
13.23–6.40 (leeward reef)			
0.33 (patch reef)			
0.36–0.93 (reef slope)	Lizard Is., Great Barrier Reef	Coral blocks; polychaetes, then replaced by sipunculans	Kiene (1985)
0.15–0.38 (reef flat)			
0.10–0.37 (lagoon)			

boreholes. By contrast, the lower intertidal zones, (notch–GB) not shaded by an epilithic cover, are dominated by endolithic cyanobacteria which are able to penetrate the substrate to a depth of 600–900 μm. This represents the light compensation depth where cyanobacterial assimilation is balanced by respiration (Le Campion-Alsumard, 1970; Golubic et al., 1975). These lower zones are also strongly influenced by interaction with other organisms on the shore (Schneider, 1976; Torunski, 1979).

Colonization of 'fresh' surfaces by micro-organisms can be extremely rapid: initial colonization may take place within 8–9 days of exposure and 90–100 per cent micro-organic cover may be achieved in less than four months (e.g., Marseille, France: Le Campion-Alsumard, 1975; Virgin Islands: Perkins and Tsentas, 1976; Jamaica: Kobluk and Risk, 1977; Aldabra Atoll, Indian Ocean and Grand Cayman Island, West Indies: Viles and Spencer, 1986), or even 3–5 weeks (Aldabra Atoll: Whitton and Potts, 1980; Potts and Whitton, 1980), leaving the disturbed substrate visually indistinguishable from the surrounding surfaces. In time, the degree of substrate modification, particularly by endoliths, can be considerable. The greatest number of boreholes are to be found in the upper intertidal zones: along the Adriatic coast (Torunski, 1979) densities reach 4,460–8,140 boreholes mm^{-2}. Up to 33 per cent by volume of the substrate may be occupied by borings (Le Campion-Alsumard, 1979). In some cases, boreholes have been shown to follow cleavage planes and elsewhere borehole position may be determined by the distribution of impurities and components of varying solubility within the host substrate (Whitton and Potts, 1982).

Boring activity: the cryptofaunal community and bivalve molluscs

Hard substrates are also weakened by the actions of larger boring organisms. Reviews by Davies (1983), Trudgill (1983b, 1985, 1987) and Hutchings (1986) point to the importance of sponges, sipunculans, polychaete worms and bivalve molluscs. On tropical coasts, boring sponges penetrate to depths of up to 80 mm (details of excavation processes are given by Pomponi (1977)). Substrate collapse and the re-distribution of rubble and detrital carbonate 'chips', 15–100 mm in size, throughout the reef environment follows (Wilkinson, 1983). Sponge chips often contribute 0.1–6.0 per cent, and exceptionally in excess of 30 per cent, of the total sediment (for silt fraction: 10–30 per cent, 98 per cent respectively) in tropical lagoons and fore-reef environments (Young and Nelson, 1985). In temperate environments, shell destruction is enhanced by sponge boring as periodically under-saturated waters (Alexandersson, 1976) can etch the increased shell surface area generated by sponge chip production (Akpan and Farrow, 1985; Young and Nelson, 1985). The wide range of sponge erosion rates (table 8.2) probably reflects the diversity of estimation techniques employed, the variety of substrates attacked and genuine

Table 8.3 Burrow extension rates, substrate removal and surface lowering by bivalve molluscs on coral reef and reef-related surfaces

Agent	Rate	Substrate	Locality	Author
Lithophaga nasuta	0.91 cm yr^{-1} (0.87 cm^3 yr^{-1})	Reef limestones	Aldabra Atoll	Trudgill (1976)
Lithophaga sp.	0.25 cm yr^{-1}	Reef limestones	Oman	Vita-Finzi and Cornelius (1973)
Lithophaga sp.	1.5 cm yr^{-1}	Beachrock	Great Barrier Reef	Otter (1937); McLean (1974)
Lithophaga lithophaga	0.4–1.3 cm yr^{-1}	Various limestones		Kleeman (1973)
Hiatella arctica	0.50–0.67 cm yr^{-1}	Limestones, mudstones, sandstones	Cumbrae, Plymouth UK	Hunter (1949)
Hiatella gallicana				
Hiatella arctica	0.125–1.00 cm yr^{-1}	Limestone	Co. Clare, Eire	Trudgill and Crabtree (1987)
Lithophaga/ Gastrochaena	0.18 kg m^{-2} yr^{-1} (0.05–4.4 cm^3 yr^{-1})	Coral reef	Florida	Hein and Risk (1975); calculation by Davies (1983)
Lithophaga/ Gastrochaena	0.13–0.15 kg m^{-2} yr^{-1} 0.01–0.02 kg m^{-2} yr^{-1} 0.04–0.05 kg m^{-2} yr^{-1}	Reef slope Flat reef Reef lagoon	Lizard Is., Great Barrier Reef	Kiene (1985)
Lithotrya sp.	0.84 cm yr^{-1} (0.78 cm^3 yr^{-1})	Reef limestones	Aldabra Atoll	Trudgill (1976)
Tridacna crocea	0.14 kg m^{-2} yr^{-1}	Reef flat	Great Barrier Reef	Hammer and Jones (1976)
Saccostrea amasa	0.02–0.38 cm yr^{-1}	Boulder rock	One Tree Island, Great Barrier Reef	Trudgill (1983a)

intra- and inter-regional differences in boring potency. What is clear, however, is that short-term experiments over-estimate boring activity, as rates slacken after 3–6 months when the initial colonization phase has been completed (Rutzler, 1975). Some rates of boring by polychaetes and worms are also now becoming available (table 8.2; see Hutchings, 1986, for species numbers and densities) but community structure and function in this area is little understood (e.g. see Ginsburg, 1983). At Lizard Island, Great Barrier Reef, the rapid colonization of fresh coral tablets by polychaete worms is subsequently replaced by more 'mature' boring communities of sipunculids and sponges (Davies and Hutchings, 1983; Hutchings and Bamber, 1985); in the Caribbean, the primary colonists are bacteria and fungi (Risk and MacGeachy, 1978; Choi, 1984).

At the macro-scale, rock-boring bivalve molluscs and barnacles use a mixture of mechanical and chemical penetration processes in attacking rock, coral and shell (for taxonomy and burrow characteristics see Evans, 1970; Warme, 1975). Unfortunately, the majority of studies on boring bivalves have been of a qualitative nature and it is not always clear than an individual present in a burrow has been responsible for the construction of that burrow (Trudgill et al., 1987). Furthermore, it is difficult to evaluate known rates of boring (table 8.3) in the absence of information on the variation in boring between individuals of different sizes and at different stages in their life histories, on the areal coverage of borings and on the percentage occupancy of these boreholes (Trudgill et al., 1987; Trudgill and Crabtree, 1987).

Grazing activity: echinoids, fish, chitons and gastropods

Substrates riddled with boreholes are also subjected to the action of surface grazers, attracted by epilithic and endolithic algae. Once again, the bulk of quantitative information on this type of biological abrasion comes from studies on living coral reefs or unconsolidated reef sediments. In such environments the principal grazing organisms are echinoids and a wide variety of reef fish. The basal teeth and spines of echinoids produce clear browsing patterns (Bromley, 1975) and, at a larger scale, hemispherical burrows and grooves in limestone substrates. Echinoids may also limit growth on live coral by grazing pressure (e.g. *Eucidaris thouarsii* in the Galapagos; Glynn et al., 1979). Species of the genus *Echinometra* are clearly capable of considerable substrate modification on tropical coasts (table 8.4), particularly on Caribbean reefs where they may reach densities of 50 individuals m^{-2} (e.g. *Echinometra*; Discovery Bay, Jamaica, Sammarco, 1982). The temperate equivalent is *Paracentrotus lividus* (table 8.4). Trudgill et al. (1987) have described a cycle of erosion by this urchin on structural benches in Co. Clare, Eire with pools deepened, widened and finally breached in the lifetime of a *P. lividus* population living for up to 10 years.

Table 8.4 Substrate removal by burrowing and grazing echinoids

Agent	Sediment removal	Substrate	Locality	Author
Tropical and sub-tropical environments				
Echinometra lucunter	24 g yr^{-1} 9.96–14.0 cm^3 yr^{-1}	Beachrock	Barbados	McLean (1967a)
Echinometra lucunter	0.77 g d^{-1} urchin^{-1} (7.0 kg m^{-2} yr^{-1})	Eolianite	Bermuda	Hunt (1969)
Echinometra lucunter	0.11 g d^{-1} urchin^{-1} (3.9 kg m^{-2} yr^{-1})	Algal ridge	Virgin Is.	Ogden (1977)
Eucidaris thouarsii	0.40–0.84 g d^{-1} urchin^{-1} 0.47–0.77 g d^{-1} urchin^{-1} (1.7 kg m^{-2} yr^{-1})	Coralline algae Live coral Overall, reef edge	Galapagos	Glynn *et al.* (1979)
Echinometra mathaei	0.11–0.13 g d^{-1} urchin^{-1}	Lagoonal coral knoll + reef platform	Enewetak Atoll	Russo (1980)
Echinostrephus aciculatus	0.18–0.40 g d^{-2} urchin^{-1}	Lagoonal coral + reef platform	Enewetak Atoll	Russo (1980)
Echinometra mathaei and *E. oblonga*	0.11–0.82 g d^{-1} urchin^{-1}	Limestone veneer on basalt platform	Hawaii	Russo (1977, 1980)
Echinometra mathaei	0.5 g d^{-1} urchin^{-1}	Dead *Acropora*	Persian Gulf	Shinn (in Hughes Clarke and Keij 1973)
Diadema antillarum	1.85 g d^{-1} urchin^{-1} 5.0–5.6 kg m^{-2} yr^{-1}	Coral reef	Barbados	Scoffin *et al.* (1980)
Diadema antillarum	2.07 g d^{-1} urchin^{-1}	Coral reef	Barbados	Lewis (1964)
Diadema antillarum	4.6 kg m^{-2} yr^{-1}	Coral reef	Virgin Islands	Ogden (1977)
Diadema antillarum	5.8 kg m^{-2} yr^{-1}	Coral reef	Barbados	Hunter (1977)
Temperate environments				
Paracentrotus lividus	0.25–1.5 cm yr^{-1} (exposed coast) 0.00–1.0 cm yr^{-1} (sheltered coast)	Limestone	Co. Clare, Eire	Trudgill *et al.*, (1987)
Paracentrotus lividus	1.1 mm yr^{-1} surface lowering 18 g yr^{-1} urchin^{-1}	Limestone	N. Adriatic	Torunski (1979)
		Limestone	N. Adriatic	Torunski (1979)

There are strong linkages through grazing between corals and coral reef fish (e.g. Patton, 1976; Reese, 1977; Neudecker, 1979), although the exact nature of these interrelationships, and their erosional consequences, is not always clear. Different feeding strategies on coral (e.g. Hiatt and Strasburg, 1960), from browsing through to coral-feeding, do differing degrees of damage to coral skeletons. These variations are part of the wider debate on the extent to which reef fish produce 'new' sediment rather than merely recycle 'old' reef sediment (table 8.5, for debate see Ogden, 1977; Frydl and Stearn, 1978; Scoffin et al., 1980; see also Cook, 1971; Howard et al., 1977; Grant, 1983; and Suchanek and Colin, 1986 on the disruption of lagoonal and intertidal sediments). What is certain however, is that some families are highly adapted to grazing and processing carbonate substrates. Pre-eminent are the parrotfish (scarids) which possess massive 'beaks' of fused teeth for scraping the surface, sets of pharyngeal bones which grind ingested particles into a fine paste and a sufficiently acidic gut (they have no stomach) to dissolve calcium carbonate and release organic matter for absorption through the gut wall. Similarly, the surgeonfish (acanthurids) have evolved a thick walled, gizzard-like stomach to break down carbonate sediments (Hutchings, 1986).

Littoral surfaces are also grazed by amphineurids and gastropods in the search for algal food sources (figure 8.7). In a classic series of experiments, McLean (1967b, 1974) established how this form of bioabrasion could be quantified by the collection of faecal pellets and the analysis of the carbonate material contained within them. The results of such experiments, allied to other estimation techniques, reveal that this group of grazers remove considerably less carbonate substrate than echinoids or fish (Table 8.6). Erosional efficiency in individual amphineurids and gastropods is related to foraging behaviour and especially to the mechanics of feeding. Siphonarian limpets, for example, have a radula with numerous fragile teeth which can only pick at the surface (Creese and Underwood, 1982) whereas the patelliform limpets possess a much stronger radula, taking the form of a self-sharpening chisel which is rapidly and repeatedly replaced (figure 8.7; e.g. Steneck and Watling, 1982; Hughes, 1985). Thus, for example, the radula of the small snail *Littorina neritoides* is not mineralized and grazing-related bioerosion in this species is limited to the removal of carbonate grains already loosened by other agencies, leaving behind a surface of shallow pits (e.g. Schneider and Torunski, 1983, figure 11c; and Trudgill et al., 1987, figure 12). Nevertheless, large numbers of littorinids, reaching densities of 18,000 individuals m^{-2} (e.g. Adriatic coast: WE zone, Schneider, 1976, figure 17) and rapid grazing by these relatively small individuals (Newell et al., 1971) partly accounts for greater biological abrasion towards mean water level. By comparison, *Patella coerulea* grazes slowly and methodically with a radula hardened by goethite and silicate (Lowenstam, 1962; Runham and Thornton, 1967). The result is a linearly-abraded surface (Schneider and Torunski, 1983, figure 11b).

Table 8.5 Sediment processing and substrate removal by grazing, feeding and browsing fish on coral reefs

Species	Location	Population density/fish biomass	Total sediment processed	Production rate or removal of 'new' sediment	Substrate	Author
Scarus croicensis (Striped parrotfish)	Panama (Caribbean)	47.0 kg ha⁻¹	0.98 kg m⁻² yr⁻¹	0.49 kg m⁻² yr⁻¹ (new sediment)	Coral reef	Ogden (1977)
Scarus vetula (Queen parrotfish)	Barbados	36.1 kg ha⁻¹				
Scarus croicensis (Striped parrotfish)						
Scarus taeniopterus (Princess parrotfish)	Barbados	4.3 kg ha⁻¹		34 ± 5 g m⁻² yr⁻¹ (66% of all sediment)	Coral reef	Scoffin et al. (1980)
Sparisoma aurofrenatum (Redband parrotfish)						
Sparisoma viride (Stoplight parrotfish)						
Sparisoma viride	Bermuda		3.0 kg m⁻² yr⁻¹	2.1 kg m⁻² yr⁻¹	Coral reef	Gygi (1975)
Scaridae	Barbados	245–347 fish ha⁻¹ 19.8–36.2 kg ha⁻¹	3.4–5.9 kg m⁻² yr⁻¹	0.4–1.7 kg m⁻² yr⁻¹	Coral reef	Frydl and Stearn (1978)
Scaridae and Acanthuridae (Surgeonfish)	Bermuda			1.3 mm yr⁻¹	Coral reef	Bromley (1978)
Grazing and browsing fish	Bermuda	55 kg ha⁻¹	0.23 kg m⁻² yr⁻¹	0.11 kg m⁻² yr⁻¹	Coral reef	Bardach (1961)
Arothron meleagris (Tetradontidae; pufferfish)	Panama (Pacific)	40 fish ha⁻¹	20.4 g indiv⁻¹ d⁻¹ Pocillopora		Coral reef	Glynn et al. (1972)
Grazing and browsing fish	Saipan Mariana Islands		0.43–0.63 kg m⁻² yr⁻¹		Coral reefs	Cloud (1959)

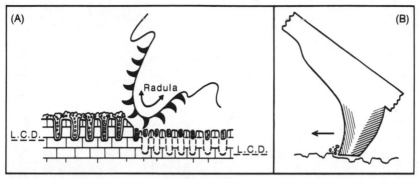

Figure 8.7 Grazing activity and substrate removal. (A) Organisms grazing on epi- and endolithic micro-organisms remove substrate, thereby encouraging endoliths to bore to a new light compensation depth (L.C.D.)
(after Torunski, 1979)
(B) radula tooth of *Patella vulgata*: orientation of crystalline growth of minerals by organic fibres provides self-sharpening for tooth
(after Runham *et al.*, 1969)

On tropical shores, the grazing tracks of the chiton *Acanthozostera (Acanthopleura)* are visible on beachrock, the product of a large, magnetite-strengthened radula. This grazing efficiency is reflected in the rates of carbonate removal achieved by chitons (table 8.6) and by the observation (Trudgill, 1985, figure 9.10) that chiton grooves may reach a depth of 0.5 mm.

Biological abrasion: a summary

Attempts to quantify bioerosional activity are still fragmentary and have yet to reach the stage where broad generalizations about the temporal and spatial patterns of substrate modification by macro-organisms can be made. At present, few rates of bioabrasion are directly comparable referring, within sites, to individuals rather than populations of particular species and, across sites, to varying coastal exposures and lithologies. These shortcomings have been documented in more detail elsewhere (Trudgill, 1985; Spencer, in press). What should be clear, however, is that there is a 'strict trophic relationship' (Schneider and Torunski, 1983, 51) between grazing organisms and endolithic cyanobacteria.

Biological corrosion and biological abrasion: ecological interactions and sedimentological and erosional consequences

This interrelationship between macro- and micro-organisms is well-illustrated at different positions on the shore: the lack of epilithic communities in the deeper zones, and the dominance of endoliths, results from the greater grazing activity at these levels, whereas the reduced

Table 8.6 Quantitative estimates of substrate removal and/or surface lowering by grazing amphineurids and gastropods

Agent	Rate of rock removal	Substrate	Locality	Author
Chiton sp.	8.0 cm^3 yr^{-1}	Beachrock	Barbados	McLean (in Trudgill, 1983b)
Acanthopleura sp.	13.0 cm^3 yr^{-1}	Beachrock	Barbados	McLean (in Trudgill, 1983b)
Acanthozostera gemmata	18.0 cm^3 yr^{-1} chiton^{-1} 0.5 mm yr^{-1}	Beachrock	Heron Is., Great Barrier Reef	McLean (1974)
Acanthozostera gemmata	0.2–2.9 mm yr^{-1}	Boulder rock	One Tree Is., Great Barrier Reef	Trudgill (1983a)
Acanthopleura brevispinosa	5.4 cm^3 yr^{-2} chiton^{-1}	Reef limestones	Aldabra Atoll	Taylor and Way (1976)
Acanthopleura granulata	38.8 cm^3 yr^{-1} chiton^{-1}	Coral rubble	Puerto Rico	Glynn (1973)
Chiton tuberculatus	11.6 cm^3 yr^{-1} chiton^{-1}	Coral rubble	Puerto Rico	Glynn (1973)
Acanthopleura granulata	36.5 cm^3 yr^{-1} chiton^{-1}	Reef limestone		
Acanthopleura granulata	38.6 cm^3 yr^{-1} chiton^{-1}	Beachrock		
Chiton squamosus	34.6 cm^3 yr^{-1} chiton^{-1}	Reef limestone		
Chiton squamosus	34.1 cm^3 yr^{-1} chiton^{-1}	Beachrock	Grand Cayman	Benn (in Spencer and Benn, in press)
Chiton marmoratus	6.8 cm^3 yr^{-1} chiton^{-1}	Reef limestone		
Chiton marmoratus	6.5 cm^3 yr^{-1} chiton^{-1}	Beachrock		
Patella coerulea	0.51–0.76 mm yr^{-1}	Limestone	N. Adriatic	Torunski (1979)
Littorina neritoides	0.07–0.13 mm yr^{-1}			
Littorina ziczac	0.40 cm^3 yr^{-1}	Beachrock	Barbados	Mclean (1967b)
Littorina meleagris	0.15 cm^3 yr^{-1}			
Nodolittorina tuberculata	0.60 cm^3 yr^{-1}			
Littorina planaxis	0.93 cm^3 yr^{-1} 2.5 mm yr^{-1}			
Littorina scutulata	3.8 mm yr^{-1}	Sandstone	California	North (1954)
Cittarium pica	1.30 cm^3 yr^{-1}			
Nerita versicolor	0.80 cm^3 yr^{-1}	Beachrock	Barbados	McLean (1967b)
Nerita tesselata	0.40 cm^3 yr^{-1}			
Acmaea sp.	154 gm^{-2} yr^{-1} 0.99 cm^3 yr^{-1} 1.5 mm yr^{-1} 2.4 g yr^{-1}	Beachrock	Barbados	McLean (in Trudgill, 1983b)
Fissurella sp.	5.0 cm^3 yr^{-1}			

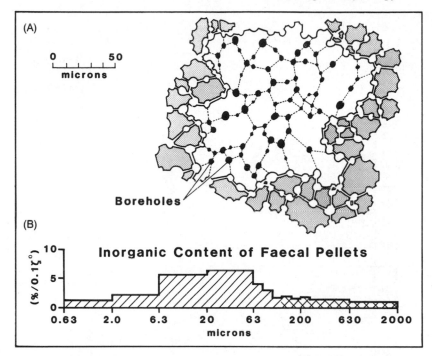

Figure 8.8 Grain size distribution of substrate particles removed by grazers is pre-determined by the pattern of boreholes of endolithic micro-organisms. (A) Lines with closely spaced boreholes act as pre-determined fracture lines (dotted lines); (B) size distribution of carbonate grains contained in faecal pellets of *Patella coerulea, Monodontia turbinata* and *Paracentrotus lividus*, N. Adriatic. Material > 125 mm (cross-hatched) composed of shell detritus and juvenile shells (after Torunski, 1979)

bioerosion from a more limited number of grazers in the upper zones allows for the development of an epilithic mat. By implication, there must be a delicate balance between the grazing behaviour of macro-organisms and the boring activity of micro-organisms. As endolithic boring is substrate penetration-limited by light, some grazing activity is required to rejuvenate cyanobacterial boring to a new compensation depth (figure 8.7). Thus, for example, Golubic and Schneider (1979) have shown experimentally that grazing by the gastropod *Littorina* is necessary for the development of the maximum density of the shallow-boring endolith *Hyella balani*.

Importantly, these ecological interactions have erosional and sedimentological consequences. Examination of rock samples by scanning electron microscopy has shown that endolithic borings are not randomly distributed but organized into intersecting trends, producing lines of preferential fracture and so pre-determining the size of grain which can be removed by a grazing organism (figure 8.8).

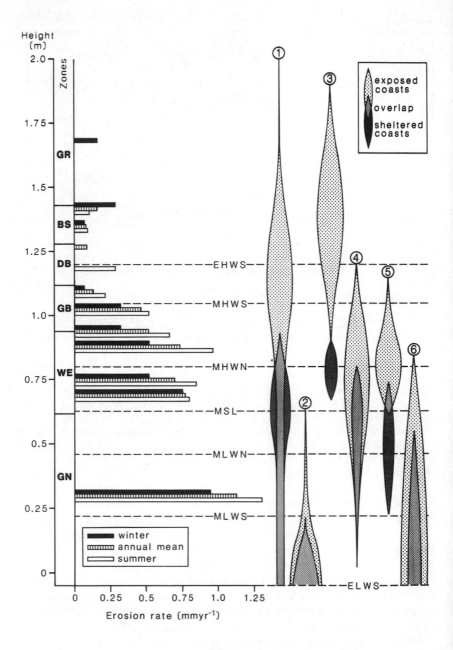

Figure 8.9 Interrelationships between erosion rates (determined by micro-erosion meter), cyanobacteria and macro-organisms with profile position and exposure, N. Adriatic coast (after Torunski, 1979) see figure 8.5 for zone key. 1 Cyanophyceae, 2 *Paracentrotus lividus*, 3 *Littorina neritoides*, 4 *Monodontia turbinata*, 5 *Patella coerulea*, 6 *Cliona* sp.

In terms of total erosion Torunski (1979) has suggested that for the northern Adriatic the balance is of the order 70–90 per cent biological abrasion to 30–10 per cent biological corrosion (figure 8.3). Studies by Trudgill (1976) at Aldabra Atoll, Indian Ocean and by Spencer (1985a) on Grand Cayman, West Indies agree over the negligible contribution of dissolution *sensu stricto* but also identify the process of physical abrasion. This is of varying importance and depends upon local conditions of exposure and position in the littoral profile. In these tropical environments grazing bioerosion accounts for 34 to 64 per cent of total erosion. By extension, within-profile variations in biological abrasion can be used to explain the form of the intertidal notch and to predict the evolution of cliffed limestone shores (figure 8.9: e.g. Trudgill, 1976; Torunski, 1979). Furthermore, changes in the characteristics and distribution of the micro-organic community under different tidal regimes and with varying levels of shoreline exposure may drive different lines of profile evolution. Such models may not be straightforward, however, and in some cases may lead to changes from bioerosion to bioconstruction with coralline and vermetid accretions (Föcke, 1978; Spencer, in press). Nevertheless, it does seem fair to claim, for tropical limestones at least, that biogeomorphological linkages from the small scale to the landform scale can be established and can make a major input to our understanding of rocky shore coastal morphologies.

Biogeomorphology of shallow subtidal sediments: algal mats and marine macrophytes

Sediment stabilization in the littoral zone is strongly mediated by biological influences, particularly in warm, shallow tropical and subtropical waters. Such effects are important at both large and small scales in the benthic boundary layer.

At the micro-level, the presence of mucus secretions – or more strictly mucopolysaccharides (Rhoads and Boyer, 1982) – are of key importance in regulating physical, chemical and biological processes at the sediment-water interface. Mucus secretions are associated not only with filamentous algae (Hommeril and Rioult, 1965; Scoffin, 1970a) and seagrasses (Kikuchi and Pérès, 1977) but also with bacteria (Webb, 1969), meiofauna, especially diatoms (Neumann et al., 1970; Holland et al., 1974; Riemann and Schrage, 1978) and macrofauna (Grenon and Walker, 1980). Besides sediment stabilization, mucus secretions influence the geometry and transport of faecal materials and near-surface decomposition reactions.

Marine macrophytes may also have a protective function near the substrate–water interface. On exposed rocky coasts holdfast-attached macro-algae are dominant while in sheltered localities and/or areas of high fine sediment supply rooted-plant communities are characteristic, with

Table 8.7 Resuspension of sediments and thresholds to resuspension in the presence and absence of diatom communities: flask experiments of Holland et al. (1974)

Experiment	Sediment mixture					
	100% sand		90% sand, 10% silt/clay		100% silt/clay	
	weight[1] (mg)	rpm[2]	weight[1] (mg)	rpm[2]	weight[1] (mg)	rpm[2]
Control - no diatoms	29.5 ± 16.10	160 ± 10	198.1 ± 88.9	100 ± 10	816.0 ± 98.0	100 ± 10
Natural diatom community	15.0 ± 2.0	160 ± 10	15.8 ± 10.0[3]	210 ± 15[3]	20.0 ± 2.0[3]	350 ± 10[3]

[1] Sediment collected after 5 mins at stirrer velocity = 160 rpm
[2] Stirrer velocity at which resuspension commenced
[3] Significant at $p = 0.05$ ('t' test)

saltmarshes and mangrove swamps in the intertidal and immediate supratidal zones and seagrasses in the subtidal environment. In tropical and subtropical seas, mangroves and seagrasses may be intimately related to coral reef constructions and their evolution (e.g. Stoddart, 1980; Woodroffe, 1982).

Shallow subtidal mats

Transport processes in intertidal and subtidal environments can be modified by the presence of biological mats in temperate and tropical waters. The majority of the literature on this subject is, however, highly qualitative in nature and there are few quantitative assessments of mat effects.

Some precise data come from the laboratory experiments of Holland *et al.* (1974) who have shown that sediment stabilization is achieved by diatom communities on temperate intertidal mudflats. Diatoms were cultured in flasks on different sediment mixtures for 12 days and then re-agitated, the degree of sediment binding being measured by the collection of resuspended sediment and by the degree of agitation required for resuspension. It is clear (table 8.7) that, apart from sand substrates where there is an absence of fine particles for resuspension, resuspension volumes were significantly reduced and threshold velocities to resuspension significantly raised when diatoms were present. Furthermore, by isolating individual species Holland *et al.* (1974) were able to show that it is the mucus-secreting diatoms in this surface-film community which prevent particle resuspension.

In tropical waters, a subtidal organic film, or 'algal mat' (Bathurst, 1975, 122) generally covers all stable sand substrates. Subtidal mats in the Bahamas are composed of assemblages of green and red algae, cyanobacteria (blue-green algae), animal-built grain tubes and diatoms. It is the latter which give the mats their characteristic pale brown to pale green colouration (Bathurst, 1967; Neumann *et al.*, 1970). Sedimentary grains are trapped and held in the mucilaginous sheaths of algal filaments (e.g. *Schizothrix* filaments: Bimini lagoon, Bahamas; Scoffin, 1970a) and by diatomaceous mucilages (Frankel and Ward, 1973). While these effects are well known, only Scoffin's (1968, 1970a) experiments with an underwater flume have provided quantitative information on how these organic binding effects are translated into a resistance to erosion: intact natural mats can survive current velocities 2 to 5 times as high as unbound laboratory sediments and 3 to 9 times greater than the maximum tidal currents (13 m s^{-1}) experienced in Caribbean lagoons. Such structures are, therefore, of considerable importance in the sedimentology of shallow marine carbonates.

Stromatolites

If sediment is supplied across a mat surface (although not at such a rate of delivery as to bury or destroy the mat), then ellipsoidal nodules, anchored to the substrate by algal filaments, may develop (Gebelein, 1969). Such forms may be regarded as incipient stromatolites, algal-laminated columnar structures which can grow into complex spheroidal or hemispheroidal forms (for geometrical classification: Logan *et al.*, 1964). The laminations are composed of sedimentary or evaporative grains interlayered with precipitated carbonate and algal filaments dominated by the cyanobacteria (blue-green algae) *Schizothrix*, *Scytonema* and *Entophysalis* (Bathurst, 1975) and locally by the green alga *Batophora* (Dravis, 1983). Diurnal growth rhythms (e.g. Monty, 1965, 1967) including heliotrophic behaviour (Awramik and Vanyo, 1986) tidal fluctuations (Gebelein and Hoffman, 1968) and variations in salinity (Black, 1933) produce alternating sediment-rich/organic-rich laminae. Simple lamination sequences are disrupted by more extreme environmental fluctuations, including periodic emersion and erosion, sediment influx and the collapse and compaction of the internal framework (Golubic, 1983). Uncemented, porous layers become hard limestone in stromatolite interiors. Although within-structure precipitation of carbonate was initially regarded as inorganic (Logan, 1961) subsequent studies have shown an intimate association between the siting of crystal growth, its mineralogy and the arrangement of algal filament sheaths and mucilage (Monty, 1967; Dravis, 1983). However, while algal-related processes may be important in ensuring the permanence of these structures, the accumulation of most marine stromatolites (as opposed to lacustrine and freshwater forms) is dominated by the mechanical trapping and binding of sediment (Black, 1933; Gebelein, 1969; Monty, 1972). Observations in the Bahamas have shown that the 'sticky flypaper' (Dill *et al.*, 1986, 55) surfaces of giant (larger than 2 m), subtidal (8 m below sea level) stromatolites instantly trap oolitic sands carried by tidal currents and so promote rapid surface accretion. The presence of an organic mat is critical: where algal filaments have been killed with formalin, accumulation has ceased (Gebelein and Hoffman, 1968). Stromatolites are both influenced by, and influence, nearshore water movements and sedimentation. In the most remarkable field of modern stromatolites at hypersaline Hamelin Pool, Shark Bay, W. Australia there is a distinctive zonation of stromatolitic types, from flat mats on protected intertidal mudflats to sinuous domes at high tide level. On headlands with strong tidal currents, discrete 'capstans' with overhanging walls are found (Bathurst, 1975). Elsewhere, the distribution of subtidal stromatolites shows a close relation to bathymetry, wave exposure and the pattern of tidal currents (Gebelein, 1969; Gebelein and Hoffman, 1968; Dill *et al.*, 1986).

Beachrock

Biological activity may also be important in the stabilization, through cementation, of intertidal beach sands in tropical and subtropical seas.

Calcium carbonate beaches of coral reef islands commonly support beachrock, a case-hardened, often multiple-generation, seaward-dipping deposit which takes on the characteristics of the intertidal slope on which it develops; cliff-foot erosion blocks, for example, may be cemented into a conglomeratic beachrock (Scoffin, 1970b). Similar lithification may also typify intertidal flats (e.g. Taylor and Illing, 1969), ramparts of coral rubble and reef-flat boulder tracts (Scoffin and Stoddart, 1983). Sand cays protected by such cemented barriers become less mobile and thus more receptive to vegetation establishment, soil development and ecological succession (Stoddart and Steers, 1977). Beachrock is usually composed of sand-sized carbonate grains cemented by micritic coatings or crusts of aragonite and Mg calcite. Explanations for the origins of these cements often involve physiochemical precipitation, either by evaporation of seawater at low tide, or by the mixing of fresh or brackish water with seawater, or by CO_2 de-gassing from seaward-flowing groundwater (Schmalz, 1971; Hanor, 1978; Scoffin and Stoddart, 1983). However, it has also been suggested that carbonate precipitation might be controlled, either directly or indirectly, by biological processes in the intertidal zone. Possible processes include photosynthetic CO_2 uptake by an intertidal microflora, increases in pH associated with denitrifying bacteria and the release of complexes promoting precipitation. Studies on the recolonization of artificially cleared beachrock surfaces have shown that the rapid (less than 6 months) development of a spongy surface crust is indeed accompanied by the concomitant development of a diverse microflora dominated by cyanobacteria (blue-green algae) (Viles and Spencer, 1986), although it has been claimed that the algae only physically trap carbonate particles and do not participate in sand cementation (Davies and Kinsey, 1973). Particularly noteworthy, however, is Krumbein's (1979) study of beachrock formation in the Gulf of Aqaba, Red Sea where rich cyanobacterial and bacterial populations appear to be processing large amounts of organic material. From field observations, beachrock mineralogy and laboratory experiments Krumbein has argued that lithification proceeds by combined biological and inorganic precipitation of carbonates with the decay of organic matter within interstitial pore spaces. Such studies at a micro-environment level, when allied to the emerging syntheses on interstitial biomass and energetics on sandy beaches (for a review see McLachlan, 1985), challenge the pre-conceived notion that beaches are devoid of organic matter (e.g. Milliman, 1974) and identify a need for further research.

Seagrasses

Subtidal environments are also stabilized by macroalgae and marine vegetation. The holdfasts of tropical green algae penetrate several centimetres below the surface and bind lagoon sediments (e.g. *Halimeda* spp.,

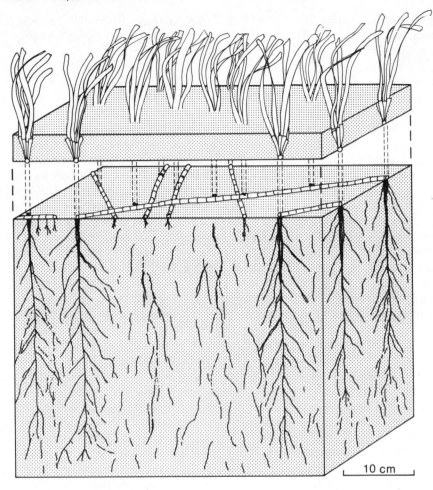

Figure 8.10 Blades and shoots, rhizome network and roots of the seagrass
Thalassia testudinum
(modified from Scoffin, 1970a)

Penicillus spp., *Batophora* spp.; Scoffin, 1970a). Of particular importance
as stabilizers, however, are the marine angiosperms, the seagrasses (for
taxonomy and distribution: den Hartog, 1970; Johnstone, 1982) which
also make a key contribution to nearshore trophic structure and productivity (figure 8.2; McRoy and McMillan, 1977; Phillips and McRoy,
1980). In temperate regions *Zostera marina*, or eelgrass, is the dominant
genus while in tropical and sub-tropical waters the best known genus is
Thalassia testudinum, or turtle grass, which is found in association with
Syringodium and *Halodule*. In the Mediterranean, *Posidonia* and
Cymodocia are important genera.

Figure 8.11 Dense seagrass meadow, Discovery Bay lagoon, N. Coast of Jamaica: water depth 5.0 m
(photograph by T. P. Scoffin)

The high productivity of seagrasses results from their ability to structure their own environments for efficient ecosystem function and maintenance. A high proportion of the annual production of plant tissue appears to be buried almost intact in bottom sediments and provides a bank of slowly decaying organic matter which is processed by bacteria and the meiofauna for detritivore populations (figure 8.2; Harrison and Mann, 1975; Thayer et al., 1975). Studies of nutrient supply within seagrass meadows have shown the importance of phosphorus supply from, and bacteria-mediated nitrogen fixation in, soft bottom sediments (Patriquin, 1972). Thus, for long-term stability, seagrasses must not only retain these nutrients but also encourage within-meadow sedimentation. Biogenic calcium carbonate is added *in situ* from a diverse leaf-encrusting community of epibionts and invertebrates (e.g. Nelsen and Ginsburg, 1986) and the physical structure of seagrass meadows encourages sedimentation.

Patterns of erosion and deposition in seagrass meadows are a function of seagrass blade type and density, tidal current velocities and, in some localities, wave action. *Thalassia* binds the substrate both through a network of overlapping rhizomes just below the sediment surface and by a more deeply-penetrating root system (figure 8.10). Flume experiments by Fonseca et al., (1982) suggest that the baffle effect from the increased surface roughness of seagrass blades (figure 8.11) and shoots results in dramatic reductions in near-bed current velocities (initial current reduction = $1.0 \, \text{cm s}^{-1}$ per $1.35 \, \text{cm}$ of *Zostera* meadow; peak current reduction = 1.0 cm s^{-1} per 2.07 cm meadow). Under such conditions deposition of suspended sediment is encouraged and seagrass beds tend to be composed of finer, more-poorly-sorted sediments than adjacent non-vegetated areas (for summary see Burrell and Schubel, 1977). Patterns of erosion are more complex. For *Thalassia*, the threshold to rhizome exposure and undercutting is a current velocity of $\simeq 50$ cm s^{-1} in areas of sparse seagrass (shoot spacing = 10–100 cm) but this threshold rises to 100 cm s^{-1} in areas of medium seagrass density (spacing = 10 cm) and to 150 cm s^{-1} where dense seagrass meadows (spacing < 2 cm) are encountered (Scoffin, 1970a). This is because seagrasses not only influence but also are influenced by current flow: even at low (below 20 cm s^{-1}) current velocities seagrass blades bend easily in order to reduce exposed surface area and to minimize drag forces and internal blade stress. Under increasing current flow, individual shoots leave the vertical to form a low-angle intermeshing canopy (Fonseca *et al.*, 1982). Thus while flow at low seagrass densities is deflected towards the bed, causing scour around individual shoots, in dense seagrass meadows there is a redirection of current flow upwards and across the top of the seagrass canopy. This effect is magnified at higher (above 40 cm s^{-1}) current velocities: in experimental flumes, bladed seagrasses (*Zostera marina*, *Halodule wrightii* and *Thalassia testudinum*) show a reduction in canopy friction with increasing current velocity because the canopy height declines with the progressive bending of individual seagrass blades (Fonseca and Fisher, 1986). Overall, therefore, laboratory experimentation suggests that seagrass meadows promote higher shear velocities (change in velocity per change in water depth) than non-vegetated areas (e.g. Fonseca *et al.*, 1982). Such a relationship has been observed in shallow water embayments characterized by low maximum current velocities (e.g. *Zostera marina* meadows at Beaufort, N. Carolina, USA, with $U_{max} \simeq 45$ cm s^{-1}; Fonseca *et al.*, 1982).

Such structural arrangements are, however, not always typical. First, they are only characteristic of pliant, wide-bladed seagrasses: the more rigid, cylindrically-bladed forms, like *Syringodium filiforme*, offer less substrate protection (Fonseca and Fisher, 1986). Secondly, spring tides and wave action may counter canopy layering and increase the likelihood of sediment resuspension between individual shoots (Ward *et al.*, 1984). Thirdly, and most fundamentally, at higher current flows (e.g. for Fonseca *et al.*, 1983: $U_{max} > 53$ cm s^{-1}) the vegetation effect is overridden and the relationships reversed: current flow dynamics structure both meadow topography and the continuity of seagrass cover. In addition, storm waves and high orbital velocities ($U_{max} > 100$ cm s^{-1}) generate 'blow-outs' in Caribbean back-reef lagoons. These are crescent-shaped, grass-free hollows with a long axis parallel to dominant wave approach and a 15–16 cm high convex scarp to seaward (e.g. Ball *et al.*, 1967; Thomas *et al.*, 1961). Under normal wave conditions (bottom orbital velocities = $U_{max} = 50$–100 cm s^{-1}) on

Barbados and in The Grenadines such blow-outs migrate seaward over a lag deposit of coral rubble such that lagoon floors are eroded and then restabilized over a 5–15 year cycle (Patriquin, 1975).

In spite of these provisos, the sediment stabilizing role of seagrasses was well illustrated in the 1930s when the die-back, or 'wasting disease' of the major temperate seagrass *Zostera marina* on N. American and European coasts resulted in the mobilization of nearshore sediments and coastline changes (e.g. Denmark: Rasmussen, 1977; Christiansen *et al.*, 1981; Devon, UK: Wilson, 1949). More recently, the possibilities of transplanting tropical seagrasses, notably *Thalassia testudinum* (turtle grass), to provide a biological means of controlling subtidal erosion following dredging have been investigated (e.g. Fonseca and Fisher, 1986).

Biogeomorphology of shallow subtidal sediments: sediment communities

Macroinvertebrates are able to produce large changes in the physical and chemical properties of marine soft sediments and there has been a great deal of work on animal–sediment relations in recent years. These studies show, however, that such relations are rarely straightforward. On a practical level, conventional bottom samplers not only produce the spatial-averaging of sediment characteristics through bulk sampling but also disturb or destroy near-surface structures and the products of particle reworking. Subsequent laboratory analyses compound these problems by generally removing the important organic bindings between sedimentary particles. More fundamentally, biology, chemistry and sediments often form a nexus of interlocking processes. This arrangement can be illustrated by three examples. Small (1–3 mm diameter) burrowers' tubes introduce isolated roughness elements into near-bottom velocity fields (e.g. Nowell and Church, 1979; Eckman *et al.*, 1981) and allow for improved local exchange, particularly of oxygen, across the sediment–water interface (Jumars and Nowell, 1984). As a result aerobic bacterial growth may be stimulated and meiofauna (e.g. Harpacticoid copepods: Thistle *et al.*, 1984) attracted, leading to the development of sediment binding by mucus secretions and raised thresholds to the entrainment of bottom sediments. This example shows that processes promoting both sediment stabilization and destabilization may be in operation at the same time. Thus, for example, gastropod tracks, through increasing the small-scale bottom roughness, may lower the threshold to sediment transport, while at the same time mucus secretions produced during gastropod locomotion may bind sedimentary particles and so decrease the probability of sediment entrainment (e.g. *Transenella tantilla*: Nowell *et al.*, 1981). In addition these interrelated physical and chemical changes which take place on the seafloor feed back into ecological consequences. Rhoads and Young (1970)

have discussed 'trophic group amenalism' in the incompatibility of deposit-
and suspension-feeding populations: intensive particle re-working by
deposit-feeders creates an easily resuspended, unstable sediment–water
interface where the delicate filtering structures of suspension feeders
become overloaded by excessive sedimentation (e.g. silt–clay substrates:
Buzzards Bay, USA: Rhoads and Young, 1970; Cape Cod, USA: Rhoads
and Young, 1971; sand substrate: Rhode Is., USA: Myers, 1977).

In spite of these complexities, some generalizations about the role of
the sediment epifauna and infauna can be made. Invertebrates can trap
and bind sediments and encourage sedimentation (e.g. Yingst and Rhoads,
1978; Nowell et al., 1981; Eckman et al., 1981). Alternatively, sediments
can also be destabilized by biological activity (e.g. Rhoads et al., 1978;
Nowell et al., 1981; Eckman et al., 1981; Grant et al., 1982). There are
also more indirect biologically-mediated controls on sediment transport.
First, animals may vertically segregate different particle sizes in the near-
bed sediment column during feeding and burrow construction and thus
alter the availability of particular grain sizes for transport. Secondly,
benthic organisms may redefine modal particle size through the creation
of faecal pellets and by the secretion of mucus which coats pellets and
sedimentary particles and binds them to one another. Thirdly, sediment
entrainment may be influenced by organism-determined alterations to the
mass or bulk geotechnical properties of sediments. These observed effects
should be set against the commonly used models of sediment transport,
such as the Yalin bedload equation and the Einstein bedload transport
model (e.g. see Smith, 1977) which rarely make any allowance for the
presence of a bottom-dwelling fauna. Indeed, hydraulic engineers generally
exclude biological activity from flume channels by the addition of growth
inhibitors to recirculating water. It is not surprising, therefore, that there
can be large discrepancies between the actual behaviour of natural
sediments and that suggested by 'abiotic' competency diagrams.

Stabilization: animal tube fields

At the meiofaunal level, a common biologically-produced roughness
element is that of fields of the tubes of polyhaetes and amphipod
crustaceans which project a few millimetres above the bed. In physically
disturbed environments, transient aggregations of tubes may rapidly reach
densities of 10^5 m^{-2} (McCall, 1977). As tube fields are often associated
with a raised bottom surface and the accumulation of fine sediments and
organic matter between the tubes (e.g. Lynch and Harrison, 1970;
Featherstone and Risk, 1977; Myers, 1977; Bailey-Brock, 1979, 1984)
it has been generally assumed that tube fields stabilize seafloor sediments.
Yet flume experiments to test this assertion have produced non-consistent
results, even when the same species of tube-builder has been used (e.g.
the polychaete Owenia fusiformis: Rhoads et al., 1978). There appear to

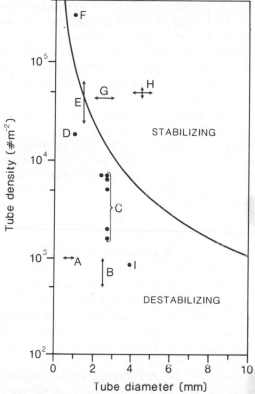

Figure 8.12   Relationship between tube diameter and tube spacing in determining noncohesive bed stability
(after Eckman *et al.*, 1981 and Rhoads and Boyer, 1982)
A   oligochaetes, B   *Owenia fusiformis*, C   *O. fusiformis*, D   *Steblospio benedict*,
E   chaetopterid polychaetes, F   *Corophium insidiosum*, G   *Ampelisca abdita*,
H   *Owenia fusiformis*, I   *Microdentopus gryllotolpa*

be two reasons for these results. First, as with seagrass blades, the spatial arrangement of specific-sized tubes is critical: at low tube densities (i.e. 1:22 ratio of unit area tubes to unit area of flume bottom) turbulent vortices are shed around each individual tube and transferred to the bed (Eckman and Nowell, 1984), whereas at high densities (i.e. 1:12–1:8) maximum turbulence is elevated away from the bed to near the top of the tube field (Rhoads and Boyer 1982; 'skimming flow' *sensu* Morris, 1955). Eckman *et al.* (1981) have shown that natural densities of tubiculous species, controlling for tube diameter, span the threshold between stable and unstable beds (figure 8.12). However, these experiments need to be confirmed by investigators at the densities and diameters of tubes predicted to give a stable bed. Second, even if the tube density falls to a level where bed scour is predicted, the bed may remain intact as a result of the presence of sediment-binding mucus secretions. The combination of tubes and exudates is not coincidental: micro-organism productivity, for example, is enhanced by the increased supply of nutrients pumped across the sediment–water inteface by tubes (Eckman, 1985; Jumars and Nowell, 1984; for complications: Luckenbach, 1986; Palmer, 1986).

At a larger scale, pioneer communities may also be characterized by opportunistic bivalve molluscs which reach densities of 10^3–10^4 m^{-2} (McCall, 1977). On death, the disarticulated valves of these colonists accumulate *in situ* and protect the sediment–water interface. In temperate estuaries and saltmarsh creeks, pavements and banks of oyster shells (*Crassotrea virginica*) similarly armour channel sides and marsh edges and trap fine sediment (e.g. Sapelo Island, USA: Salazar-Jimenez *et al.*, 1982; Howard and Frey, 1985).

Sediment destabilization by invertebrates

Benthic organisms directly destabilize natural substrates by varying the boundary properties of the flow through an increase in the surface roughness of the bed. Laboratory experiments with both animal tubes (e.g. Eckman *et al.*, 1981) and clam tracks (e.g. Nowell *et al.*, 1981) have shown the generation of a surface micro-relief and a lowering of the critical entrainment velocity. The macrofauna also alters the probability of sediment entrainment more indirectly as a result of feeding and burrowing activity.

The feeding strategies of most benthic invertebrates are based either on suspended organic particles (both living and non-living) and inorganic detritus or on deposited detritus. Suspension-feeders trap mobile food resources on ciliate- or mucus-covered surfaces, aggregate the ingested non-food materials in the gut and void them as part of discrete pellets or faecal strings. Deposit-feeders, either at the surface or at a certain depth below it, similarly achieve particle aggregation in the process of faecal pellet production; however, they are capable of ingesting larger particle sizes than suspension-feeders and are thought to be quantitatively more important in 'pelletizing' sea floor sediments (Rhoads, 1967). Styles of sediment re-working are shown in figure 8.13 and quantitative rates in table 8.8.

The actual pattern of re-working can be seen as describing a series of 'advective loops' (Aller, 1982, 57) to a characteristic depth around individual deposit-feeders. Such simple patterns are, however, complicated by the presence of individuals of different sizes or at different stages of the life cycle; by the mobility of individual feeders; by local burrow infilling; and by the interaction with solely surface deposit-feeders. In fact, for many deposit-feeders the notion of homogeneous, random sediment mixing is misleading. A common feeding pattern involves the preferential ingestion of fine material at depth (leaving pockets of coarse particles) and its

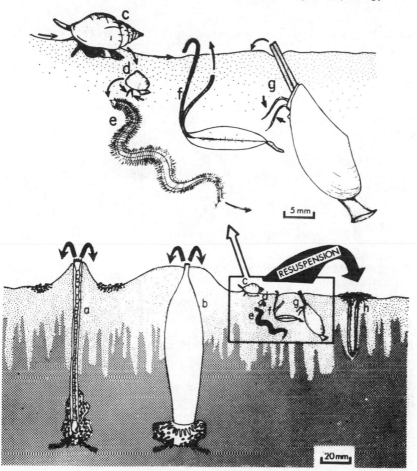

Figure 8.13 Methods of mixing and recycling of sediment by deposit-feeders: (A) maldanid polychaete, (B) holothurian, (C) gastropod (*Nassarius*), (D) nuclid bivalve (*Nucula* sp.), (E) errant polychaete, (F) tellinid bivalve (*Macoma* sp.), (G) nuclid bivalve (*Yoldia* sp.), (H) anemone (*Cerianthus* sp.). Oxidized mud lightly stippled, reduced mud densely stippled
(from Rhoads, 1974)

subsequent ejection at the surface. Such behaviour is characteristic of many polychaete worms (the 'conveyor belt' species of Rhoads (1974)) whose vertically-orientated tubes function not only as efficient sediment pathways but also as sediment trapping and binding structures at the surface (e.g. *Arenicola marina*: Wells, 1945; Baumfalk, 1979; figure 8.14). Other organisms selectively remove coarser material from the surface or defecate fines at depth. Particular particle sizes deposited at the surface by one organism may be removed by neighbouring species, either directly or from

Table 8.8   Some rates of sediment re-working by macroinvertebrates

Species	Habitat and feeding strategy[1]	Burrow depth (cm)	Sediment transport (g indiv.$^{-1}$ d^{-1})	Density (indiv. m^{-2})
Bivalves				
Macoma baltica	IMD	4-6	0.043	100-1000
Yoldia limatula	IMD	2-4	0.28	10-100
Nucula annulata	IMD	1-2	0.0033	100-1000
Polychaetes				
Pectinaria gouldii	IMD	1-6	1.6	10
Arenicola marina	IMD	15-60	4.5	10-100
Clymenella torquata	ISD	20-30	0.9	100-1000
Amphitrite ornata	ISD	30	4.5	1-10
Holothurians				
Leptosynapta tenuis	IMD	10	5-14	10-100
Parastichopus parvimensis	EMD	0	22.0	0.4
Holothuria atra	EMD	0	86.5	0.44
Holothuria vitiensis	EMD	0	73.0	0.072
Holothuria arenicola	IMD	15-20	105.0	1.2
Paracaudina chilensis	IMD	20	158.0	1-10

[1] I = infaunal; E = epifaunal; M = mobile; S = sedentary; D = deposit-feeder
Source: Aller in Carney, 1981

resuspensions (e.g. Powell, 1977; Aller and Yingst, 1978; for review see Aller, 1982).

Whatever the exact pattern of cycling, it is clear that the formation and deposition of faecal pellets may both change the grain size-frequency distribution of bottom sediments (e.g. holothurians: figure 8.15; Massin and Doumen, 1986) and vertically and horizontally segregate particular grain-size classes. Detritus-feeders tend to target sediments with a high proportion of particles in the silt–clay range (less than 0.0625 mm in diameter) as potential food sources: the high surface-area to volume ratios of these fine particle sizes provide an attractive micro-environment for the attachment and growth of algal, diatom and microbial communities (Dale 1974; figure 8.16). By comparison, most faecal pellets fall within the fine sand size fraction (0.125–0.25 mm in diameter). Over time, therefore, pelletization should convert a well-sorted, dominantly fine-grained deposit, through a bimodal stage with pellets and fine silt–clay, to an accumulation of sand-sized pellets. Rates of 'biodeposition' are documented by Haven and Morales-Alamo (1972), Kraeuter (1976) and Smith and Frey (1985).

Some pellets move as discrete particles. Risk and Moffat (1977) have described the flood tide transport of the pellets and pseudofaeces of *Macoma balthica* in Nova Scotia and Wanless *et al.* (1981) have studied the hydrodynamics of carbonate faecal pellets in S. Florida and the Bahamas. In

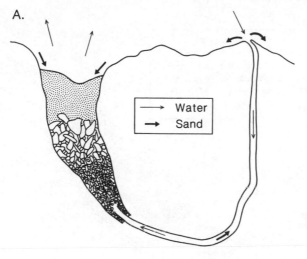

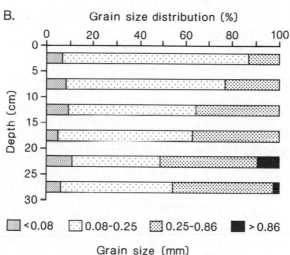

Figure 8.14 Feed strategy and segregation of particle sizes: (A) movements of water and sand as a result of feeding by the lugworm, *Arenicola marina* (worm lives at lowest point of burrow)
(after Wells, 1945)
(B) vertical grain size distribution from *Arenicola* tidal flats, Dutch Wadden Sea
(after Baumfalk, 1979)

the latter case, considerable variability in settling velocities was observed: 'hardened' pellets, characterized by internal, intragranular micritization, settled at rates similar to those expected of same diameter quartz spheres but 'soft' pellets, particularly when containing fine aragonite needle mud, settled at rates equivalent to quartz spheres of 10–25 per cent the diameter

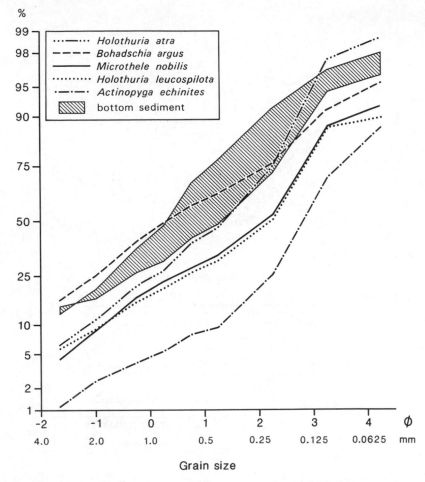

Figure 8.15 Particle size distribution (cumulative percentage, probability ordinate) of gut sediment from five holothurians compared to bottom sediment, Laing Island reef flat, Papua New Guinea
(from data in Massin and Doumen, 1986)
Sediment processing may result in a finer particle size distribution

of the pellets, thus explaining the wide dispersal of these pellets across shallow carbonate banks (figure 8.17). Similar or slower settling rates are characteristic of large pellets with high water contents and levels of organic matter but a low silt-sized mineral grain component (e.g. Haven and Morales-Alamo, 1968). Many of these pellets support transport theories: they suggest that faecal pellets should be more easily transported than unpelletized sediment as the critical shear velocity for fine sand entrainment is lower than that for the cohesive silt–clay fraction. Flume experiments by Nowell *et al.* (1981) have, however, produced conflicting results.

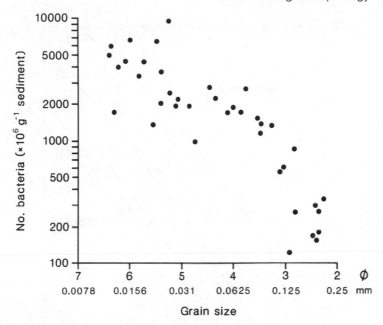

Figure 8.16 Relationship between bacterial numbers per grain of sediment to mean grain diameter
(after Dale, 1974)

On the one hand, 1 mm diameter pellets of the polychaete *Amphicteis scaphobranchiata* were found to be transported as bedload at a bed shear velocity of $1.8\ \mathrm{cm\ s^{-1}}$ whereas the critical erosion velocity for fine silt suspension is $6\ \mathrm{cm\ s^{-1}}$. On the other hand, pellets on sandy silt to fine sand substrates were not entrained before the ambient sediment. In this case it appears that faecal mounds are held together, and bound to the bed, by mucus secretions. Such biological bindings, established over periods as short as three days on initially sterile glass beads, have been shown to increase critical erosion velocities by over 60 per cent (figure 8.18; Rhoads *et al.*, 1978). Under such conditions, there appear to be three velocity thresholds to transport: first, intermittent shaking of faecal mounds; secondly the initiation of sediment motion; and thirdly, the removal of faecal accumulations in the form of large aggregates, 1–2 mm across, or as fibrous strings of mucus-bound particles (Nowell *et al.*, 1981). Subsequent transport is then complicated by invertebrate-faecal-pellet–sediment-transport interactions as classically demonstrated by Newell (1965); recently voided pellets of little nutritional value are likely to be removed intact from the feeding area whereas older pellets, recolonized by micro-organisms but with degraded mucus bindings, are prone to local disintegration, providing a food source of suitably-sized particles to a deposit-feeder (Taghon *et al.*, 1984; see also Hargrave, 1976).

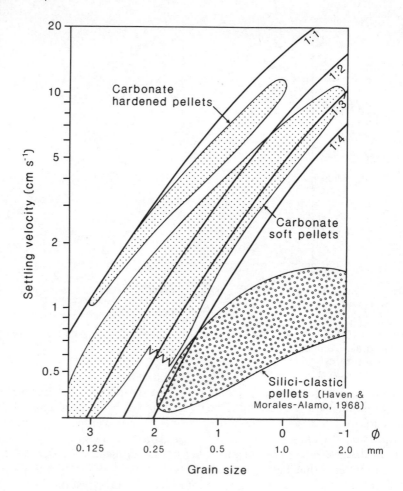

Figure 8.17   Settling behaviour of carbonate pellets and silici-clastic pellets relative to the settling velocity of quartz spheres
(after Wanless *et al.*, 1981)

Benthic organisms and the mass properties of sediments

As well as controlling sediment sizes and their distribution, bioturbating organisms also exert a strong influence on sediment water content, inter-particle porosity and sediment cohesion. These effects are greater in silts and clays which compact less readily than cohesionless sands.

Micro-profiles of water content obtained by freeze-core techniques show that high levels are associated with re-worked layers (figure 8.19). Intensely bioturbated fine-grained sediments exhibit water contents in excess

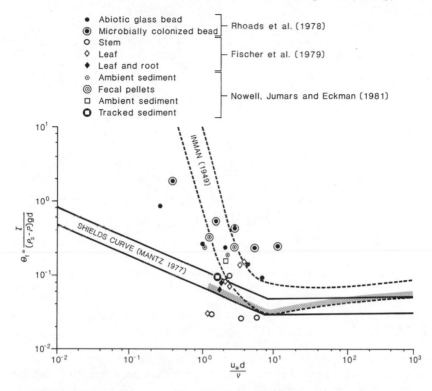

Figure 8.18 Competency diagram showing the effects of faecal pellets, tracks and micro-organisms on critical threshold velocities
(from Nowell *et al.*, 1981)
Shields diagram with non-dimensional shear stress plotted against boundary Reynolds number. Hatched area from Shields; solid lines = extensions from Mantz (1977). Dashed lines = calculation of dimensionless shear and Reynolds number for quartz-density material at 20°C
(Inman, 1949)

of 50 per cent and commonly over 70 per cent. High water contents are further reflected in the low bulk strength properties of the uppermost levels of sublittoral muds and in the susceptibility of bottom sediments to erosion. Inverse correlations between erosion resistance and sediment water content have been reported by several authors (e.g. Postma, 1967; Southard *et al.*, 1971; Lonsdale and Southard, 1974) and Young and Southard (1978) have shown that threshold bottom shear velocities can be decreased on muddy substrates from 12 cm s^{-1} to 5 cm s^{-1} with the introduction and establishment of deposit feeders (and see also Grant *et al.*, 1982). Seasonal changes in sediment water content can be explained by the annual, temperature-controlled cycle of macrofaunal activity (Rhoads, 1970). In Long Island Sound, where the annual range of bottom sediment

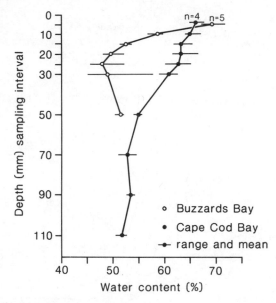

Figure 8.19 Water content profiles in mud and the bioturbation layer
(from Rhoads, 1974)

temperature exceeds 15°C, these changes result in the bottom sediments
being most easily eroded in the spring and early summer and most resistant
to erosion in winter (figure 8.20; Rhoads *et al.*, 1978). Similar controls
on estuarine mudflat levels, largely through seasonal algal growth, have
been quantified by Frostick and McCave (1979).

Benthic organisms and the
chemical properties of marine sediments

The sediment–water interface is not only biologically active but also
chemically reactive. The metabolism of organisms in the benthic boundary
layer gives rise to an oxygen demand, but dissolved oxygen from the surface
penetrates depths of only centimetres (on coarse sand surf beaches) or even
millimetres (fine-grained sediments). Beneath the oxidized layer, the
sediment is anaerobic and chemically reducing. Here, most of the
important reactions taking place are associated with the decomposition
of organic matter. It is generally assumed that these decomposition
reactions are vertically stratified below the sediment–water interface (figure
8.21). Under such conditions, and in the absence of 'biological' or physical
disturbance, solutes consumed or generated within the sediment column
diffuse along concentration gradients between an upper, well-mixed water
column reservoir and within-sediment diagenetic reactions. Models of this
type, using one-dimensional transport-reaction equations, have been widely

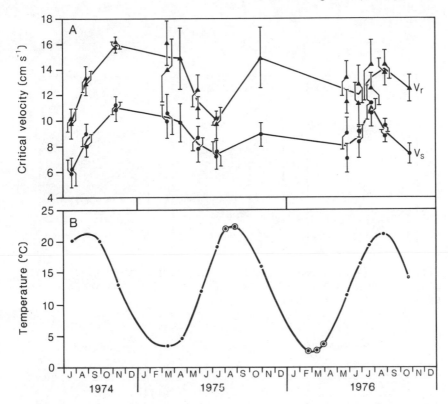

Figure 8.20 Seasonal changes in seafloor erodibility, central Long Island Sound: (A) seasonal change in mean critical rolling velocity (V_r) and mean suspension velocity (V_s), vertical bars = one SD; (B) seasonal change in bottom sediment temperature
(from Rhoads *et al.*, 1978)

applied (e.g. Lerman, 1979; Berner, 1980). Bioturbation clearly upsets this simple pattern and the presence of burrows and faecal pellets generates a complex mosaic of biogeochemical microenvironments (figure 8.21).

Where tubes and burrows are present, interstitial solutes can diffuse laterally into these cavities as well as vertically towards the sediment–water interface (and the converse is true for solutes whose source is the overlying water column). The addition of these pathways to the classic models (e.g. Aller, 1980a) suggests transport rates 10 to 100 times greater than in sediment undisturbed by macrofaunal activity (e.g. with burrows: effective solute diffusion coefficient, $D_s' \approx 10^{-5}$ to $\approx 10^{-4}$ cm s^{-1}; no burrows: $D_s \approx 10^{-6}$ to $\approx 10^{-5}$ cm s^{-1}; Aller, 1982) and a much closer correspondence between model predictions and measured concentration profiles (figure 8.22). These models also show that concentrations of pore

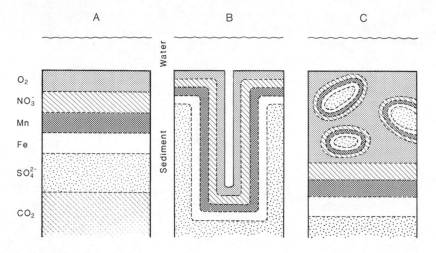

Figure 8.21 (A) Classically assumed vertical zonation in sediments; (B) reaction zonation around irrigated burrow micro environments; (C) reaction geometries associated with faecal pellet micro-environments
(from Aller, 1982)

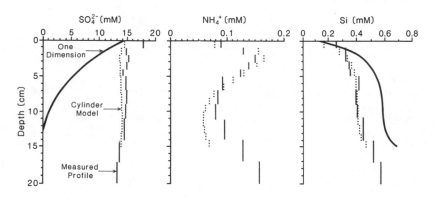

Figure 8.22 SO_4^{2-}, NH_4^+ and Si concentration profiles in pore water, Mud Bay, S. Carolina. Note improved fit of cylinder model (modelling presence of burrows) over one-dimensional model. The one-dimensional NH_4^+ profile is off scale and not plotted
(from Aller, 1980a)

water solutes, and the magnitude of solute flux, are not only related to reaction rate (itself stimulated by feeding activity and particle re-working) but also to the size and abundance of the burrowing organisms (figure 8.23; Aller, 1980a; Aller and Yingst, 1985). Not surprisingly, therefore, early opportunistic colonizers of chemically hostile substrates crowd together in patches of high abundance (Aller, 1980b).

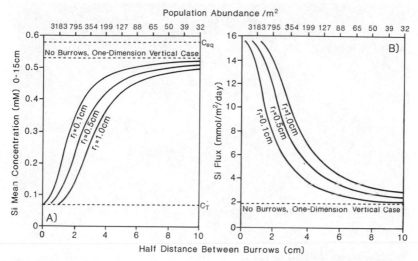

Figure 8.23 Concentration and flux of solutes and the size and abundance of infauna: (A) Expected average concentration of Si in pore water from 0–15 cm as a function of burrow size (r_1) and abundance, C_{eq} = apparent solubility; (B) the total flux of dissolved silica from sediment as a function of burrow size (r_1) and abundance
(from Aller, 1980a)

These patterns may be further complicated by variations in the permeability of burrow linings as solute concentrations build up around relatively impermeable burrow walls. Interestingly, however, it appears that different solute groups have very different diffusive permeabilities: NH_4 and $Si(OH)_4$, for example, respond to the presence of burrow walls, whereas humic acids are distributed as if burrow structures were absent from sediments (Aller, 1983). The burrow lining may also be the site of important chemical–microbiological interactions. Thus, for example, King (1986) has shown that the secretion of a bromophenol (2, 4-dibromophenol or DBP) in the burrow-lining mucus of the hemicordate *Saccoglossus kowalewskii* inhibits the microbial processes which would normally attack the burrow lining and encourages the deposition of a 2 mm thick layer of iron oxyhydroxides on the burrow walls, thereby increasing the longevity of the burrow and maintaining the presence of burrows within sediments.

Applied coastal biogeomorphology:
the case of *Callianassa*

In concluding this account, it is useful to take the sediment-processing activities of callianassid shrimps as a case study. *Callianassa* illustrates not

Figure 8.24 *Callianassa* mounds, Davies Reef, Great Barrier Reef, Australia: water depth 20 m, knife = 30 cm
(photograph by T. P. Scoffin)

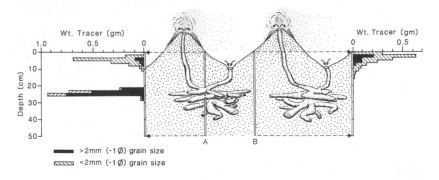

Figure 8.25 Sediment re-working of callianassid shrimps: expected distribution of original surface tracer as a result of burrowing activity
(after Tudhope and Scoffin, 1984; burrow geometry after Shinn, 1968)

only many of the processes of sediment modification, and their ecological consequences, discussed in this chapter but also the applied value of bioturbation studies.

Tropical lagoon sands are often characterized by remarkable fields of episodically-active volcano-like mounds, up to 30 cm high and reaching densities of 10 mounds m^{-2} (figure 8.24; e.g. Florida reef tract: Shinn, 1968; Pacific Ocean: de Vaugelas et al., 1986; Indian Ocean: Farrow, 1972, Braithwaite and Talbot, 1972; Red Sea: de Vaugelas and de Saint Laurent, 1984). These mounds, and the funnel depressions between them, are the surface expression of the deposit-feeding decapod crustacean Callianassa. Callianassids strip organic matter from sediments which arrive at the burrow entrances in the depressions and, in doing so, actively sort the different grain sizes. Field excavation and resin casting techniques have revealed that the gravel fraction and gravel-grade disarticulated bivalve shells are stacked in a system of indurated ramifying chambers, often about 50 cm below the sediment–water interface but exceptionally to a depth of 2.0 m (Shinn, 1968; Tudhope and Scoffin, 1984; Suchanek, 1985). On Enewetak Atoll, introduced Halimeda flakes, 4.0–5.8 mm in diameter, were found in such storage galleries within 24 hours of introduction at burrow entrances (Suchanek et al., 1986). By comparison, the finer (1.0–2.0 mm in diameter) sediments are expelled to the mound apices and into the water column by an episodic pumping action; effluent volumes of 0.46–1.95 and 1.55–4.45 l $burrow^{-1}$ day^{-1} have been recorded at Enewetak Atoll and Bikini Atoll respectively (Colin et al., 1986). Seeding experiments on St. Croix, US Virgin Islands, have shown that sediments finer than 0.4 mm are preferentially expelled, with a modal grain size of 0.13 mm (Suchanek, 1983). The larger sediments are deposited almost immediately and contribute to mound-building, but finer-grained sediments are carried in suspension away from the mound top. This is a significant transport process, particularly in environments of generally low energy. Studies on St. Croix, for example, have demonstrated that 216 kg sediment day^{-1} can be fluxed through a shallow back-reef lagoon in this way and under ambient current flows (i.e. $5–15 cm s^{-1}$) which would not normally entrain sand-sized particles (Roberts et al., 1982). In addition, mound sand may be resuspended after deposition: mean bottom resuspension rates of 19 mg cm^{-2} day^{-1} have been reported from Jamaica (Aller and Dodge, 1974). The result of Callianassa bioturbation is, therefore, a well-sorted, virtually gravel-free upper layer overlying a patchy distribution of sub-surface gravel (figure 8.25; Miller, 1984; Tudhope and Scoffin, 1984). The maintenance of the upper layer is, however, highly dynamic and rates of sediment re-working can be considerable (table 8.9). While the rapid burial of the gravel-sized fraction leads to good preservation, the repeatedly reworked sand-grade sediments are subjected to an increased rate of breakdown from microborer infestation (Tudhope and Scoffin, 1984). Furthermore, the continued mobility of the surface sediments severely limits the occurrence of algal mats, prevents the development of seagrass beds (Roberts et al., 1982; Suchanek, 1983) and Halimeda meadows (Colin, 1986) and limits colonization by burrowing organisms (e.g. Brenchley, 1982; Posey, 1986).

Table 8.9 Sediment processing by callianassid shrimps in tropical and sub-tropical lagoons

Species	Density (indiv. m^{-2})	Total sediment processed (kg m^{-2} d^{-1})	Substrate	Location	Author
Callianassa spp.	6.7 ± 2.4 2.1 ± 1.4	3.395 0.819	Coarse-fine sand Seagrass bed	St. Croix, Caribbean	Roberts et al. (1982)
Callianassa rathbunae	5.0–7.0 0.02–0.15	2.590 0.004	Coarse-fine sand Seagrass bed	St. Croix, Caribbean	Suchanek (1983)
Callianassa laurae	0.2–0.3 0.2–0.3	0.009 0.005	Sand Seagrass	Gulf of Aqaba Red Sea	de Vaugelas and de Saint Laurent (1984) and de Vaugelas (1985)
Callichirus armatus	0.4–0.5 2.0–3.0	0.019–0.50 0.115–0.27	Fine sand Carbonate mud	Tahiti Mataiva Atoll, N. W. Tuamotus	de Vaugelas et al. (1986)
Callianassids (various species)	0.1 ± 0.1 (3 m) –54.8 ± 2.1 (17 m) for large mounds (>5 cm high)	240.1 ± 206.7 cm^3 m^{-2} d^{-1} for large mounds (>5 cm high); 56.0 ± 52.9 cm^3 m^{-2} d^{-1} for small mounds (<5 cm high)	Medium sand	Enewetak Atoll Marshall Is.	Suchanek and Colin (1986) and Suchanek et al. (1986)

As a result of 43 nuclear weapons tests between 1948 and 1958 the atoll rim islands and extensive areas of the lagoon floor at Enewetak Atoll, northern Marshall Islands, were severely contaminated by a diverse suite of fallout radionuclides. Subsequent radiological surveys and associated clean-up campaigns have viewed the lagoon as having been a passive, featureless collecting dish for these radionuclides and have only considered burial of fission products to a depth of 20 cm. Recent sediment coring to – 2.0 m below the lagoon surface has, however, indicated considerable radionuclide concentrations below the top 20 cm and greatly in excess of the concentrations at the lagoon surface and in intertidal and subaerial environments (table 8.10; McMurtry et al., 1985, 1986; Suchanek et al., 1986). In particular, a heavily contaminated layer of fine-grained carbonate containing a radioactive iron-rich (magnetite and pyrite) non-carbonate residue has been encountered at a depth of 48–66 cm. Cores have shown that callianassid burrows intersect this layer and are associated with anomalously high levels of radioactivity. This may indicate bio-accumulation of radionuclides: indurated burrow walls are rich in humic substances which are well known for their absorptive capacities for heavy metals and transuranic elements (de Vaugelas, 1985). Modelling and coastal sediment studies (e.g. Santschi et al., 1983) have shown that radionuclide transport is controlled by molecular diffusion through sediment pore waters and that, as predicted (Aller, 1980a; and see this chapter) this transport is enhanced by the presence of burrows. Pore water advection by callianassid burrowing and pumping has been demonstrated from Bermuda (Waslenchuk et al., 1983) and it seems certain that shrimp bioturbation enhances the mobility of at least the more soluble radio-nuclides (e.g. ^{137}Cs). The presence of callianassid burrows probably also promotes the initial release of radionuclides absorbed into the fallout particles at Enewetak. Oxidation of the magnetite and pyrite particles by relatively oxygenated pore waters introduced by the animals has been suggested as one possible means of radionuclide release (e.g. McMurtry et al., 1986).

These findings have considerable implications for both the atoll radionuclide inventory and its future mobilization. New evidence for bioturbation to depths of at least 1.5 m (Suchanek et al., 1986) requires a re-calculation of inventory size; it seems likely that previous estimates have been 3 to 10 times too low (McMurtry et al., 1985). Furthermore, former calculations, based on sediment-surface–water-column reactions, predicted that 50 per cent of the present inventory of $^{239+240}Pu$ would have been remobilized in solution and discharged from the lagoon over the next 250 years. A much larger inventory will obviously require a considerably longer period of time for the same level of dissipation – unless, of course, the enhancement of radionuclide remobilization and transport by callianassid activity overrides the presence of a much larger store (McMurtry et al., 1986).

Table 8.10 Radionuclide concentrations ($pCig^{-1}$) by environment, Runit Island, Enewetak Atoll

Environment	Radionuclide ^{60}Co	^{90}Sr	^{125}Sb	^{137}Cs	^{155}Cs	^{207}Bi	^{238}Pu	$^{239+240}Pu$	^{241}Am
Atoll soils	0.64 (0.01–20)	1.7 (0.09–20)	—	0.40 (0.02–3.6)	—	—	—	3.2 (0.02–50)	—
Beaches	0.13 (0.03–1.6)	6.4 (1.2–3.0)	—	0.30 (0.03–9.0)	—	—	—	2.7 (0.34–18)	—
Lagoon surface	1.6 (0.19–18)	—	—	—	5.5 (0.82–54)	0.47 (0.17–7.9)	6.2 (1.1–96)	26 (3.1–190)	2.2 (0.09–31)
Lagoon cores	180 (39–4662)	—	33 (4.8–1420)	15 (2.3–204)	132 (46–1447)	0.58 (0.12–4.4)	8.2 (0.75–39)	82 (8.1–360)	12 (1.2–56)

Source: McMurtry et al., 1986

Clearly, any such re-evaluations have important repercussions for marine food chains (e.g. *Halimeda* may selectively uptake $^{239+240}$Pu at concentration factors of up to 32×10^4; Noskin, 1972) and the long-term health of the Marshallese, who consume a strongly marine-based diet. Such implications are not restricted to Enewetak but must also apply to the bomb test islands of Rongelap and Bikini. And no doubt similarly bioturbated radionuclide inventories characterize the atoll lagoons of Hao and Mururoa in French Polynesia.

Note

[1]This chapter is concerned with the effects of *benthos* (bottom-dwelling organisms) on the sedimentology and geomorphology of shallow subtidal environments. The benthos is commonly sorted according to size into *macrobenthos*, *meiobenthos* and *microbenthos* with dividing lines at about 1.0 mm and 100 μm in diameter or 10^{-4} and 10^{10} g wet weight (Mann, 1982). Macrobenthos can be further classified according to life habits: *epifauna* spend the majority of their time on the substratum surface; *infauna* live primarily beneath this surface. *Mobile* benthos move actively within or on the substratum, *sedentary* benthos do not, and *attached* organisms are incapable of movement along the bottom. The terms filter feeder, deposit feeder and herbivore are explained in the main text.

Acknowledgements

Production of this chapter in the Department of Geography, University of Manchester, was greatly assisted by the graphical skills of Mr Graham Bowden and Mr Nick Scarle and, in particular, by the word-processing expertise of Miss Rachel Creasey.

References

Acker, K. L. and Risk, M. J. 1985: Substrate destruction and sediment production by the boring sponge *Cliona caribbea* on Grand Cayman Island. *Journal of Sedimentary Petrology* 55, 705–11.

Ackpan, E. B. and Farrow, G. E. 1985: Shell bioerosion in high-latitude low-energy environments: Firths of Clyde and Lorn, Scotland. *Marine Geology* 67, 139–50.

Alexandersson, E. T. 1976: Actual and anticipated petrographic effects of carbonate undersaturation in shallow seawater. *Nature* (Lond.) 262, 653–5.

Aller, R. C. 1980a: Quantifying solute distributions in the bioturbated zone of marine sediments by defining an average micro-environment. *Geochimica et cosmochimica acta* 44, 1955–65.

—— 1980b: Relationships of tube-dwelling benthos with sediment and overlying water chemistry. In Tenore, K. R. and Coull, B. C. (ed), *Marine Benthic Dynamics*, pp. 285–308. Columbia: Univ. S. Carolina Press.

—— 1982: The effects of macrobenthos on chemical properties of marine sediment and overlying water. In McCall, P. L. and Tevesz, M. J. S. (eds), *Animal-sediment relations*, pp. 53–102. New York: Plenum Press.

—— 1983: The importance of diffusive permeability of animal burrow linings in determining marine sediment chemistry. *Journal of Marine Research* 41, 299–322.

—— and Dodge, R. E. 1974: Animal–sediment relations in a tropical lagoon, Discovery Bay, Jamaica. *Journal of Marine Research* 32, 209–32.

—— and Yingst, J. Y. 1978: Biogeochemistry of tube-dwellings: A study of the sedentary polychaete *Amphitrite ornata* (Leidy). *Journal of Marine Research* 36, 201–54.

—— and Yingst, J. Y. 1985: Effects of the marine deposit-feeders *Heteromastus filiformis* (Polychaeta), *Macoma balthica* (Bivalvia) and *Tellina texana* (Bivalvia) on averaged sedimentary solute transport, reaction rates and microbial distributions. *Journal of Marine Research* 43, 615–45.

Awramik, S. M. and Vanyo, J. P. 1986: Heliotropism in modern stromatolites. *Science* 231, 1279–81.

Bailey-Brock, J. H. 1979: Sediment trapping by chaetopterid polychaetes on a Hawaiian fringing reef. *Journal of Marine Research* 37, 643–56.

—— 1984: Ecology of the tube-building polychaete *Diopatra leuckarti* Kinberg, 1865 (Onuphidae) in Hawaii: community structure and sediment stabilizing properties. *Zoological Journal of the Linnean Society* 80, 191–9.

Bak, R. P. M. 1976: The growth of coral colonies and the importance of crustose coralline algae and burrowing sponges in relation with carbonate accumulation. *Netherlands Journal of Sea Research* 10, 285–337.

Ball, M. M., Shinn, E. A. and Stockman, K. W. 1967: The geological effects of hurricane Donna in South Florida. *Journal of Geology* 75, 583–97.

Bardach, J. E. 1961: Transport of calcareous fragments by reef fishes. *Science* 133, 98–9.

Bathurst, R. G. C. 1967: Sub-tidal gelatinous mat, sand stabilizer and food, Great Bahama Bank. *Journal of Geology* 75, 736–8.

—— 1975: *Carbonate Sediments and Their Diagenesis* (2nd edn). Amsterdam: Elsevier.

Baumfalk, Y. A. 1979: Heterogeneous grain size distribution in tidal flat sediment caused by bioturbation activity of *Arenicola marina* (Polychaeta). *Netherlands Journal of Sea Research* 13, 428–40.

Berner, R. A. 1980: *Early Diagenesis – A Theoretical Approach*. Princeton: Princeton Univ. Press.

Black, M. 1933: The algal sediments of Andros Island, Bahamas. *Philosophical Transactions, Royal Society of London* 222B, 165–92.

Braithwaite, C. J. R. and Talbot, M. R. 1972: Crustacean burrows in the Seychelles, Indian Ocean. *Palaeogeography, Palaeoclimatology and Palaeoecology* 11, 265–85.

Brenchley, G. A. 1982: Mechanisms of spatial competition in marine soft-bottom communities. *Journal of Experimental Marine Biology and Ecology* 60, 17–33.

Brock, R. E. 1979: An experimental study of the effects of grazing by parrotfishes and the role of refuges in benthic community structure. *Marine Biology* 51, 381–8.

Bromley, R. G. 1975: Comparative analysis of fossil and recent echinoid bioerosion. *Palaeontology* 18, 725–39.

—— 1978: Bioerosion of Bermuda reefs. *Palaeogeography, Palaeoclimatology, Palaeoecology* 23, 169–97.

Burrell, D. C. and Schubel, J. R. 1977: Seagrass ecosystem oceanography. In McRoy, C. P. and Helfferich, C. (eds), *Seagrass ecosystems: a scientific prospective*, pp. 195–232. New York: Marcel Dekker.

Carney, R. S. 1981: Bioturbation and biodeposition. In Boucot, A. J., (ed.), *Principles of Benthic Marine Palaeoecology*, pp. 357–400. New York: Academic Press.

Carr, N. C. and Whitton, B. A. 1982: *The Biology of Cyanobacteria. Botanical Monographs* 19. Oxford: Blackwell Scientific Publications.

Chapman, V. J. and Trevarthen, C. B. 1953: General schemes of classification in relation to marine coastal zonation. *Journal of Ecology* 41, 198–204.

Chapman, V. J. (ed.) 1977: *Wet Coastal Ecosystems (Ecosystems of the World, 1)*. Amsterdam: Elsevier.

Choi, D. R. 1984: Ecological succession of reef cavity dwellers (Coelobites) in coral rubble. *Bulletin of Marine Science* 35, 72–80.

Christiansen, C., Christofferson, H., Dalsgaard, J. and Nornberg, P. 1981: Coastal and near-shore changes correlated with dieback in eelgrass (*Zostera marina* L.) *Sedimentary Geology*, 28, 163–73.

Coles, S. M. 1979: Benthic microalgal populations on intertidal sediments and their role as precursors to salt marsh development. In Jefferies, R. L. and Davy, A. J. (eds), *Ecological processes in coastal environments*, pp. 25–42. Oxford: Blackwell Scientific Publications.

Colin, P. L. 1986: Benthic community distribution in the Enewetak lagoon, Marshall Islands. *Bulletin of Marine Science* 38, 129–43.

——, Suchanek, T. L. and McMurtry, G. 1986: Water pumping and particulate resuspension by callianassids (Crustacea: Thalassinidea) at Enewetak and Bikini Atolls, Marshall Islands. *Bulletin of Marine Science* 38, 19–24.

Cook, D. O. 1971: Depressions in shallow marine sediments made by benthic fishes. *Journal of Sedimentary Petrology* 41, 577–8.

Cloud, P. E. 1959: Geology of Saipan, Mariana Islands. Part 4. Submarine topography and shoal water ecology. *United States Geological Survey Professional Paper* 280-K, 361–445.

Creese, R. G. and Underwood, A. J. 1982: Analysis of inter- and intra-specific competition amongst intertidal limpets with different methods of feeding. *Oecologia* (Berl.) 53, 337–46.

de Vaugelas, J. 1985: Sediment reworking by callianassid mud-shrimp in tropical lagoons: A review with perspectives. *Proceedings, 5th International Coral Reef Congress*, Tahiti 6, 617–22.

——, Delesalle, B. and Monier, C. 1986: Aspects of the biology of *Callichirus armatus* (A. Milne Edwards 1870) (Decapoda, Thallasinidea) from French Polynesia. *Crustaceana* 50, 204–16.

—— and de Saint Laurent, M. 1984: Premières données sur l'écologie de *Callichirus laurae* de Saint Laurent sp. nov. (Crustacé, Decapodé, Thallasinide): son action bioturbatrice sur les formations sédimentaires du Goffe d'Aqaba (Mer Rouge). *Compte rendu hebdomadaire des séances de l'Academie des Sciences* Paris 298 (III), 147–52.

Dale, N. 1974: Bacteria in intertidal sediments: factors related to their distribution. *Limnology and Oceanography* 19, 509–18.

Davies, P. J. 1983: Reef growth. In Barnes, D. J. (ed.), *Perspectives on Coral Reefs*, pp. 69–106. Townsville/Canberra: Australia Institute of Marine Science/B. Clouston.

—— and Hutchings, P. A. 1983: Initial colonization, erosion and accretion on coral substrate: experimental results. Lizard Island, Great Barrier Reef. *Coral Reefs* 2, 27–35.

—— and Kinsey, D. W. 1973: Organic and inorganic factors in Recent beach rock formation, Heron Island, Great Barrier Reef. *Journal of Sedimentary Petrology* 43, 59–81.

Den Hartog, C. 1970: *The Seagrasses of the World*. London: North Holland Publishing Company.

Dill, R. F., Shinn, E. A., Jones, A. T., Kelley, K. and Steinen, R. P. 1986: Giant subtidal stromatolites forming in normal salinity waters. *Nature* (Lond.) 324, 55–8.

Dolan, R. 1972: Barrier dune systems along the Outer Banks of North Carolina: a reappraisal. *Science* 176, 286–8.

Dravis, J. J. 1983: Hardened subtidal stromatolites. *Science* 219, 385–6.

Eckman, J. E. 1985: Flow disruption by an animal-tube mimic affects sediment bacterial colonization. *Journal of Marine Research* 43, 419–35.

—— and Nowell, A. R. M. 1984: Fluid and sediment dynamic effects on marine benthic community. *Sedimentology* 31, 851–862.

——, Nowell, A. R. M. and Jumars, P. A. 1981: Sediment destabilization by animal tubes. *Journal of Marine Research* 39, 361–74.

Evans, J. W. 1970: Palaeontological implications of a biological study of rock boring clams (Family Pholadidae). In Crimes, T. P. and Harper, J. C. (eds), *Trace Fossils*, pp. 127–41. Liverpool: Seel House Press.

Farrow, G. E. 1972: Back-reef and lagoonal environments of Aldabra Atoll distinguished by their crustacean burrows. *Symposia of the Zoological Society of London* 28, 455–500.

Featherstone, R. P. and Risk, M. J. 1977: Effects of tube-building polychaetes on intertidal sediments of the Minas Basin, Bay of Fundy. *Journal of Sedimentary Petrology* 47, 446–50.

Fischer, J. S., Pickral, J. and Odum, W. E. 1979: Organic detritus particles: initiation of motion criteria. *Limnology and Oceanography* 24, 529–32.

Föcke, J. W. 1978: Limestone cliff morphology on Curaçao (Netherlands Antilles) with special attention to the origin of notches and vermetid/coralline algal surf benches ('cornices', 'trottoirs'). *Zeitschrift für Geomorphologie* 22, 329–49.

Fonseca, M. S. and Fisher, J. S. 1986: A comparison of canopy friction and sediment movement between four species of seagrass with reference to their ecology and restoration. *Marine Ecology – Progress Series* 29, 15–22.

Fonseca, M. S., Fisher, J. S., Zieman, J. C. and Thayer, G. W. 1982: Influence of the seagrass, *Zostera marina* L., on current flow. *Estuarine, Coastal and Shelf Science* 15, 351–64.

Fonseca, M. S., Zieman, J. C., Thayer, G. W. and Fisher, J. S. 1983: The role of current flow in structuring seagrass meadows. *Estuarine Coastal and Shelf Science* 17, 367–80.

Fosberg, F. R. 1971: Mangroves v. tidal waves. *Biological Conservation* 4, 38–9.

Frankel, J. and Ward, D. J. 1973: Mucilagenous matrix of some estuarine sands in Connecticut. *Journal of Sedimentary Petrology* 43, 1090–5.

Fremy, P. 1945: Contributions à la physiologie des thallophytcs marines perforant et cariant les roches calcaires et les coquilles. *Annales de l'Institut océanographique*, Monaco/Paris 22, 107–42.

Frey, R. W. and Basan, P. B. 1978: Coastal salt marshes. In Davis, R. A. Jr. (ed.), *Coastal Sedimentary Environments*, pp. 101–69. New York: Springer-Verlag.

Frostick, L. E. and McCave, I. N. 1979: Seasonal shifts of sediment within an estuary mediated by algal growth. *Estuarine and Coastal Marine Science* 9, 569–76.

Frydl, P. and Stearn, C. W. 1978: Rate of bioerosion by parrotfish in Barbados reef environments. *Journal of Sedimentary Petrology* 48, 1149–57.

Gallagher, E. D., Jumars, P. A. and Trueblood, D. D. 1983: Facilitation of soft-bottom benthic succession by tube builders. *Ecology* 64, 1200–16.

Gebelein, C. D. 1969: Distribution, morphology and accretion rate of recent subtidal algal stromatolites, Bermuda. *Journal of Sedimentary Petrology* 39, 49–69.

—— and Hoffman, P. 1968: Intertidal stromatolites and associated facies from Cape Sable, Florida. *Geological Society of America, Special Papers* 121, 109.

Gill, E. D. and Lang, J. G. 1983: Micro-erosion meter measurement of rock wear on the Otway coast of Southeast Australia. *Marine Geology* 52, 141–56.

—— 1953: Intertidal erosion on the Florida Keys. *Bulletin of Marine Science of the Gulf and Caribbean* 3, 59–69.

—— 1983: Geological and biological cavities in coral reefs. In Barnes, D. J. (ed.), *Perspectives on Coral Reefs*, pp. 148–53. Townsville/Canberra. Australian Institute of Marine Sciences/B. Clouston.

—— and Lowenstam, H. A. 1958: The influence of marine bottom communities on depositional environment of sediments. *Journal of Geology* 66, 310–18.

Glynn, P. W. 1973: Aspects of the ecology of coral reefs in the western Atlantic region. In Jones, O. A. and Endean, R. (eds), *Biology and Geology of Coral Reefs*, Volume II, Biology 1, pp. 271–324. New York: Academic Press.

—— Stewart, R. H. and McCosker, J. E. 1972: Pacific coral reefs of Panama: Structure, distribution and predators. *Geologische Rundschau* 61, 483–519.

——, Wellington, G. M. and Birkeland, C. 1979: Coral reef growth in the Galapagos: limitation by sea urchins. *Science* 203, 47–9.

Golubic, S. 1983: Stromatolites, fossil and recent: a case history. In Westbrook, P. and de Jong, E. W., (eds), *Biomineralization and biological metal accumulation*, pp. 313–26. Dordrecht: D. Reidel Publishing Company.

—— and Schneider, J. 1979: Carbonate dissolution. In Trudinger, P. A. and Swaine, D. J. (eds), *Biochemical cycling of mineral forming elements*, pp. 107–29. New York: Elsevier.

——, Perkins, R. S. and Lukas, K. J. 1975: Boring micro-organisms and microborings in carbonate substrates. In Frey, R. W., (ed.), *The Study of Trace Fossils*, pp. 229–59. Berlin: Springer-Verlag.

——, Friedmann, I. and Schneider, J. 1981: The lithobiontic ecological niche with special reference to microorganisms. *Journal of Sedimentary Petrology* 51, 475–9.

Grant, J. 1983: The relative magnitude of biological and physical sediment reworking in an intertidal community. *Journal of Marine Research* 40, 659–77.

Grant, W. D., Boyer, L. F. and Sandford, L. P. 1982: The effects of bioturbation on the initiation of motion of intertidal sands. *Journal of Marine Research* 40, 659–77.

Grenon, J.-F. and Walker, G. 1980: Biochemical and rheological properties of pedal mucus of the limpet *Patella vulgata* L. *Comparative Biochemistry and Physiology* 66B, 451–8.

Gygi, R. A. 1975: *Sparisoma viride* (Bonnaterre) the stoplight parrotfish, a major sediment producer on coral reefs of Bermuda? *Eclogae Geologicae Helvetiae* 68, 327–59.

Hamner, W. M. and Jones, M. S. 1976: Distribution, burrowing and growth rates of the clam *Tridacna crocea* on interior reef flats. *Oecologia* (Berl.) 24, 207–27.

Hanor, J. S. 1978: Precipitation of beachrock cements: mixing of marine and meteoric waters. *Journal of Sedimentary Petrology* 48, 489–501.

Hargrave, B. T. 1976: The central role of invertebrates faeces in sediment decomposition. In Anderson, J. M. and MacFadyen, A. (eds), *The Role of Terrestrial and Aquatic Organisms in Decomposition Processes*, pp. 301–21. Oxford: Blackwell Scientific Press.

Harrison, P. G. and Mann, K. H. 1975: Detritus formation from eelgrass (*Zostera marina* L.): the relative effects of fragmentation, leaching and decay. *Limnology and Oceanography* 20, 924–34.

Hatcher, B. G. 1983: Grazing in coral reef ecosystems. In Barnes, D. J. (ed.), *Perspectives on Coral Reefs*, pp. 164–79. Townsville/Canberra: Australian Institute of Marine Science/B. Clouston.

Haven, D. S. and Morales-Alamo, R. 1968: Occurrence and transport of faecal pellets in suspension in a tidal estuary. *Sedimentary Geology* 2, 141–51.

—— and —— 1972: Biodeposition as a factor in sedimentation of fine suspended solids in estuaries. *Geological Society of America Memoir* 133, 121–130.

Hein, F. J. and Risk, M. J. 1975: Bioerosion of coral heads: inner patch reefs, Florida reef tract. *Bulletin of Marine Science* 25, 133–8.

Hiatt, R. W. and Strasburg, D. W. 1960: Ecological relationships of the fish fauna on coral reefs in the Marshall Islands. *Ecological Monographs* 30, 65–127.

Highsmith, R. C. 1981: Lime-boring algae in hermatypic coral skeletons. *Journal of Experimental Marine Biology and Ecology* 55, 267–81.

Hodgkin, E. P. 1964: Rate of erosion of intertidal limestone. *Zeitschrift für Geomorphologie* 8, 385–92.

—— 1970: Geomorphology and biological erosion of limestone coasts in Malaysia. *Geological Society of Malaysia Bulletin* 3, 27–51.

Holland, A. F., Zingmark, R. G. and Dean, J. M. 1974: Quantitative evidence concerning the stabilization of sediments by marine benthic diatoms. *Marine Biology* 27, 191–6.

Hommeril, P. and Rioult, M. 1965: Etude de la fixation des sédiments meubles par deux algues marines: *Rhodothamniella floridula* (Dillwyn) J. Feldm. et *Microcoleus chtonoplastes* Thur. *Marine Geology* 3, 131–55.

Hopley, D. 1982: *The Geomorphology of the Great Barrier Reef*. New York: Wiley-Interscience.

Howard, J. D. and Frey, R. W. 1985: Physical and biogenic aspects of backbarrier sedimentary sequences, Georgia Coast, U.S.A. *Marine Geology* 63, 77–127.

Howard, J. D., Mayou, T. V. and Heard, R. W. 1977: Biogenic sedimentary structures formed by rays. *Journal of Sedimentary Petrology* 47, 339–46.

Hudson, J. H. 1977: Long-term bioerosion rates on a Florida reef: new method. *Proceedings of the Third International Coral Reef Symposium*, Miami 2, 491–8.

Hughes, R. N. 1985: *A functional biology of marine gastropods*. London: Croom-Helm.

Hughes Clarke, M. W. and Keij, A. J. 1973: Organisms as producers of carbonate sediment and indicators of environment in the southern Persian Gulf. In Purser, B. H. (ed.), *The Persian Gulf: Holocene carbonate sedimentation and diagenesis in a shallow epicontinental sea*, pp. 33–56. Berlin: Springer.

Hunt, M. 1969: A preliminary investigation of the habits and habitat of the rock-boring urchin *Echinometra lucunter* near Devonshire Bay, Bermuda. In Ginsburg, R. N. and Garrett, P. (eds), *Seminar on organism sediment relationships* Bermuda Biological Research Station, Special Publications 2, 35–40.

Hunter, I. G. 1977: Sediment production of *Diadema antillarum* on a Barbados fringing reef. *Proceedings of the Third International Coral Reef Symposium*, Miami 2, 105–9.

Hunter, W. R. 1949: The structure and behaviour of *Hiatella gallicana* (Lamarck) and *H. artica* (L), with special reference to the boring habit. *Proceedings, Royal Society of Edinburgh* B63, III, 271–89.

Hutchings, P. A. 1983: Cryptofaunal communities of coral reefs. In Barnes D. J. (ed.), *Perspectives on Coral Reefs*, pp. 200–8. Townsville/Canberra: Australian Institute of Marine Science/B. Clouston.

—— 1986: Biological destruction of coral reefs. A review. *Coral Reefs* 4, 239–52.

—— and Bamber, L. 1985: Variability of bioerosion rates at Lizard Island, Great Barrier Reef: preliminary attempts to explain these rates and their significance. *Proceedings of the Fifth International Coral Reef Congress*, Tahiti 5, 333–8.

Inman, D. L. 1949: Sorting of sediments in the light of fluid mechanics. *Journal of Sedimentary Petrology* 19, 51–70.

Johnstone, I. M. 1982: Ecology and distribution of seagrasses. *Monographiae Biologicae* 42, 497–512.

Jumars, P. A. and Nowell, A. R. M. 1984: Fluid and sediment dynamic effects on marine benthic community structure. *American Zoologist* 24, 45–55.

Kaye, C. A. 1959: Shoreline features and Quaternary shoreline changes in Puerto Rico. *United States Geological Survey Professional Paper* 317, 1–140.

Kiene, W. E. 1985: Biological destruction of experimental coral substrates at Lizard Island, Great Barrier Reef, Australia. *Proceedings of the Fifth International Coral Reef Congress*, Tahiti 5, 339–44.

Kikuchi, T. and Pérès, J. M. 1977: Consumer ecology of seagrass beds. In McRoy, C. P. and Helfferich, C. (eds), *Seagrass Ecosystems*, pp. 148–93. New York: Marcel Dekker.

King, G. M. 1986: Inhibition of microbial activity in marine sediments by a bromophenol from a hemicordate. *Nature* (Lond.) 323, 257–9.

Kirk, R. M. 1977: Rates and forms of erosion on intertidal platforms at Kaikoura peninsula, South Island, New Zealand. *New Zealand Journal of Geology and Geophysics* 20, 571–613.

Kleeman, K. 1973: *Lithophaga lithophaga* (L.) (Bivalvia) in different limestones. *Malacologia* 14, 345–7.

Kobluk, D. R. and Risk, M. J. 1977: Rate and nature of infestation of carbonate substrates by a boring alga *Ostreobium* sp. *Journal of Experimental Marine Biology and Ecology* 27, 107–15.

Kohlmeyer, J. 1969: The role of marine fungi in the penetration of calcareous substrates. *American Zoologist* 9, 741–6.

Kraeuter, J. N. 1976: Biodeposition by salt-marsh invertebrates. *Marine Biology* 35, 215–23.

Krumbein, W. E. 1979: Photolithotrophic and chemoorganotrophic activity of bacteria and algae as related to beachrock formation and degradation (Gulf of Aqaba, Sinai). *Geomicrobiology Journal* 1, 134–203.

Le Campion-Alsumard, T. 1970: Cyanophycées marines endoliths colonisant les surfaces rocheuses denudées (Etages supralittoral et mediolittoral de la region de Marseille). *Schweizerische Zeitschrift für Hydrologie* 32, 552–8.

—— 1975: Etude expérimentale de la colonisation d'éclats de calcite par les cyanophycées endoliths marines *Cahiers de biologie marine* 12, 177–85.

—— 1979: Les cyanophycées endoliths marines. Systematique, ultrastructure et biodestruction. *Oceanologia Acta* 2, 143–56.

Lee, H., II and Schwartz, R. C. 1980: Biological processes affecting the distribution of pollutants in marine bioturbation. In Baker, R. A., (ed.), *Contaminants and Sediments*, volume 2, pp. 555–605. Ann Arbor: Ann Arbor Scientific Publishers.

Lerman, A. 1979: *Geochemical Processes: Water and Sediment Environments.* New York: John Wiley.

Lewis, J. R. 1964: *The ecology of rocky shores.* London: English Universities Press.

Logan, B. W. 1961: Cryptozoan and associated stromatolites from the Recent, Shark Bay, Western Australia. *Journal of Geology* 69, 517–553.

——, Rezak, R. and Ginsburg, R. N. 1964: Classification and environmental significance of algal stromatolites. *Journal of Geology* 72, 68–83.

Lonsdale, P. and Southard, J. B. 1974: Experimental erosion of North Pacific red clay. *Marine Geology* 17, M51–M60.

Lowenstam, H. A. 1962: Magnetite in denticle capping in Recent chitons (Polyplacophora). *Bulletin of the Geological Society of America* 73, 435–8.

Luckenbach, M. W. 1986: Sediment stability around animal tubes: The roles of hydrodynamic processes and biotic activity. *Limnology and Oceanography* 31, 779–87.

Lynch, M. and Harrison, W. 1970: Sedimentation caused by a tube-building amphipod. *Journal of Sedimentary Petrology* 40, 434–5.

McCall, P. L. 1977: Community patterns and adaptive strategies of the infaunal benthos of Long Island Sound. *Journal of Marine Research* 35, 221–6.

McLachlan, A. 1985: Sandy beach ecology – a review. In McLachlan, A. and Erasmus, T. (eds), *Sandy beaches as ecosystems*, pp. 321–80. The Hague: W. Junk.

McLean, R. F. 1967a: Erosion of burrows in beachrock by the tropical sea urchin *Echinometra lucunter. Canadian Journal of Zoology* 45, 586–8.

—— 1967b: Measurement of beachrock erosion by some tropical marine gastropods. *Bulletin of Marine Science* 17, 551–61.

—— 1974: Geologic significance of bioerosion of beachrock. *Proceedings of the Second International Coral Reef Symposium*, Brisbane 2, 401–8.

McMurtry, G. M., Schneider, R., Colin, P. L., Buddemeier, R. W. and Suchanek, T. H. 1985: Redistribution of fallout radionuclides in Enewetak lagoon sediments by callianassid (Crustacea: Thalassinidea) bioturbation. *Nature* (Lond.) 313, 674–7.

McMurtry, G. M., Schneider, R., Colin, P. L., Buddemeier, R. W. and Suchanek, T. H. 1986: Vertical distribution of fallout radionuclides in Enewetak lagoon sediments: Effects of burial and bioturbation on the radionuclide inventory. *Bulletin of Marine Science* 38, 35–55.

McRoy, C. P. and McMillan, C. 1977: Production ecology and physiology of seagrasses. In McRoy, C. P. and Helfferich, C., (eds), *Seagrass Ecosystems*, pp. 53–87. New York: Marcel Dekker.

Mann, K. H. 1982: *Ecology of Coastal Waters*. Oxford: Blackwell Scientific Publications.

Mantz, P. A. 1977: Incipient transport of fine grains and flakes by fluids – an extended Shields diagram. *American Society of Civil Engineers, Journal of the Hydraulics Division* 103, 601–15.

Massin, C. and Doumen, C. 1986: Distribution and feeding of epibenthic holothuroids on the reef flat at Laing Island (Papua New Guinea). *Marine Ecology: Progress Series* 31, 185–95.

Miller, M. F. 1984: Bioturbation of intertidal quartz-rich sands: A modern example and its sedimentologic and palaeoecologic implications. *Journal of Geology* 92, 201–16.

Milliman, J. D. 1974: *Recent sedimentary carbonates. Part I: Marine Carbonates*. Berlin: Springer.

Molinier, R. and Picard, J. 1952: Recherches sur les herbiers de phanérogames marine du littoral mediterraneen français. *Annales de l'Institut océanographique Monaco/Paris* 27, 157–234.

Monty, C. 1965: Recent algal stromatolites in the windward lagoon, Andros Island, Bahamas. *Annales de la Société géologique de Belgique* 88, 269–276.

—— 1967: Distribution and structure of recent stromatolitic algal mats, eastern Andros Island, Bahamas. *Annales de la Société géologique de Belgique* 90, 55–100.

—— 1972: Recent algal stromatolitic deposits, Andros Island, Bahamas. Preliminary report. *Geologische Rundschau* 61, 742–83.

Moore, C. H. and Shedd, W. W. 1977: Effective rates of sponge bioerosion as a function of carbonate production. *Proceedings of the Third International Coral Reef Symposium*, Miami 2, 499–505.

Morris, H. M. 1955: A new concept of flow in rough conduits. *Transactions of the American Society of Civil Engineers* 120, 373–98.

Mottershead, D. N. 1982: Coastal spray weathering of bedrock in the supratidal zone at East Prawle, South Devon. *Field Studies* 5, 663–84.

Myers, A. C. 1977: Sediment processing in a marine subtidal sandy bottom community. I. Physical aspects. *Journal of Marine Research* 35, 609–32.

Nelsen, J. E. Jr. and Ginsburg, R. N. 1986: Calcium carbonate production by epibionts on *Thalassia* in Florida Bay. *Journal of Sedimentary Petrology* 56, 622–8.

Neudecker, S. 1979: Effects of grazing and browsing fishes on the zonation of corals in Guam. *Ecology* 60, 666–72.

Neumann, A. C. 1966: Observations on coastal erosion in Bermuda and measurements of the boring rate of the sponge *Cliona lampa*. *Limnology and Oceanography* 11, 92–108.

—— 1968: Biological erosion of limestone coasts. In Fairbridge, R. W. (ed.), *Encyclopedia of Geomorphology*. New York: Reinhold, 75–81.

——, Gebelein, C. P. and Scoffin, T. P. 1970: The composition, structure and erodibility of subtidal mats, Abaco, Bahamas. *Journal of Sedimentary Petrology* 40, 274–97.

Newell, R. C. 1965: The role of detritus in the nutrition of two marine deposit feeders, the prosobranch *Hydrobia ulvae* and the bivalve *Macoma balthica*. *Proceedings of the Zoological Society of London* 144, 25–45.

——, Pye, V. I. and Ahsanullah, M. 1971: Factors affecting the feeding rate of the winkle *Littorina littorea*. *Marine Biology* 9, 138–44.

North, W. J. 1954: Size, distribution, erosive activities and gross metabolic efficiency of the marine intertidal snails *Littorina planaxis* and *L. scutulata*. *Biological Bulletin* 106, 185–97.

Noskin, V. E. 1972: Ecological aspects of plutonium dissemination in aquatic environments. *Health Physics* 22, 537–550.

Nowell, A. R. M. and Church, M. 1979: Turbulent flow in a depth-limited boundary layer. *Journal of Geophysical Research* 84, 4816–24.

Nowell, A. R. M., Jumars, P. A. and Eckman, J. E. 1981: Effects of biological activity on the entrainment of marine sediments. *Marine Geology* 42, 133–53.

Odum, E. P. 1971: *Fundamentals of Ecology* (3rd edn). Philadelphia: W. B. Saunders.

Ogden, J. 1977: Carbonate sediment production by parrot-fish and sea urchins on Caribbean reefs. In Frost, S. H. J., Weiss, M. P. and Saunders, J. B. (eds), *Reefs and related carbonates – ecology and sedimentology*, American Association of Petroleum Geologists, Studies in Geology 4, 281–7.

Otter, G. W. 1937: *Rock-destroying organisms in relation to coral reefs*. *Scientific Reports, Great Barrier Reef Expedition 1928–29*, British Museum (Natural History) 1, 323–52.

Palmer, M. A. 1986: Hydrodynamics and structure: interactive effects on meiofauna dispersal. *Journal of Experimental Marine Biology and Ecology* 104, 53–68.

Patriquin, D. G. 1972: The origin of nitrogen and phosphorus for growth of the marine angiosperm *Thalassia testudinum*. *Marine Biology* 15, 5–46.

—— 1975: 'Migration' of blowouts in seagrass beds at Barbados and Carriacou, West Indies and its ecological and geological implications. *Aquatic Botany* 1, 163–89.

Patton, W. K. 1976: Animal associates of living coral reefs. In Jones, O. A. and Endean, R. E. (eds), *Biology and Geology of Coral Reefs*, volume III, pp. 1–36. New York: Academic Press.

Perkins, R. D. and Tsentas, I. 1976: Microbial infestation of carbonate substrates planted on the St Croix shelf, West Indies. *Bulletin of the Geological Society of America* 87, 1615–28.

Phelps, D. K. 1966: Partitioning of the stable elements, Fe, Zn, Se, Sm, within a benthic community. Anasco Bay, Puerto Rico. In Aberg, B. and Hungate, F. P. (eds), *Radioecological Concentration Processes*, pp. 721–34. New York: Pergamon Press.

Phillips, R. C. and McRoy, C. P. (eds) 1980: *Handbook of seagrass biology*. New York: Garland.

Pomeroy, L. R., Darley, W. M., Dunn, L., Gallagher, J. L., Haines, E. B. and Whitney, D. M. 1981: Primary production. In Pomeroy, L. R. and Wiegert, R. G. (eds), *The Ecology of a Salt Marsh*, pp. 39–67. New York: Springer-Verlag.

Pomponi, S. A. 1977: Etching cells of boring sponges: an ultrastructural analysis. *Proceedings of the Third International Coral Reef Symposium*, Miami 2, 485–90.

Posey, M. H. 1986: Changes in a benthic community associated with dense beds of a burrowing deposit feeder, *Callianassa californiensis*. *Marine Ecology: Progress Series* 31, 15–22.

Postma, H. 1967: Sediment transport environment. In Lauff, G. H. (ed.), *Estuaries*, pp. 158–79. Washington DC: American Association for the Advancement of Science.

Potts, M. and Whitton, B. A. 1980: Vegetation of the intertidal zone of the lagoon of Aldabra, with special reference to the photo-synthetic prokaryotic communities. *Proceedings of the Royal Society* 208B, 13–55.

Powell, E. N. 1977: Particle size selection and sediment reworking in a funnel feeder, *Leptosynapta tenuis* (Holothuriodea, Synoptidae). *Internationale Revue der gesamten Hydrobiologie* 62, 385–403.

Psuty, N. P. 1965: Beach-ridge development in Tabasco, Mexico. *Annals, Association of American Geographers* 55, 112–24.

Purdy, E. G. and Kornicker, L. S. 1958: Algal disintegration of Bahamian limestone coasts. *Journal of Geology* 66, 97–9.

Rasmussen, E. 1977: The wasting disease of eelgrass (*Zostera marina*) and its effects on environmental factors and fauna. In McRoy, C. P. and Helferrich, C. (eds), *Seagrass Ecosystems*, pp. 1–51. New York: Marcel Dekker.

Reese, E. S. 1977: Coevolution of corals and coral feeding fishes of the family Chaetodont. *Proceedings of the Third International Coral Reef Symposium*, Miami 1, 267–74.

Reise, K. 1985: *Tidal flat ecology*. Berlin: Springer-Verlag.

Revelle, R. and Emery, K. O. 1957: Chemical erosion of beach rock and exposed reef rock. *United States Geological Survey, Professional Papers* 260-T, 699–709.

Revelle, R. and Fairbridge, R. W. 1957: Carbonates and carbon dioxide in Hedgpeth, W. (ed.), *Treatise of marine ecology and paleoecology*, Geological Society of America Memoir 67, volume 1, 239–96.

Rhoads, D. C. 1967: Biogenic reworking of intertidal and subtidal sediments in Barnstable Harbor and Buzzards Bay, Massachusetts. *Journal of Geology* 75, 461–76.

—— 1970: Mass properties, stability and ecology of marine muds related to burrowing activity. In Crimes, T. P. and Harper, J. C. (eds), *Trace Fossils*, pp. 391–406. Liverpool: Seel House Press.

—— 1974: Organism–sediment relations on the muddy sea floor. *Oceanography and Marine Biology, An Annual Review* 12, 263–300.

—— and Boyer, L. F. 1982: Effect of marine benthos on sediment physical properties. In McCall, P. L. and Tevesz, M. J. S. (eds), *Animal–sediment relations*, pp. 3–52. New York: Plenum Press.

—— and Young, D. K. 1970: The influence of deposit-feeding organisms on sediment stability and trophic structure. *Journal of Marine Research* 28, 150–78.

—— and Young, D. K. 1971: Animal–sediment relations in Cape Cod Bay, Massachusetts. II: Reworking by *Molpadia oolitica* (Holothuroidea). *Marine Biology* II, 255–61.

—— , Yingst, J. Y. and Ullman, W. 1978: Seafloor stability in central Long Island Sound. Part I: Temporal changes in erodibility of fine-grained sediment. In Wiley, M. L. (ed.), *Estuarine Interactions*, pp. 221–4. New York: Academic Press.

Riemann, F. and Schrage, M. 1978: The mucus-trap hypothesis on feeding of aquatic nematodes and implications for biodegradation and sediment texture. *Oecologia* (Berl.) 34, 75–88.

Risk, M. J. and MacGeachy, J. K. 1978: Aspects of bioerosion of modern Caribbean reefs. *Revista de biologica tropicale* 26 (Supplement 1), 85–105.

Risk, M. J. and Moffat, J. S. 1977: Sedimentological significance of fecal pellets of *Macoma balthica* in the Minas Basin, Bay of Fundy. *Journal of Sedimentary Petrology* 47, 1425–36.

Roberts, H. H., Suchanek, T. H. and Wiseman, W. J. Jr. 1982: Lagoon sediment transport: the significant effect of *Callianassa* bioturbation. *Proceedings of the Fourth International Coral Reef Symposium*, Manila 1, 459–65.

Robinson, L. A. 1977: Marine erosive processes at the cliff foot. *Marine Geology* 23, 257–71.

Rooney, W. S. J. and Perkins, R. D. 1972: Distribution and geologic significance of microboring organisms within sediments of the Arlington Reef Complex, Australia. *Bulletin of the Geological Society of America* 83, 1139–50.

Rowe, G. T. 1974: The effects of the benthic fauna on the physical properties of deep-sea sediments. In Inderbitzen, A. L. (ed.), *Deep-Sea Sediments: Physical and Mechanical Properties*, pp. 381–400. New York: Plenum Press.

Runham, W. W. and Thornton, P. R. 1967: Mechanical wear of the gastropod radula: a scanning electron microscope study. *Journal of Zoology* 153, 445–52.

Runham, W. W., Thornton, P. R., Shaw, D. A. and Wayte, R. C. 1969: The mineralization and hardness of the radular teeth of the limpet *Patella vulgata* L. *Zeitschrift für Zellforschung und microskopische Anatomie* 99, 608–26.

Russo, A. R. 1977: Water flow and the distribution and abundance of echinoids (Genus: *Echinometra*) on an Hawaiian reef. *Australian Journal of Marine and Freshwater Research* 28, 693–702.

—— 1980: Bioerosion by the two rock boring echinoids (*Echinometra mathaei* and *Echinostrephus aciculatus*) on Enewetak Atoll, Marshall Islands. *Journal of Marine Research* 38, 99–110.

Rützler, K. 1975: The role of burrowing sponges in bioerosion. *Oecologia* (Berl.) 19, 203–16.

Salazar-Jimenez, A., Frey, R. W. and Howard, J. D. 1982: Concavity orientations of bivalve shells in estuarine and nearshore shelf sediments, Georgia. *Journal of Sedimentary Petrology*, 52, 565–86.

Sammarco, P. W. 1982: Echinoid grazing as a structuring force in coral communities: whole reef manipulations. *Journal of Experimental Marine Biology and Ecology* 61, 33–55.

Santschi, P. H., Li, Y.-H., Adler, D. M., Amdurer, M., Bell, J. and Nyffeler, U. P. 1983: The relative mobility of natural (Th, Pb and Po) and fallout (Pu, Am, Cs) radionuclides in the coastal marine environment: results from model ecosystems (MERL) and Narragansett Bay. *Geochemica et cosmochimica acta* 47, 201–10.

Schiechtl, H. 1980: *Bioengineering for land reclamation and conservation*. Edmonton: University of Alberta Press.

Schmalz, R. F. 1971: Formation of beachrock at Eniwetok Atoll: In Bricker, O. P. (ed.), *Carbonate sediments*, pp. 17–24. Baltimore: John Hopkins Univ. Press.

Schneider, J. 1976: Biological and inorganic factors in the destruction of limestone coasts. *Contributions to Sedimentology* 6, 1–112.

—— and Torunski, H. 1983: Biokarst on limestone coasts, morphogenesis and sediment production. *Marine Ecology* 4, 45–63.

Schwartz, R. C. and Lee, H., III. 1980: Biological processes affecting the distribution of pollutants in marine sediments. Part I: Accumulation, trophic transfer, biodegradation and migration. In Baker, R. E. (ed.), *Contaminants and Sediments*, volume 2, pp. 533–53. Ann Arbor: Ann Arbor Scientific Publishers.

Scoffin, T. P. 1968: An underwater flume. *Journal of Sedimentary Petrology* 38, 244–6.

—— 1970a: The trapping and binding of subtidal carbonate sediments by marine vegetation in Bimini Lagoon, Bahamas. *Journal of Sedimentary Petrology* 40, 249–73.

—— 1970b: A conglomeratic beachrock in Bimini, Bahamas. *Journal of Sedimentary Petrology* 40, 756–9.

—— and Stoddart, D. R. 1983: Beachrock and intertidal cements. In Goudie, A. S. and Pye, K., (eds), *Chemical sediments and geomorphology*, pp. 401–25. London: Academic Press.

—— , Stearn, C. W., Boucher, D., Frydl, P., Hawkins, C. M., Hunter, I. G. and MacGeachy, J. K. 1980: Calcium carbonate budget of a fringing reef on the west coast of Barbados. *Bulletin of Marine Science* 30, 475–508.

Shinn, E. A. 1968: Burrowing in recent lime sediments. *Journal of Palaeontology* 42, 879–94.

Smith, J. D. 1977: Modeling of sediment transport on continental shelves. In Goldberg, E. D., McCave, I. N., O'Brien, J. J. and Steele, J. M. (eds), *The Sea*, volume 6, pp. 539–77. New York: J. WIley.

Smith, J. M. and Frey, R. W. 1985: Biodeposition by the ribbed mussel *Genkensia demissa* in a salt marsh, Sapelo Island, Georgia. *Journal of Sedimentary Petrology* 55, 817–28.

Southard, J. B., Young, R. A. and Hollister, C. D. 1971: Experimental erosion of calcareous ooze. *Journal of Geophysical Research* 76, 5903–9.

Southward, A. J. 1964: Limpet grazing and the control of vegetation on rocky shores. In Crisp, D. J. (ed.), *Grazing in terrestrial and marine environments*, pp. 265–73. Oxford: Blackwell Scientific Publications.

Spencer, T. 1983: *Limestone erosion rates and microtopography: Grand Cayman Island, West Indies*. Unpublished Ph.D.thesis, University of Cambridge.

—— 1985a: Marine erosion rates and coastal morphology of reef limestones on Grand Cayman Island, West Indies. *Coral Reefs* 4, 59–70.

—— 1985b: Weathering rates on a Caribbean reef limestone: results and implications. *Marine Geology* 69, 195–201.

—— in press. Limestone coastal morphology: The biological contribution. *Progress in Physical Geography*.

—— and Benn, J. R. in press. Limestone topography and rock weathering in the Cayman Islands. In Stoddart, D. R., Davies, J. E. and Brunt, M. A. (eds), *The Biogeography and Ecology of the Cayman Islands*. The Hague: W. Junk.

Steers, J. A. 1957: The coast as a field for physiographic research. *Transactions and Papers, Institute of British Geographers*, 22, 1–13.

Steneck, R. S. and Watling, L. 1982: Feeding capabilities and limitations of herbivorous molluscs: a functional group approach. *Marine Biology* 68, 299–319.

Stephenson, W. 1961: Experimental studies on the ecology of intertidal environments at Heron Island. II: The effects of substratum. *Australian Journal of Marine and Freshwater Research* 12, 164–76.

Stoddart, D. R. 1964: Storm conditions and vegetation in equilibrium of reef islands. *Proceedings, Ninth Conference on Coastal Engineering*, 893–906.

—— 1980: Mangroves as successional stages, Inner Reefs of the Northern Great Barrier Reef. *Journal of Biogeography* 7, 269–84.

—— and Steers, J. A. 1977: The nature and origin of reef islands. In Jones, O. A. and Endean, R. E. (eds), *Biology and Geology of Coral Reefs*, volume IV, pp. 60–102. New York: Academic Press.

Suchanek, T. H. 1983: Control of seagrass communities and sediment distribution by *Callianassa* (Crustacea, Thalassinidea) bioturbation. *Journal of Marine Research* 41, 281–98.

—— 1985: Thalassinid shrimp burrows: ecological significance of species-specific architecture. *Proceedings of the Fifth International Coral Reef Congress*, Tahiti 5, 205–10.

—— and Colin, P. L. 1986: Rates and effects of bioturbation by invertebrates and fishes at Enewetak and Bikini Atolls. *Bulletin of Marine Science* 38, 25–34.

——, Colin, P. L., McMurtry, G. M. and Suchanek, C. S. 1986: Bioturbation and redistribution of sediment radionuclides in Enewetak Atoll lagoon by callianassid shrimp: biological aspects. *Bulletin of Marine Science* 38, 144–54.

Taghon, G. L., Nowell, A. R. M. and Jumars, P. A. 1984: Transport and breakdown of fecal pellets: Biological and sedimentological consequences. *Limnology and Oceanography* 29, 64–72.

Taylor, J. C. M. and Illing, L. V. 1969: Holocene intertidal calcium carbonate sedimentation, Qatar, Persian Gulf. *Sedimentology* 12, 69–107.

Taylor, J. D. and Way, K. 1976: Erosive activities of chitons at Aldabra Atoll. *Journal of Sedimentary Petrology* 46, 974–7.

Teas, H. J. 1977: Ecology and restoration of mangrove shorelines in Florida. *Environmental Conservation* 4, 51–8.

Thayer, G. W., Adams, S. M. and La Croix, M. W. 1975: Structural and functional aspects of a recently established *Zostera marina* community. In Cronin, L. E. (ed.), *Estuarine Research* I, pp. 518–40. New York: Academic Press.

Thistle, D., Reidenauer, J. A., Findlay, R. H. and Waldo, R. 1984: An experimental investigation of enhanced harpacticoid (Copepoda) abundances around isolated seagrass shoots. *Oecologia* (Berl.) 63, 295–9.

Thom, B. G. 1967: Mangrove ecology and deltaic geomorphology: Tabasco, Mexico. *Journal of Ecology* 55, 301–43.

—— 1982: Mangrove ecology: A geomorphological perspective. In Clough, B. (ed.), *Mangrove Ecosystems in Australia: Structure, Function and Management*, pp. 3–17. Canberra: ANU Press.

—— 1984: Coastal landforms and geomorphic processes. In Snedaker, S. and Snedaker, J. G. (eds), *The mangrove ecosystem: research methods*, pp. 3–17. Paris: UNESCO.

——, Wright, L. D. and Coleman, J. M. 1975: Mangrove ecology and deltaic estuarine geomorphology: Cambridge Gulf – Ord River, Western Australia. *Journal of Ecology* 63, 203–32.

Thomas, L. P., Moore, D. R. and Work, R. C. 1961: Effects of hurricane Donna on the turtle grass beds of Biscayne Bay, Florida. *Bulletin of Marine Science of the Gulf and the Caribbean* 11, 191–7.

Torunski, H. 1979: Biological erosion and its significance for the morphogenesis of limestone coasts and for nearshore sedimentation. *Senckenbergiana maritima* 11, 193–265.

Trudgill, S. T. 1976: The marine erosion of limestones on Aldabra Atoll, Indian Ocean. *Zeitschrift für Geomorphologie, Supplementband* 26, 164–200.

—— 1983a: Preliminary estimates of intertidal limestone erosion, One Tree Island, Southern Great Barrier Reef, Australia. *Earth Surface Processes and Landforms* 8, 189–93.

—— 1983b: Measurements of rates of erosion of reefs and reef limestones. In Barnes, D. J. (ed.), *Perspectives on Coral Reefs*, pp. 256–62. Townsville/Canberra: Australian Institute of Marine Science/B. Clouston.

—— 1985: *Limestone Geomorphology*, Geomorphology Texts 8. London: Longman.

—— 1987: Bioerosion of intertidal limestone, Co. Clare, Eire – 3: Zonation, process and form. *Marine Geology* 71, 111–21.

—— and Crabtree, R. W. 1987: Bioerosion of intertidal limestone, Co. Clare, Eire – 2: *Hiatella artica. Marine Geology* 74, 99–109.

——, High, C. J. and Hanna, F. K. 1981: Improvements to the micro-erosion meter (MEM). *British Geomorphological Research Group, Technical Bulletin* 29, 3–17.

——, Smart, P. L., Friederich, H. and Crabtree, R. W. 1987: Bioerosion of intertidal limestone, Co. Clare, Eire – 1: *Paracentrotus lividus. Marine Geology* 74, 85–98.

Tudhope, A. W. and Risk, M. J. 1985: Rate of dissolution of carbonate sediments by microboring organisms, Davies Reef, Australia. *Journal of Sedimentary Petrology* 55, 440–7.

Tudhope, A. W. and Scoffin, T. P. 1984: The effects of *Callianassa* bioturbation on the preservation of carbonate grains in Davies Reef lagoon, Great Barrier Reef, Australia. *Journal of Sedimentary Petrology* 54, 1091–6.

Van den Hoek, C., Colijn, F., Cortel-Breeman, A. M. and Wanders, J. B. W. 1972: Algal-vegetation types along the shores of inner bays and lagoons of Curaçao and of the Lagoon Lac (Bonaire), Netherlands Antilles. *Verhandelingen der K. nederlandsche akademie van wetenschappen* 61, 3–72.

Van den Hoek, C., Breeman, A. M., Bak, R. P. M. and Van Buurt, G. 1978: The distribution of algae, corals and gorgonians in relation to depth, light attenuation, water movement and grazing pressure in the fringing coral reef of Curaçao, Netherlands Antilles. *Aquatic Botany* 5, 1–46.

Viles, H. A. and Spencer, T. 1986: 'Phytokarst', blue-green algae and limestone weathering. In Paterson, K., and Sweeting, M. M. (eds), *New Directions in Karst*, pp. 115–40. Norwich: Geobooks.

Viles, H. A. and Trudgill, S. T. 1984: Long term remeasurements of micro-erosion meter rates, Aldabra Atoll, Indian Ocean. *Earth Surface Processes and Landforms* 9, 89–94.

Vita-Finzi, C. and Cornelius, P. F. S. 1973: Cliff sapping by molluscs in Oman. *Journal of Sedimentary Petrology* 43, 31–2.

Wanders, J. B. W. 1977: The role of benthic algae in the shallow reef of Curaçao (Netherlands Antilles). III: The significance of grazing. *Aquatic Botany* 3, 356–90.

Wanless, H. R., Burton, E. A. and Dravis, J. 1981: Hydrodynamics of carbonate fecal pellets. *Journal of Sedimentary Petrology* 51, 27–36.

Ward, L. G., Boynton, W. R. and Kemp, W. M. 1984: The influence of waves and seagrass communities on suspended particulates in an estuarine embayment. *Marine Geology* 59, 85–103.

Warme, J. E. 1975: Borings as trace fossils, and the processes of marine bioerosion. In Frey, R. W. (ed.), *The Study of Trace Fossils*, pp. 181–229. Berlin: Springer.

Warthin, A. S. Jr. 1959: Ironshore on some W. Indian islands. *Transactions, New York Academy of Sciences* 2/21, 649–52.

Waslenchuk, D. G., Matson, E. A., Zajac, R. N., Dobbs, F. C. and Tramontano, J. M. 1983: Geochemistry of waters vent by a bioturbating shrimp in Bermudian sediments. *Marine Biology* 72, 219–25.

Webb, L. E. 1969: Biologically significant properties of submerged marine sands. *Proceedings of the Royal Society of London* 174B, 355–402.

Wells, G. P. 1945: The mode of life of *Arenicola marina* L. *Journal of the Marine Biological Association of the United Kingdom* 26, 170–207.

Whitton, B. A. and Potts, M. 1980: Blue-green algae (Cyanobacteria) of the oceanic coast of Aldabra. *Atoll Research Bulletin* 238, 1–8.

—— and —— 1982: Marine littoral. In Carr, N. G. and Whitton, B. A. (eds), *The Biology of Cyanobacteria. Botanical Monographs* 19, pp. 515–42. Oxford: Blackwell Scientific Publications.

Wilkinson, C. R. 1983: Role of sponges in coral reef structural processes. In Barnes, D. J. (ed.), *Perspectives on Coral Reefs*, pp. 263–74. Townsville/Canberra: Australian Institute of Marine Sciences/B. Clouston.

Wilson, D. P. 1949: The decline in *Zostera marina* L. at Salcombe and its effects on the shore. *Journal of Marine Biological Association of the United Kingdom* 28, 395–412.

Woodroffe, C. D. 1982: Geomorphology and development of mangrove swamps, Grand Cayman Island, West Indies. *Bulletin of Marine Science* 32, 381–98.

Yingst, J. Y. and Rhoads, D. C. 1978: Seafloor stability in Central Long Island Sound. Part II: Biological interactions and their potential importance for seafloor erodibility. In Wiley, M. N. (ed.), *Estuarine Interactions*, pp. 245–60. New York: Academic Press.

Young, H. R. and Nelson, C. S. 1985: Biodegradation of temperate-water skeletal carbonates by boring sponges on the Scott shelf, British Columbia, Canada. *Marine Geology* 65, 33–45.

Young, R. A. and Southard, J. B. 1978: Erosion of fine-grained marine sediments: Sea-floor and laboratory experiments. *Bulletin of the Geological Society of America* 89, 663–72.

9 Organisms and karst geomorphology

H. A. Viles

Introduction

Scattered observations of biological influences upon weathering and landform development in karst areas have appeared in the literature since the nineteenth century. Sollas (1880), for example, observed the denuding effect of a lichen upon a limestone surface and speculated upon the mechanisms involved. So far, however, there has been little consensus on the overall geomorphological importance of such organic influences, although some workers have suggested specific process–form relationships with organic processes leading to *biokarst* features. In many cases, organic influences have been regarded merely as interesting oddities, producing small-scale features of localized importance. There have been few rigorous scientific studies of biological effects upon geomorphological processes, and most theories of karst landform development ignore any organic role.

There is a clear need for a considered assessment of the role of biological communities in shaping the geomorphology of the terrain upon which they grow; which would link organic and inorganic parts of the ecosystem. This chapter aims to go part of the way towards fulfilling this goal through a review of recent work. Firstly, some information is presented upon the nature and extent of karst environments and the general influence of organisms on erosion rates. Secondly, the occurrence of biokarst features is discussed and some examples given of ecosystems developed upon karst terrain. Thirdly, recent examples of biological influences upon weathering/erosion and deposition/consolidation of limestone surfaces are presented. Finally, some conclusions are drawn upon the importance of organisms to explanation in karst geomorphology.

The overall impact of organisms on dissolution rates

The word 'karst' is a germanicized form of the Slavian word 'krs', which was originally used as a name for the large area of limestone with distinctive surficial and underground geomorphology found in Yugoslavia and known as the Dinaric karst (Sweeting, 1972; Roglic, 1972). The usage of the word has, since then, been expanded to cover areas similar to this type site on carbonate and other relatively easily soluble rocks, and in the process has become the subject of much debate. The problems of the expanded usage of the term 'karst' are familiar ones in geomorphology and relate to difficulties of taxonomy within the subject. Karst may be defined as terrain produced mainly by solutional erosion, which usually produces an important underground drainage system. Karst areas are found in all climatic zones, and their distribution in the northern hemisphere is documented in Herak and Stringfield (1972).

Many data have been collected upon dissolution rates in different karst areas which can be related to runoff (Atkinson and Smith, 1976; Jennings, 1985), although there are problems of accuracy with some of the data (Gunn, 1986, p. 380). Figure 9.1 presents a compilation of observations of erosion rates from a range of karst areas plotted against runoff. It can be seen that runoff variations explain much of the variability in erosion rates. The remainder of the variability in erosion rates is ascribed to differences in solute concentrations. These differences are a function of temperature (through its control of reaction kinetics, CO_2 production and CO_2 solubility) and the evolution system (i.e. whether reactions take place under open or closed equilibrium conditions).

Most points which plot above the line in figure 9.1 are from vegetated areas, whereas those falling below the line are from non-vegetated or sparsely vegetated arctic and alpine areas. Bearing in mind that the data set may be inadequate (see Gunn, 1986, p. 380), it can still be seen that there is some correlation between the presence of soil and vegetation and elevated levels of dissolution of $CaCO_3$. Microbial decomposition and

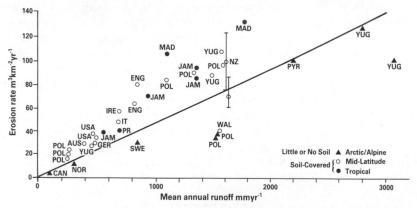

Figure 9.1 Graph of limestone erosion rate against mean annual runoff for selected sites (AUS = Australia, CAN = Canada, ENG = England, GER = West Germany, IRE = Ireland, IT = Italy, JAM = Jamaica, MAD = Madagascar, NZ = New Zealand, NOR = Norway, POL = Poland, PR = Puerto Rico, PYR = Pyrenees, SWE = Sweden, USA = United States, WAL = Wales, YUG = Yugoslavia (from Jennings, 1985)

other processes lead to higher levels of CO_2 in soil atmospheres. Similar microbial processes may act within the upper zone of weathered bedrock (the epikarstic zone) leading to the production of ground-air CO_2 (Atkinson, 1977) which may be an important agent of limestone dissolution.

Some specific studies confirm the role of biogenic CO_2 in increasing dissolution rates. Woo and Marsh (1977), for example, evaluated the effects of tundra vegetation on limestone solution on Ellesmere Island, North West Territories, Canada. Input and output waters were sampled along a test stream reach, as was soil water at the base of an adjoining vegetated slope. Results showed higher values of water hardness and bicarbonate concentration throughout the growing season than those recorded for non-vegetated areas in the Arctic. Woo and Marsh ascribed these findings to the biogenic production of CO_2 within the study area.

Biogenically produced CO_2, therefore, seems to play a general role in influencing overall rates of solutional erosion in karst landscapes. Other factors are also important, especially the dominant control of runoff, and conversely other organically produced acids may also be involved in limestone denudation. Although much work has gone into explaining the nature and variability of rates of solutional erosion in karst terrains, it is still unclear in many cases how such generalized rates may be related to the sculpture of karst landforms.

Karst landforms can be classified, fairly simply, into those occurring on bare rock, those on soil covered surface, and those formed underground. All these occur at a range of scales, from minor rock sculpture to large hills (see table 9.1). Contemporary geomorphological theory aims to explain the genesis of this range of features, as well as how the overall system operates. A largely hydrological perspective has been taken by karst theorists, as the genesis and development of karst landforms can be related to water flow.

Table 9.1   A classification of karst forms according to size and situation

Situation	Max. dimension	Examples
Subaerial surfaces	<10 m	Karren
	10–1,000 m	limestone pavements, tufas
Soil-covered surfaces	<10 m	subsoil karren
	10–1,000 m	closed depressions, residual hills, cones and towers
Subsurface	<10 m	scallops, small speleothems
	>10 m	cave passages and galleries

The question of biokarst

Karst geomorphology is full of terminology, much of which has not been defined rigorously, or whose use has been extended beyond its original meaning and application. Many of the terms used in karst geomorphology come from local names for specific features. They are, therefore, described in a wide variety of languages, leading to some terminological confusion. There has been a long-term trend in karst geomorphology to name forms *before* any rigorous investigation of the processes that cause or influence them. The proliferation of terms for small-scale sculpting on limestone surfaces (karren), for example, obscures the fact that little is known about how such forms originate, and whether there is any justification for separating them on morphometric criteria alone.

Although observations of organisms influencing surface weathering were made in the nineteenth century and probably earlier, a direct relationship with landforms was not noted until more recently. Jones (1965) made an early suggestion that limestone pavements were largely sculpted by biological processes under a soil cover, and several workers have recognized the importance of biological processes in the formation of coastal karren and notch features (e.g. Neumann, 1968; Hodgkin, 1970).

Folk *et al.* (1973) made a more explicit connection between organic action and landforms in their description of 'phytokarst' at Hell, Grand Cayman Island. They observed a complexly dissected area of pinnacles up to 3 m high in an inland, swampy area. Dissection was not gravitationally oriented, and led to a random, spongy appearance. The pinnacles were covered in a black coating (green in more shady spots) which turned out to be a layer of blue-green algae (Cyanobacteria).

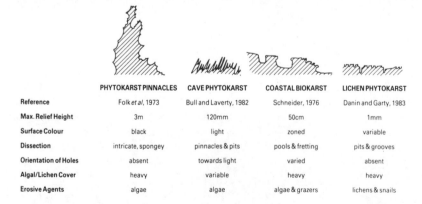

	PHYTOKARST PINNACLES	CAVE PHYTOKARST	COASTAL BIOKARST	LICHEN PHYTOKARST
Reference	Folk *et al*, 1973	Bull and Laverty, 1982	Schneider, 1976	Danin and Garty, 1983
Max. Relief Height	3m	120mm	50cm	1mm
Surface Colour	black	light	zoned	variable
Dissection	intricate, spongey	pinnacles & pits	pools & fretting	pits & grooves
Orientation of Holes	absent	towards light	varied	absent
Algal/Lichen Cover	heavy	variable	heavy	heavy
Erosive Agents	algae	algae	algae & grazers	lichens & snails

Figure 9.2 Characteristics of phytokarst/biokarst observed from different areas (compiled with data from Folk *et al.*, 1973; Bull and Laverty, 1982; Schneider, 1976; and Danin and Garty, 1983)

Filamentous forms of these algae were observed penetrating into the substrate (see figure 9.2). Folk *et al.* concluded from this juxtaposition of unusually fretted pinnacles and algal coating that the algal filaments boring into the rock caused the non-gravitationally oriented hollows and, indeed, the pinnacle forms themselves. Hence, they named the forms 'phytokarst', and such forms have since been identified elsewhere (Strecker *et al.*, 1984; Palmer, 1984). Subsequent research has suggested that Folk *et al.* overstated their case. Certainly, to date no convincing process–form links have been established between algal boring and large-scale terrestrial landforms (Viles, 1987a). Viles and Spencer (1986) suggest that structural control is of major importance in shaping the gross morphology of the pinnacles at Hell. Similarly, a *suite* of weathering processes, acting upon heterogeneous rock surfaces, is more likely to be responsible for the minor complex sculpting, rather than algal boring alone.

The term 'phytokarst' has been applied to other, supposedly similar, features from very different environments. Waltham and Brook (1980) and Bull and Laverty (1982) observed bed rock sticks, angled towards the light and covered with algae, at the entrance to caves in Mulu, Borneo (see figure 9.2). They concluded that these features were a product of phototrophic algal erosion. Similar features, although on a smaller scale, have been observed in Australia (Jennings, 1982) and Austria (D. Gebauer, pers. comm.). This 'phytokarst' is covered with an algal coating, like that recognized by Folk *et al.* (1973), but again process–form links are difficult to establish. The term phytokarst has also been used to describe micro-scale grooves and pits produced by lichen thalli upon Mediterranean calcrete surfaces (Klappa, 1979; and see figure 9.2).

Work from intertidal limestone platforms, where a suite of organic processes has been observed to produce assemblages of karren features, has led to the term 'biokarst' being adopted (Schneider, 1976; Torunski, 1979; Le Campion-Alsumard, 1979). The term 'biokarst' implies that a wider range of organic processes are involved (i.e. grazing, abrasion, and boring) than in 'phytokarst' which is limited to plant action. The term 'biokarst' has been applied to other, possibly analogous features (Viles, 1984) where rock sculpture is presumed to be carried out by suite of microorganisms, plants and animals. Depositional landforms, as well as erosional features, may be classed as biokarst forms, although in both cases more process–form investigations need to be carried out to elucidate important links. Work from Aldabra Atoll, Indian Ocean, suggests that no one distinctive terrestrial landform can be directly related to biological action, although there is evidence to indicate that blue-green algae (Cyanobacteria) have a general impact upon surface weathering (Viles, 1987a). Process interactions occur, so in some cases the effects of biological processes are outweighed by 'inorganic' processes, and vice versa.

Two important general questions emerge from a consideration of biokarst:

1   What processes/conditions lead to the production of features already identified as biokarst?
2   Are there more general links between organic weathering processes and karst landforms?

Observations of biokarst features and biological weathering processes have been made by researchers in a range of disciplines, e.g. geology, geomorphology, ecology, and microbiology, and different perspectives emerge from these. An ecological perspective would seem to provide the best framework for future studies, capable of recognizing the interrelatedness of organic and inorganic processes.

Characteristics of ecosystems on limestone

Ecosystems developed in limestone areas are as varied as the limestone landscapes themselves, as they are both heavily influenced by climatic and historical factors, and lately by the 'human impact' (Goudie, 1986). However, some general features have been recognized which distinguish these ecosystems from those found upon other rock types. According to Jennings (1985), root systems penetrate deeper and productivity is lower in vegetation on karst as opposed to non-karst areas. This in part reflects the lack of surface water in karst landscapes and the characteristically nutrient-poor, thin mineral soils found there. Generalizations about ecosystems upon limestone really depend upon *soil type*. Characteristically, limestone soils are thin, rich in calcium and high in pH when formed primarily from the parent rock material. Where more allochthonous material is present, however, organic-rich and sometimes highly acidic soils are found. Weathering rates beneath soils vary according to soil acidity (Trudgill, 1976a). The type of vegetation present in a limestone area is related to soil type (and itself influences the development of soils) with calcicole types developing on calcareous soils and calcifuge species preferring more acidic substrates.

Several detailed studies have been made upon ecosystems (or selected components of these) in various limestone areas which are of interest here. Analyses have focused upon such parameters as the species present, the biomass, rates of initial colonization and succession, as well as links between parts of the ecosystem (due to mineral and energy flows). Examples presented below are divided into three categories: ecosystems upon bare, subaerial limestone surfaces; those on soil covered limestone surfaces; and those within cave systems.

Subaerial exposures of limestone without soil cover have relatively simple ecosystems developed upon them. Communities upon such substrates are primarily composed of pioneering organisms, i.e. cryptogams and

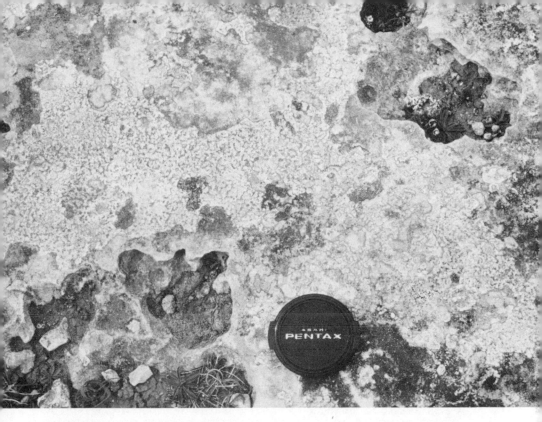

Figure 9.3 A community of endolithic and epilithic lichens growing on a calcrete surface in Mallorca (lens cap diameter c 5 cm)

microorganisms. Such organisms occupy a range of microhabitats within the rock surface (Golubic *et al.*, 1981). A simple classification can be made into those which live *upon* the rock surface (epiliths), those which live in preformed cavities within the surface (chasmoliths), and those which actively bore into the substrate (endoliths). A slightly more complex classification of niches is presented in figure 8.4 in chapter 8 of this book. Common types of organisms found in these microhabitats upon limestone substrates include bacteria, fungi, algae (including cyanobacteria), lichens and mosses. An example of a lichen community on an exposed carbonate surface is shown in figure 9.3.

McCarthy (1983) described lichen and algal communities upon limestone pavements in the Burren, south-east Ireland, using simple quantitative techniques. Four distinctive communities were recognized occupying clints (flat pavement surfaces); grikes (runnels) and shaded areas; kamenitzas (solution pools); and erratic boulders, respectively. Clint surfaces were commonly covered by blue-green algae (Cyanobacteria), and endolithic and sub-epilithic lichens. *Verrucaria baldensis* was the most abundant lichen upon clints. A low diversity of lichen species was found on the most xeric of clint surfaces, and well-developed lichen thalli were relatively rare. In grikes and shaded clint areas a nondescript growth of blue-green algae and endolithic lichens was common. In very shaded sites bryophytes became more common. In kamenitzas a zonation was found, which could be related to frequency of wetting. On erratic boulders and drystone walling the lichen flora was influenced to an extent by the availability of nutrients from bird droppings. Where bird droppings were common (bird perch sites), well developed lichen thalli of *Aspicilia calcarea* were found. A great difference was noted between the lichen flora on clints and that found on erratic boulders, with epilithic species more common on the boulders.

McCarthy (1983) found no evidence, in the Burren sites studied, of a successional sequence involving a flora dominated by endolithic and sub-epilithic species being superseded by epiliths, as had been suggested by previous workers. Instead, McCarthy explained differences in floras in terms of microenvironmental parameters. Similar studies have been carried out in Israel by Danin and Garty (1983). They investigated the distribution of cyanobacteria and lichens over limestone hillslopes in the Negev desert and concluded that the major control over the distribution of these organisms is the cumulative imbibition time, or CIT. The CIT is the length of time that water can usefully be absorbed by organisms and is controlled by the temperature and moisture received by the substrate. In the Negev sites studied the CIT was often influenced by dew. North-facing rocks were found to be inhabited by epilithic lichens, and south-facing ones by endolithic cyanobacteria. Stones upon the south-facing slopes were inhabited by endolithic lichens, and were found to receive more dew. These findings show that there is a clear microenvironmental control upon community characteristics. Some of the communities studied in these investigations were found to leave a characteristic weathering 'imprint' upon the rock surface. Endolithic lichens, for example, create holes in the rock surface where their fruiting bodies are, and grooves upon the rock surface where two or more thalli meet.

Studies of the microflora upon limestone surfaces on Aldabra Atoll, Indian Ocean, indicate a certain correlation between blue-green algal communities and environmental factors (Whitton, 1971; Viles, 1987a). Blue-green algae are the most abundant type of rock surface microorganisms as indicated in figure 9.4. Lichens are relatively rare upon the atoll, and occur mainly on nutrient-enriched bird perch sites. Endolithic, chasmolithic and epilithic types of blue-green algae are present, as well as certain unidentified fungi and bacteria. Endolithic species are most common in frequently wetted sites, e.g. in kamenitzas and on intertidal surfaces. Distinctive zonations of species are found down the sides of kamenitzas, related to wetting frequency. On drier, unshaded surfaces, epilithic species are common, although a patchy chasmolithic layer is also found.

Colonization experiments carried out on Aldabra Atoll revealed gross differences in the speed of colonization upon cleared rock surfaces between

Figure 9.4 Blue-green algal cover on limestone surfaces from Aldabra Atoll: the kamenitza bottoms are white due to the absence of thick epilithic algal cover (Geological hammer is *c*. 30 cm long)

frequently wetted and dry sites (see figure 9.5). Wetter sites characteristically experienced more rapid colonization, whereas dry sites were often only sparsely covered with blue-green algae after one year. Epilithic species were found to colonize first, and there was little evidence of a chasmolithic layer after one year—suggesting that some kind of succession must occur. Similar recolonization experiments carried out upon intertidal surfaces on Mediterranean and Jamaican coasts reveal a much faster recolonization rate there, with endoliths the dominant component of the flora (Le Campion-Alsumard, 1979; Kobluk and Risk, 1977). At some sites in the Aldabra study blue-green algal colonization was observed to occur preferentially in topographic hollows.

The studies of lower plant communities upon bare limestone surfaces discussed above reveal some potentially important findings in terms of biological weathering. Communities of cyanobacteria and lichens are common on calcareous surfaces in nearly all terrestrial environments. Their spatial and temporal patterning is highly influenced by microclimatic and substrate characteristics which interact to control microenvironmental conditions. Competition and succession also influence community

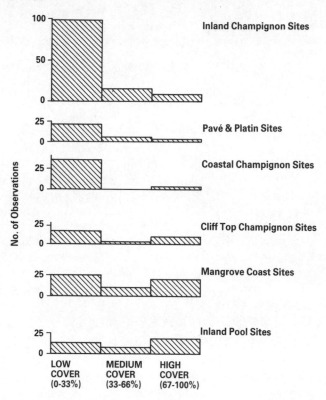

Figure 9.5 Frequency distributions of percentage cover of algal growths on cleared squares of limestone after one year, from Aldabra Atoll

characteristics. There is an autogenic component to some successions whereby the organisms directly weather and alter the surface, making conditions more favourable for other species. These features of community patterning will control the occurrence and importance of biological weathering on any one surface.

More complex ecosystems develop where a soil cover is present, as higher plants can colonize. Several studies have been carried out on the nature of ecosystems on soil-covered limestones and comparisons made with non-limestone communities in similar areas. Crowther (1982, 1987a), for example, studied tropical rain forest vegetation on limestones in Malaysia. The overall tree flora was found to be diverse, with many endemic species, but the cover was sparse and less luxuriant than on adjacent lithologies. Crowther studied thirteen 30 m × 30 m plots and found that soil cover within these ranged from 9.3 to 96 per cent. Soil cover and tree size generally increased with decreasing slope angle towards the base of slopes. Some lithological control was evident with impure limestones giving rise to smoother meso-topography with deeper and more continuous soil cover (and more luxuriant tree cover) than purer limestone sites. Estimates were made of nutrient cycling within six plots and it was found that there was very active cycling of calcium and magnesium (Crowther, 1987a). Crowther (1987a, p. 154) concludes that 'There is therefore much ecological diversity within tropical karst terrain, much of which is related to the depth and nutrient status of the soil'.

Similar studies have been carried out in four contrasting rain forests in Sarawak, which provide comparisons between alluvial forest, Dipterocarp forest, heath forest and forest on limestone (Proctor et al., 1983). The limestone forest here was found to have very shallow soils in comparison with the other forests studied (mean depths on limestone 11 cm, range 0–55 cm), with little development of soil horizons. Bare rock accounted for 9 per cent of the ground surface area in the limestone forest. The limestone forest was also the least species-rich, although the soil was found to be rich in nutrients. In terms of biomass, the limestone forest ranked below the Dipterocarp and heath forests, but above the alluvial forest.

The vegetation and soil characteristics in many limestone areas have been seriously altered by human occupation, with potential consequent effects upon limestone weathering. In the Dinaric karst areas, for example, there has been extensive forest clearance and replanting (Herak, 1972), and these changes will have had a great impact upon both hydrology and geomorphology. Evidence has been collected to suggest that during human occupation of the Burren area, Ireland, in prehistoric times there was catastrophic soil erosion (Drew, 1983). This human-induced soil erosion exhumed the pavement surfaces which had developed under soil cover, and exposed them to a different weathering regime. In the very different environment of the Yucatan peninsular, Central America, human occupation since the early days of the Mayan civilization has been reflected in major changes to the vegetation of this limestone area. The Mayan civilization was based upon agriculture, including orchards and terraced farming (Turner, 1986). Large areas of the peninsula are now covered by sisal and other cash crops, and the area of original forest vegetation is shrinking. Annual burning and other farming practices influence soil erosion, and may affect subsoil limestone weathering in some parts of the peninsula.

Soil and vegetation systems upon limestone substrates are highly influenced by climate, lithology and human interference. In many cases the limestone substrate causes the vegetation to be less rich than on other rocks in the same area, and bare rock patches are more common in limestone areas, as shown in figure 9.6. As in other environments vegetation, animals and microorganisms within limestone areas will influence soil characteristics, and through their influence on soil and hydrology will affect geomorphology.

Caves support very different ecosystems to those found on terrestrial surfaces, the obvious control being the lack of sunlight in most parts of

Figure 9.6 Soil and vegetation-covered limestone, interspersed with exposed limestone covered with epilithic lichens, from Mallorca

cave systems. Organisms depending upon light as an energy source (phototropic organisms) are therefore largely absent from such ecosystems, and their place is taken by organisms which can obtain energy directly from minerals (chemolithotrophic organisms). Heterotrophic fungi and bacteria are also common in caves which break down organic waste within the ecosystem. A range of animals, with special adaptations to cave environments, is also found. A clear cycling of nutrients and energy is found within cave systems, as cave animals consume heterotrophic microorganisms. Although some organisms live permanently in cave environments, others feed outside and only use caves for shelter, e.g. bats and birds, thus providing a link with surface ecosystems. Water entering cave systems from the ground surface provides another link with surface ecosystems, bringing in organic debris, minerals and microorganisms.

Various studies have been made of components of cave ecosystems, and special attention has been paid to enumerating the types of microorganisms present. Caumartin (1963), for example, reviewed the microbial flora of caves and described a range of organisms including chemolithotropic iron and sulphur bacteria, heterotrophic bacteria, actinomycetes and fungi. Several species of cyanobacteria have been discovered in caves with special adaptations to the very low light intensities experienced (Cox, 1977; Cox and Marchant, 1977). Many of these microorganisms (especially chemo-lithotropic species) are involved in mineral transformations, and through these processes influence weathering in the cave environment.

From the review of ecosystems on bare, soil-covered and cave surfaces presented in the preceding paragraphs it can be seen that the range of communities encountered is vast, and any generalizations are difficult to uphold. Some general points emerge, however. In many karst areas a complex pattern of bare and covered rock surfaces is present, forming a more or less well-linked ecosystem, which may also have an underground component. Nutrient cycling in any one of these components affects solutional and other weathering processes both locally and throughout the karst system. Important features of ecosystems to consider in evaluating their impact on limestone weathering are the spatial patterning of vegetation and lower plant communities and their links with grazing animals; and also successional and other changes in communities over time.

Biological influences upon weathering and erosion in limestone areas

Having established the general characteristics of limestone topography and associated ecosystems in a range of areas, some specific examples of the weathering and erosional interface between the two can be examined. A brief examination of the pertinent literature reveals that four main types of biological weathering and erosion can be recognized: those due to (a) lower plants; (b) higher plants; (c) animals; and (d) a combination of two or three of (a) to (c).

Many recent studies have focused upon microorganic and cryptogamic influences upon limestone weathering (see table 9.2). This focus of work has probably resulted because microorganisms have direct, small-scale links with limestone substrates, which may be preserved in the geological record. These small-scale links can be studied more easily now due to recent advances in microscopic and microanalytical techniques.

The major forms of lower plant life responsible for limestone weathering appear to be various types of algae (mainly blue-green algae), fungi, bacteria and lichens. Observations of microorganic weathering have been made in caves, on intertidal platforms, calcrete surfaces, limestone pavements, desert and raised reef surfaces, and on buildings in lightly polluted urban areas. There is still much controversy over the reality and nature of weathering instigated by these organisms. Two major hypotheses have been put forward:

1 a microorganic layer serves to protect the surface of rocks and stone from weathering and erosion; and
2 a microorganic layer actively promotes surface weathering through a range of chemical and physical mechanisms.

Table 9.2 Some examples of the contributions of microorganisms and lower plants to terrestrial weathering and erosion of carbonates

Organism	Process/Location	Reference
sulphur bacteria	calcite weathering, thermal spring	Kieft and Caldwell, 1984
	building-stone weathering	Paine et al., 1933
bacteria	building-stone weathering	Pochon and Jaton, 1968
fungi	dolomitic and calcitic building-stone weathering	Koestler et al., 1985
fungi/algae	limestone and dolomite weathering	Degelius, 1962
blue-green algae	corrosion, lake Krustensteine	Schneider, 1977
	weathering of limestone walls, Jerusalem	Danin, 1983
	limestone weathering, Aldabra Atoll	Viles, 1987a
	corrosion of tufa	Pentecost, 1978
blue-green and red algae	weathering of limestone cave pinnacles, Mulu, Borneo	Bull and Laverty, 1982
blue-green algae and lichens	building-stone weathering, Italy	Del Monte and Sabbioni, 1986
blue-green algae fungi & lichens	limestone weathering, deserts	Krumbein, 1972
lichens	calcite and dolomite weathering	Muxart and Blanc, 1979
	limestone weathering	Ascaso et al., 1982
	calcrete weathering, Mediterranean	Klappa, 1979
	marble weathering, Venice	Lloyd, 1974
	building-stone weathering	Bech-Anderson et al, 1983

Observations from buildings and various bare rock surfaces have been used to add support to the hypothesis that microorganisms protect substrates from other weathering processes (Jennings, 1985, p. 71; Lallemant and Deruelle, 1978). A thick layer of microorganisms, it is suggested, absorbs rainfall and salts from the atmosphere, preventing them from attacking the underlying rock and leading to the production of relatively smooth surfaces. On the other hand, much evidence has been collected to back up the second hypothesis, suggesting that microorganisms actively weather rock surfaces. Krumbein (1972), for example, suggests that a community of blue-green algae, fungi and bacteria form a 'biological weathering front' which progressively lowers the surface. The two hypotheses are not necessarily mutually exclusive. Different types of microorganisms may have different effects, and there may be a general climatic or environmental control upon biological processes. It has been suggested, for example, that lichens protect rock surfaces under humid climates and weather the surface under arid conditions. There may be a measure of autocorrelation as well, as the occurrence of different species of microorganism is to some extent controlled by environmental factors. The scale factor may also be important to biological activity, with weathering occurring on the micro-scale under individual microorganisms, but the overall community protecting the surface. Growth and succession within rock surface microorganism communities may lead to changes over time in their effects upon the underlying rock.

The mechanisms by which microorganisms weather underlying rock substrates have been a matter of debate for many years. Table 9.3 presents a compilation of some suggested processes and effects. Mechanical and/or a variety of chemical processes have been suggested as direct mechanisms, but indirect effects may also be important. Indirect effects occur when, for example, rainwater running over lichen communities becomes acidified, leading to increased solutional erosion downslope (Jones, 1965). Suggested direct mechanisms range from chemical production of boreholes and acid etching of minerals to cracking due to swelling and shrinking. Recent reviews of the mechanisms suggested for the main types of microorganisms involved are given in Lukas (1979) and Viles (1984).

Some species of lichens and cyanobacteria, especially endolithic forms, have been found to produce holes within rock surfaces leading to recognizable features, as illustrated by figures 9.7 and 9.8. Several workers (notably Folk et al., 1973) have concluded from this that such species are dominantly responsible for weathering, and by implication are also responsible for surface lowering and small-scale rock sculpture. The situation is undoubtedly more complex than this in most terrestrial situations. Although recognizable holes may be produced, it is unclear whether this itself produces an important amount of weathering, overriding other processes. Work carried out on lichens from exposed limestone surfaces on the Mendips, south west England (Viles, 1987b) suggests that depths and densities of lichen effects are very variable and the resultant influence on weathering is difficult to elucidate from field evidence. The formation of holes in the surface by some species of lichens is an established fact, but it is unclear from field evidence whether this process is accompanied by a general surface lowering under the lichen community or whether the lichens protect the weakened surface from erosion until they are themselves removed by grazing or death.

Studies by Klappa (1979) indicated that endolithic lichens growing on Mediterranean calcrete surfaces produce cavities in the surfaces, leading to important textural and fabric changes which are preserved in the stratigraphic record. In this case, therefore, although lichens are involved in weathering there is little allied surface lowering. Thin section and electron microscope studies of field material from Aldabra Atoll also

Table 9.3 Mechanisms and effects of carbonate weathering by microorganisms and lower plants

| Mechanisms | | | Effects | | | |
| | | | | Microtopographic feature-producing | | |
Direct	*Indirect*	*Protective*	*Surface-weakening*	*Feature*	*Organism responsible*	*Dimensions*
Chelation by epiliths and chasmoliths	Acidification of runoff waters passing over epilithic communities	Absorption of water/acids from atmosphere, preventing them interacting with substrate	Direct micro-organic weathering weakens the substrate, encouraging erosion	Boreholes	fungi, blue-green algae	c. 1–25 μm diam. up to 12 mm deep
Increased CO_2 during respiration				Circular pits	lichen fruiting bodies	c. 100–200 μm diam. c. 0.5–1 mm deep
Production of acidic metabolic by-products		Insulation of substrate from temperature changes, frost and wind.		Grooves	lichens (boundary between two thalli)	c 0.5–2 mm wide c. 50–500 μm deep.
Boring by chemical and/or physical means						

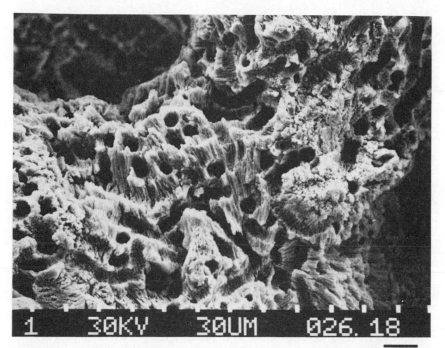

Figure 9.7 Scanning electron microscope photograph of blue-green algal boreholes from Aldabra Atoll
(scale bar = 30 µm)

Figure 9.8 Detailed view of the blue-green algal boreholes shown in figure 9.7
(scale bar = 10 µm)

indicate a complicated pattern of influences of blue-green algae on surface lowering (Viles, 1987a). Cyanobacteria upon terrestrial surfaces on Aldabra Atoll produce etching of the substrate, sometimes forming boreholes and troughs. In some cases this seems to be accompanied by surface lowering, but in others a clear 'altered zone' is formed in the top few millimetres of substrate. There is no clear link between these different microscale features and larger scale topographic styles on Aldabra so, for example, cyanobacterial boreholes and associated surface lowering are found both on highly dissected topography and also on flatter surfaces.

More field observations from different areas, combined with laboratory investigations of process mechanisms, are necessary to establish a general picture. The fact that most seemingly bare limestone surfaces are covered, at least patchily, by a range of lower plants implies that any influence they have upon weathering will be of general applicability to subaerial surfaces, and the lighter parts of cave entrances.

Higher plants have also been observed to influence weathering and erosion of limestone, although the effects have often been discounted as of no general importance–largely because they act at an intermediate scale and are difficult to monitor. Direct and indirect effects (the latter mainly mediated through soil cover) of a chemical and physical nature have been observed for many years, but only recently have any quantitative estimates of their importance been obtained. Stephens (1843) gives an early report of tree roots splitting limestone, in this case limestone walls of the Mayan ruins at Kabah in the Yucatan peninsula. Figure 9.9 illustrates the scene, which Stephens (1843, p. 242) describes as follows:

> the rankness of tropical vegetation is hurrying to destroy these interesting remains. The tree is called the alamo, or elm . . . Springing up beside the front wall, its fibres crept into cracks and crevices, and became shoots and branches which, as the trunk rose, in struggling to rise with it, unsettled and overturned the wall, and still grew, carrying up large stones fast locked in their embraces, which they now hold aloft in the air . . . no sketch can convey a true idea of the ruthless gripe in which these gnarled and twisted roots encircle sculptured stones.

More recent observations of root effects on limestone surfaces have suggested that both physical and chemical processes are important in producing 'root karst'. Several examples of root karst phenomena are given in Esteban and Klappa (1982) and Wall and Willford (1966), and an example from tropical, coastal limestone is shown in figure 9.10. Few studies have been made upon the question of the spatial frequency of root/rock contacts within limestone areas and how great an overall contribution to weathering might therefore be made by tree root processes.

Figure 9.9 Tree roots attacking limestone blocks from the Mayan ruins at Kabah, Mexico
(from Stephens, 1843)

Figure 9.10   'Root karst' from a tropical coast: roots of the mangrove *Rhizophora mucronata* and associated highly dissected limestone, Aldabra Atoll

Crowther (1987b) has produced some very interesting estimates of the magnitude of root action on tropical forest soil-covered limestones, using information from nutrient cycling studies. This work indicates that in the areas studied in Peninsular Malaysia, the combined uptake of calcium and magnesium by plants exceeds 400 kg ha^{-1} yr^{-1}. This is equivalent to a surface lowering rate of more than 40 mm 1,000 yr^{-1}, compared with the rates obtained by Crowther for chemical denudation loss of 57.5–73.4 mm 1,000 yr^{-1}. These results suggest that the soil–root–rock zone is an important locus of limestone solution within tropical forest environments. Other effects have also been observed when trees make direct contact with limestone surfaces. For example, rainwater running down tree bark (stem flow) becomes acidified leading to increased solution at the tree base. In some cases pronounced micro-runnels and polished surfaces are produced (see Trudgill, 1985, p. 37 for a clear example).

Animals have been recorded as contributing to weathering and erosion of limestone in certain areas, as well as higher and lower plant forms. Animal influences are particularly noticeable in the intertidal zone (see Trudgill, 1985 and chapter 8 of this volume), where large populations of vagile and sessile invertebrates are found. Such organisms have been found to be involved in the formation of notches in tropical limestone cliffs, as well as in the development of coastal karren assemblages.

Some scattered observations have been made upon animal-induced weathering in terrestrial areas, although the situation is more complicated than that found in coastal areas. Faunal effects can be split into three categories: the production of boreholes or pits; abrasion; and the action of excrement. Faunal boreholes are relatively common in the intertidal zone, but have been reported only rarely from terrestrial sites. Crowther (1979) recorded circular depressions of 2–5 mm diameter and 2–4 mm depth on bare limestone surfaces in West Malaysia which seemed to be produced by snails (of the genus *Alycaeus*). These circular pits were found to occur at densities sometimes exceeding $1.5 \, cm^{-2}$, and were often occupied by living snails. Crowther studied water chemistry over limestone micro-catchments in the area, and found that total hardness was greater in waters running off snail-covered sites than from sites where snail holes were absent. Crowther ascribes this increase in hardness to the increased surface area produced by pitting, but over a longer time span snail action will itself influence weathering rates.

Stanton (1984) has identified potentially similar snail holes in limestone exposures on the Mendip Hills, southwest England. Tubular holes of *c*. 20 mm diameter and up to 50 mm deep have been found, with complex forms also present. These holes are normally only found rising vertically into overhanging surfaces, but sometimes occur on vertical surfaces. Stanton (1984) notes that these forms are found most commonly in the low cliff complexes and rock exposures which occur on the slopes of closed depressions in the the Mendip Hills. A developmental sequence can be recognized from shallow scrapes through simple holes to complex holes and rock honeycombs. Snail holes are also found in quarries only 200 years old on the Mendip Hills, suggesting that they can form at rates of up to $0.15 \, mm \, yr^{-1}$. Many of the holes here are occupied by snails (*Cepaea nemoralis*), which are presumed to have formed them. Undoubtedly, more observations are needed of such holes and the mechanisms of their formation before any general comments can be made about the extent and importance of snail action on terrestrial limestone surfaces.

Some animals abrade rock surfaces through the action of claws (through walking) and jaws (through grazing). In the intertidal zone species of molluscs, echinoderms, crabs and fish graze rock surfaces feeding upon endolithic and other surface-dwelling microorganisms (see Trudgill, 1985, pp. 138–42; and Chapter 8, this volume). Grazing on terrestrial surfaces is more rare, probably due to the increased availability of more easily obtained food sources, and a smaller proportion of bare rock areas. On Aldabra Atoll giant tortoises (*Geochelone gigantea*) and land crabs are found to abrade surface rocks, producing characteristic grooves. Tortoise-produced grooves are often found in groups of 4 or 5, with widths of 1–5 mm and may be up to 1 m long. Land crabs produce shorter, smaller grooves in larger groups which may reach high densities,

Figure 9.11 Crab-produced grooves upon limestone surfaces covered with epilithic algae from Aldabra Atoll
(scale bar = 10 cm)

as shown in figure 9.11. The distribution of these features is very patchy on Aldabra Atoll, with tortoise-produced grooves concentrated around major drinking holes. Giant tortoises also abrade rough rock surfaces with the underside of their shells (the plastron) which causes rounding of small rock pinnacles. Similar claw groove features have been recorded by Splettstoesser (1985) from sandstone outcrops in the Falkland Islands, caused by the feet of rockhopper penguins (*Eudyptes crestatus*). There have also been suggestions that bats produce groove features near their roosts in caves, although this has been the subject of some debate (King-Webster and Kenny, 1958; Hooper, 1958).

Faunal excrement becomes deposited upon limestone outcrops and in caves and, depending upon its chemical composition or the composition of leachates from its microbial decomposition, may cause or encourage surface weathering. Some work has been carried out on the role of pigeon droppings in the weathering of calcareous building stones in urban areas. Bassi and Chiatante (1976) showed, from studies on Milan Cathedral, Italy, that fungal deterioration of marble was stimulated by pigeon excrement. It is commonly observed that bird droppings stimulate lichen growth upon 'bird perch' sites, and may lead to nitrate enrichment of stone surfaces. It is unclear, however, whether bird droppings play any direct role in stone weathering. Jennings (1963) recognized that bird guano corrodes bedrock and speleothems within caves. Bird droppings and mammalian excrement also influence water chemistry in solution pools upon limestone surfaces, which may encourage blue-green algal growth and corrosion of the pool sides.

It is clear from the preceding discussion that various components of ecosystems upon bare and soil-covered limestone surfaces, and in caves, *do* affect weathering and erosion. The interrelatedness of ecosystems and the presence of several different weathering processes, however, makes the situation complicated and difficult to unravel, as has been found in coastal areas. Interrelations of biological processes may occur. A key component of the model of coastal bioerosion put forward by Schneider (1976), for example, is the *synergistic* association of various biological processes which together lead to surface lowering. Microboring algae weaken the surface of the limestone, through the production of a network of boreholes, which grazing animals can easily remove. 'Inorganic' processes (e.g. solution and abrasion) can be ruled out on the section of the Istrian coast which Schneider studied, and therefore associations of biological processes can be directly related to coastal karren features and sediment production (Schneider and Torunski, 1983; Figure 8.5, this volume).

A few studies have indicated that similar process associations may act in terrestrial areas. From work in the Negev desert, for example, Danin and Garty (1983) produced a model involving lichens, cyanobacteria and snails (*Echnodrulus albulus*) to explain the production of microscale topographic features, as shown in figure 9.2. In this example, endolithic lichens and algae produce microscale pits in the surface, as well as grooves where two lichen thalli meet. Grazing snails feed on the lichens and produce deeper and wider grooves in the process. A synergistic association of processes exists, therefore, leading to recognizable micro-karren features. Desert and intertidal environments are similar in some respects, being stressed environments with little or no soil cover, which makes them of particular biogeomorphological interest. In other environments a soil/no-soil mosaic is usually present, complicating any synergistic process associations.

Biological influences upon deposition and consolidation of limestone

Dissolution of $CaCO_3$, $MgCO_3$ and $CaMg(CO_3)_2$ in carbonated water is a reversible process. Therefore, under certain conditions precipitation of these minerals occurs leading to the production of constructive landforms. This process is not wholly limited to areas of carbonate bedrocks as calcrete, for example, forms in a variety of situations, but it is nevertheless an important phenomenon in karst landscapes. Precipitation of carbonate

Table 9.4 Contributions of microorganisms and lower plants to terrestrial precipitation and consolidation of carbonates

Feature	Associated organisms	Reference
Freshawater		
travertine	bacteria	Julia, 1983; Chafetz and Folk, 1984
	bacteria and algae	Casanova, 1981
	algae and mosses	Golubic, 1969
tufa	blue-green algae	Pentecost, 1978, 1985
	blue-green algae and other microorganisms	Adolphe, 1981
oncolites	algae	Fritsch and Pantin, 1946
	blue-green algae	Golubic and Fisher, 1975; Riding, 1983
lake micro-reefs	blue-green algae	Schneider, 1977; Schneider *et al.*, 1983
Krustensteine	*Chara globularis*	Pentecost, 1984
Subaerial		
desert stromatolites	blue-green algae	Krumbein and Giele, 1979
laminar calcretes	lichens	Klappa, 1979
calcareous crusts	algae	Bunting and Christensen, 1980
case-hardening	blue-green algae	Trudgill, 1976b; Viles, 1987a
biodiagenesis	algae	Jones and Kahle, 1985, 1986
aussen Stalactiten	algae and mosses	W. H. Monroe, pers. comm.
Caves		
stalactites	fungi	Went, 1969
sinter crusts	blue-green algae	Friedmann, 1955
	fungi	Schneider, 1977
moonmilk	bacteria	Caumartin and Renault, 1958
	bacteria and fungi	Pochon *et al.*, 1964

material leads to the formation of tufas, travertines, oncolites, lake stromatolites, speleothems, calcretes and case-hardened surfaces in terrestrial environments. These forms can all be regarded as chemical sediments, consolidated by diagenetic processes. As with limestone weathering and erosion processes, organisms often influence, and in some cases may play a key role in, such precipitational processes (see table 9.4 for some examples). The landforms produced by precipitation of carbonate material have sometimes been regarded as forms of biokarst (Viles, 1984), due to this biological influence, and the term 'biochemical sediments' may also be appropriate for them. The exact role of organisms in the genesis of these features is still debated, and a variety of mechanisms have been proposed. In the following pages a few examples of the roles played by organisms (mainly microorganisms) in precipitational processes will be given for four important zones within the terrestrial karst system: subaerial surfaces, freshwaters, subsoil areas, and caves.

There have been several recent studies of precipitation on subaerially exposed limestone surfaces, which leads to the development of altered, often case-hardened, layers. These phenomena are of interest to geomorphologists (for their influence on landform development) and geologists (as distinctive case-hardened exposure surfaces may be preserved in the stratigraphical record and provide useful indicators of subaerial exposure). Case-hardening is itself an elusive phenomenon, with great variation in recorded thicknesses of case-hardened layers from a few centimetres (e.g. on Aldabra, as reported by Trudgill, 1976b) to a few metres (e.g. in Puerto Rico, as recorded by Ireland, 1979). The global distribution of case-hardening is still relatively unknown, although it is presumed that it is essentially a feature of semi-arid and sub-humid areas, being influenced by evaporation. Case-hardened surfaces are in many ways analogous to calcrete layers, and similar processes may be involved in their formation.

On Aldabra Atoll, where case-hardened surfaces are relatively common, there seems to be some correlation between the occurrence of case-hardening and the development of endolithic, epilithic and chasmolithic microflora. Some blue-green algae and other microorganisms (notably bacteria) are capable of precipitating $CaCO_3$ directly, and they may also act as efficient nuclei for the growth of inorganically precipitated $CaCO_3$, and the trapping of particulate $CaCO_3$. Some or all of these processes have been found to operate in tufas, and may well operate in the formation of case-hardening. As Klappa has demonstrated, lichen action (both precipitational and erosional) causes textural and fabric changes on calcrete surfaces, leading to the production of laminar calcretes (Klappa, 1979). Another example of subaerial precipitation of carbonates are the 'aussen Stalactiten' found in humid tropical areas. These stalactite-like forms develop in sheltered areas, often at the mouth of caves, and on cliff faces at the base of limestone towers. They commonly have a spongy texture and contain moss and algae covered with $CaCO_3$ (W. H. Monroe, pers. comm.).

In calcium carbonate-rich surface waters (streams and lakes) precipitation often occurs leading to the formation of stromatolites, oncolites and other coated grains, tufas and travertines and lake chalks and marls. Precipitation of $CaCO_3$ occurs when the water becomes supersaturated, which happens when CO_2 is removed. Organisms influence the CO_2 balance of surface waters through respiration and photosynthesis, and in some parts of stream reaches can control whether precipitation of $CaCO_3$ can occur. Recent work on tufas from a wide range of environments suggests that the role played by blue-green algae in tufa formation (through direct precipitation, and influence on CO_2 levels) has been overstressed (Pentecost, 1978, 1985). Actual precipitation by blue-green algae may account for only a small percentage of annual accretion of $CaCO_3$. Evidence further suggests that blue-green algae associated with tufas and with lake Krustensteine (i.e. carbonate crusts and furrows found in lakes, see Schneider, 1977) are active in eroding these deposits through biochemical boring. Undoubtedly, precipitation of $CaCO_3$ can occur without the participation or presence of organisms, but *in vitro* and *in vivo* studies both indicate that many organisms found in association with carbonate deposits are capable of precipitating $CaCO_3$ (Krumbein, 1979; Riding, 1977).

Many studies have been made of precipitation and sedimentation of calcium carbonate within lakes, which are not necessarily found within karst areas, but are of interest here. Calcium carbonate deposits (e.g. chalks and marls) are a common constituent of lake sediments and may either be produced from precipitation within the lake, or come from inwash. Several species within lakes are capable of forming extracellular $CaCO_3$ due to photosynthesis, and some benthic faunal species have calcareous skeletons which contribute to sedimentation of $CaCO_3$. Planktonic photosynthesis within lakes also plays an active role in the precipitation of $CaCO_3$, through the removal of CO_2. Various studies indicate that such biological influences play an important role in carbonate sedimentation in lakes, although there are obviously also other controls on water chemistry (Kelts and Hsu, 1978; Last, 1982).

Precipitation of $CaCO_3$ and other minerals is an important component of cave environments, leading to the development of speleothems, moonmilk etc. Bacteria and fungi have been found to play a potentially important role in such precipitational processes (see Viles, 1984, for a general review). Danielli and Edington (1983), for example, cultured bacteria from samples of moonmilk from Ogof Ffynnon Ddu cave in Wales, and found these to be capable of precipitating $CaCO_3$. It is difficult to transfer the results of *in vitro* experiments to natural situations, but it is clear that microorganisms in the cave environment are associated with chemical sediments there.

The preceding brief review of the range of influences of organisms upon the precipitation and consolidation of carbonate minerals indicates a very different situation to that of biological weathering and erosion. Microorganisms dominate biological influences upon precipitation, and synergistic associations may involve the interaction of degradational and constructional processes.

Some conclusions

The work presented in the preceding sections of this chapter provides much evidence of biological involvement in several earth surface processes within karst areas. Many complexities remain which can obscure any simple link between organisms and geomorphology. Firstly, there are many different mechanisms through which organisms influence particular geomorphological processes. In the case of dissolution of calcium carbonate, for example, microorganisms may play both direct and indirect roles. Secondly, many different organisms may be involved in any one earth surface process, and conversely, some species (especially microorganisms) may participate in a range of processes. Environmental factors (such as temperature, humidity and grazing pressure) often control whether particular biological processes occur, and how important they are. So, for example, a particular species of cyanobacteria may adopt the endolithic habit in stressed environments (e.g. desert and intertidal surfaces) with associated effects on weathering, but may be epilithic in more favourable environments. All these complications mean that it is difficult to establish *rates* of operation of biological weathering etc. to compare with estimates of dissolution rates and other earth surface process rates.

Further complexities arise when the impact of whole communities upon karst geomorphology is studied. Synergistic associations of processes are often encountered. Different organisms may affect geomorphological processes at very different scales and trying to estimate all effects in one study would be very difficult. Plant and animal communities also change over time (due to succession, human interference etc) and such changes may also be difficult to include in geomorphological studies.

The effects of organisms upon karst geomorphology can be classified according to scale if it is assumed that they contribute to the association of earth surface processes present. On the large scale, disturbances to organism communities lead to changes in process associations and subsequent geomorphological effects. On the medium scale, biological processes are an important component of process associations in certain areas, and as such influence the overall rates of denudation. On the small scale, in certain circumstances specific biological processes may dominate process associations and produce recognizable biokarst features.

References

Adolphe, J.-P. 1981: Examples de contributions microorganiques dans les constructions carbonatées continentales. *Bulletin de l'association de géographie française* 479–480, 194–5.

Ascaso, C. Galvin, J. and Rodriguez-Pascual, C. 1982: The weathering of calcareous rocks by lichens. *Pedobiologia* 24, 219–29.

Atkinson, T. C. 1977: Carbon dioxide in the atmosphere of the unsaturated zone; an important control of groundwater hardness in limestones. *Journal of Hydrology* 35, 111–23.

—— and Smith, D. I. 1976: The erosion of limestones. In Ford D. T. and Cullingford, C. H. D. (eds), *The Science of Speleology*, pp. 151–77. London: Academic Press.

Bassi, M. and Chiatante, D. 1976: The role of pigeon excrement in stone biodeterioration. *International Biodeterioration Bulletin* 12, 73–9.

Bech-Andersen, J., Christensen, P., Oxley, T. A. and Barry, S. 1983: Studies of lichen growth and deterioration of rocks and building materials using optical methods. In Oxley, T. A. and Barry, S. (eds), *Biodeterioration*, pp. 568–72. 5 Chichester: J. Wiley & Sons.

Bull, P. A. and Laverty, M. 1982: Observations on phytokarst. *Zeitschrift für Geomorphologie* NF 26, 437–57.

Bunting, B. T. and Christensen, L. 1980: Micromorphology of calcareous crusts from the Canadian High Arctic. *Geologiska Foreningens i Stockholm Forhandlingar* 100, 361–7.

Casanova, J. 1981 Morphologie et biolithogenèse des barrages de travertines. *Bulletin de l'Association de geographie française* 479–480, 192–3.

Caumartin, V. 1963: Review of the microbiology of underground environments. *National Speleological Society Bulletin USA* 25, 1–14.

—— and Renault, P. 1958: La corrosion biochimique dans un reseau karstique et la genese de mondmilch. *Notes biospéleologiques* 13, 87–109.

Chafetz, H. S. and Folk, R. L. 1984: Travertines: Depositional morphology and the bacterially controlled constituents. *Journal of Sedimentary Petrology* 54, 289–316.

Cox, G. 1977: A 'living fossil' in the twilight zone: A cave-wall deposit of unique ultrastructure. *Proceedings, 7th International Speleological Congress, Sheffield*, 129–31.

—— and Marchant, H. 1977: Photosynthesis in the deep twilight zone: Microorganisms with extreme structural adaptations to low light. *Proceedings, 7th International Speleological Congress, Sheffield*, 31–3.

Crowther, J. 1979: Limestone solution on exposed rock outcrops in West Malaysia. In Pitty, A. F. (ed.), *Geographical Approaches to Fluvial Hydrology*, pp. 31–50. Norwich: Geobooks.

—— 1982: Ecological observations in a tropical karst terrain, West Malaysia. I: Variations intopography, soils and vegetation. *Journal of Biogeography* 9, 65–78.

—— 1987a: Ecological observations in tropical karst terrain, West Malaysia. II: Rainfall interception, litterfall and nutrient cycling. *Journal of Biogeography* 14, 145–55.

—— 1987b: Ecological observations in tropical karst terrain, West Malaysia. III: Dynamics of the vegetation–soil–bedrock system. *Journal of Biogeography* 14, 157–64.

Danielli, H. M. C. and Edington, M. A. 1983: Bacterial calcification in limestone caves. *Geomicrobiological Journal* 3, 1–16.

Danin, A. 1983: Weathering of limestone in Jerusalem by cyanobacteria. *Zeitschrift für Geomorphologie* NF 27, 413–21.

—— and Garty, J. 1983: Distribution of cyanobacteria and lichens on hillsides of the Negev Highlands and their impact on biogenic weathering. *Zeitschrift für Geomorphologie* NF 27, 423–44.

Degelius, G. 1962: Uber Verwitterung von Kalk- und Dolomitgestein durch Algen und Flechten. In Hedval, U. A. (ed.), *Chemie im Dienst der Archaologie, Bautechnik und Denkmalpflege*, pp. 156–163. Goteborg: Akademieforlaget-Gumperts.

Del Monte, M. and Sabbioni, C. 1986: Chemical and biological weathering of an historical building: Reggio Emilia cathedral. *The Science of the Total Environment* 50, 165–82.

Drew, D. P. 1983: Accelerated soil erosion in a karst area: the Burren, Western Ireland. *Journal of Hydrology* 61, 113–26.

Esteban, M. and Klappa, C. F. 1983: Subaerial exposure. In Scholle, P. A., Bebout, D. G. and Moore, C. H. (eds), *Carbonate Depositional Environments* (Tulsa, Oklahoma; AAPG). AAPG Memoir 33, 1–54.

Folk, R. L., Roberts, H. H. and Moore, C. H. 1973: Black phytokarst from Hell, Cayman Islands, British West Indies. *Bulletin, Geological Society of America* 84, 2351–60.

Friedmann, J. 1955: Geitleria calcarea n.gen.n.sp. a new athmophytic lime-encrusting blue-green algae. *Botanisker Notiser* 108, 439–45.

Fritsch, F. E. and Pantin, C. F. A. 1946: Calcareous concretions in a Cambridgeshire stream. *Nature* 157, 397–9.

Golubic, S. 1969: Cyclic and non-cyclic mechanisms in the formation of travertine. *Verhandlungen der Internationale Vereinigung für Limnologie* 17, 956–61.

—— and Fisher, A. G. 1975: Ecology of calcareous nodules forming in the Little Connestoga Creek near Lancaster, Pennsylvania. *Verhandlungen der Internationale Vereinigung für Limnologie* 19, 2315–23.

——, Friedmann, I. and Schneider, J. 1981: The lithobiontic ecological niche with special reference to microorganisms. *Journal of Sedimentary Petrology* 57, 475–9.

Goudie, A. S. 1986: *The Human Impact* (2nd edn). Oxford; Basil Blackwell.

Gunn, J. 1986: Solute processes and karst landforms. In Trudgill, S. T. (ed.), *Solute Processes* pp. 262–437. London: J. Wiley & Sons.

Herak, M. 1972: Karst of Yugoslavia. In Herak, M. and Stringfield, V. T., *Karst: Important Karst Areas of the Northern Hemisphere*, pp. 25–84. Amsterdam: Elsevier.

—— and Stringfield, V. T. 1972: *Karst: Important Karst Areas of the Northern Hemisphere*. Amsterdam: Elsevier.

Hodgkin, E. P. 1970: Geomorphology and biological erosion of limestone coasts in Malaysia. *Bulletin, Geological Society of Malaysia* 3, 27–51.

Hooper, J. H. D. 1958: Bat erosion as a factor in cave formation. *Nature* 182, 1464.

Ireland, P. 1979: Geomorphological variations of 'case-hardening' in Puerto Rico. *Zeitschirft für Geomorphologie* NF Supplementband 32, 9–20.

Jennings, J. N. 1963: Geomorphology of the Dip Cave, Wee Jasper, New South Wales. *Helictite* 2, 57–80.

—— 1982: Karst of northeastern Queensland reconsidered. In *Tower Karst*, Chillagoe Caving Club Occasional Paper No. 4, pp. 13–52.

—— 1985: *Karst Geomorphology* Oxford: Basil Blackwell.

Jones, B. and Kahle, C. F. 1985: Lichen and algae: agents of biodiagenesis in karst breccia from Grand Cayman Island. *Bulletin, Canadian Society of Petroleum Geologists* 32, 446–61.

—— and —— 1986: Dendritic calcite crystals formed by calcification of algal filaments in a vadose environment. *Journal of Sedimentary Petrology* 56, 217–27.

Jones, R. J. 1965: Aspects of the biological weathering of limestone pavements. *Proceedings, Geologists' Association* 76, 421–33.

Julia, R. 1983: Travertines. In Scholle, P. A., Bebout, D. G. and Moore, C. H. (eds), *Carbonate Depositional Environments* (Tulsa, Oklahoma; AAPG). AAPG Memoir 33, 64–72.

Kelts, K. and Hsu, K. J. 1978: Freshwater carbonate sedimentation. In Lerman A. (ed.), *Lakes, Chemistry, Geology and Physics*, pp. 295–323. New York: Springer-Verlag.

Kieft, T. L. and Caldwell, D. E. 1984: Weathering of calcite, pyrite and sulphur by *Thermothrix thiopara* in a thermal spring. *Geomicrobiological Journal* 3, 201–16.

King-Webster, W. A. and Kenny, J. S. 1958: Bat erosion as a factor in cave development. *Nature* 181, 1813.

Klappa, C. F. 1979: Lichen stromatolites: Criterion for subaerial exposure and a mechanism for the formation of laminar calcretes (caliche). *Journal of Sedimentary Petrology* 49, 387–400.

Kobluk, D. R. and Risk, M. J. 1977: Rate and nature of infestation of a carbonate substratum by a boring alga. *Journal of Experimental Marine Biology and Ecology* 27, 107–15.

Koestler, R. J., Charola, A. E., Wypysky, M. and Lee, J. J. 1985: Microbiologically induced deterioration of dolomitic and calcitic stone as viewed by scanning electron microscopy. *Proceedings, 5th International Conference on Deterioration and Preservation of Stone*. Lausanne, Switzerland; Presses Polytechniques Romandes, 617–26.

Krumbein, W. E. 1972: Rôle des microorganismes dans la genèse, la diagenèse et la degradation des roches en place. *Revue d'écologie et Biologie du Sol* 3, 283–319.

—— 1979: Calcification by bacteria and algae. In Trudinger, P. A. and Swaine, D. J. (eds), *Biogeochemical cycling of mineral-forming elements*, pp 47–68. Amsterdam: Elsevier.

—— and Giele, G. 1979: Calcification in a coccoid cyanobacterium associated with the formation of desert stromatolites. *Sedimentology* 26, 593–604.

Lallemant, R. and Deruelle, S. 1978: The presence of lichens on stone monuments: nuisance or protection? *Proceedings, International Symposium on deterioration and Protection of Stone Monuments*. UNESCO/RILEM, Paris, Paper No. 4.6, 6 pp.

Last, W. M. 1982: Holocene sedimentation in Lake Manitoba, Canada. *Sedimentology* 29, 691–704.

Le Campion-Alsumard, T. 1979: Les cyanophycées endolithes marines. Systematique, ultrastructure, écologie et biodestruction. *Oceanologia Acta* 2, 143–56.

Lloyd, A. O. 1974: Lichen attack on marble at Torcello, Venice. *Petrolio e Ambiante* 1974, 221–4.

Lukas, K. J. 1979: The effects of marine microphytes on carbonate substrata. *Scanning Electron Microscopy* 1980/II, 447–56.

McCarthy, P. M. 1983: The composition of some calcicolous lichen communities in the Burren, western Ireland. *Lichenologist* 15, 231–48.

Muxart, T. and Blanc, P. 1979: Contribution a l'étude de l'alteration differentielle de la calcite et de la dolomite dans les dolomies sous l'action des lichens. Premières observations au microscope optique et au M.E.B. *Proceedings, International Symposium on Karst Erosion*. Union Internationale de Spéleologie, Aix en Provence-Marseille-Nimes, 165–74.

Neumann, A. C. 1968: Biological erosion of limestone coasts. In Fairbridge, R. W. (ed.), *Encyclopedia of geomorphology*, pp. 75–81. New York: Reinhold.

Paine, S. G., Linggood, F. V., Schimmer, F. and Thrupp, T. C. 1933: The relationship of microorganisms to the decay of stone. *Philosophical Transactions, Royal Society* 22B, 97–127.

Palmer, A. N. 1984: Geomorphic interpretation of karst features. In LaFleur, R. G. (ed.), *Groundwater as a Geomorphological Agent*, pp. 173–209. Boston: Allen and Unwin.

Pentecost, A. 1978: Blue-green algae and freshwater carbonate deposits. *Proceedings, Royal Society* 200B, 43–61.

—— 1984: The growth of Chara globularis and its relationship to calcium carbonate deposition in Malham Tarn. *Field Studies* 6, 53–8.

—— 1985: Association of cyanobacteria with tufa deposits: Identity, enumeration and nature of the sheath material revealed by histochemistry. *Geomicrobiological Journal* 4, 285–98.

Pochon, J. and Jaton, D. 1968: Biological factors in the alteration of stone. In Walters, H. and Elphick, J. J. (eds), *Biodeterioration of Materials*, pp. 258–68. Amsterdam: Elsevier.

——, Chalvignac, M. and Krumbein, W. E. 1964: Recherches biologiques sur le mondmilch. *Compte rendu hebdomadaire des séances de l'Academie des Sciences, Paris* 258, 5113–15.

Proctor, J., Anderson, J. M., Chai, P. and Vallack, H. W. 1983: Ecological studies in four contrasting lowland rain forests in Gunung Mulu National Park, Sarawak. I: Forest environment, structure and floristics. *Journal of Ecology* 71, 237–60.

Riding, R. 1977: Calcified Plectonema (blue-green algae): a recent example of Girvanella from Aldabra Atoll. *Paleontology* 20, 33–46.

—— 1983: Cyanoliths (cyanoids): Oncoids formed by calcified cyanophytes. In Peryt, T. M. (ed.), *Coated Grains*, pp. 276–83. Berlin: Springer-Verlag.

Roglic, J. 1972: Historical review of morphologic concepts. In Herak, M. and Stringfield, V. T. (eds), *Karst: Important karst areas of the northern hemisphere*, pp. 1–18. Amsterdam: Elsevier.

Schneider, J. 1976: Biological and inorganic factors in the destruction of limestone coasts. *Contributions to Sedimentology* 6, 112 pp.

—— 1977: Carbonate construction and decomposition by epilithic and endolithic microorganisms in salt- and fresh-water. In Flugel, E. (ed.), *Fossil Algae*, pp. 248–60. Berlin: Springer-Verlag.

—— and Torunski, H. 1983: Biokarst on limestone coasts, morphogenesis and sediment production. *Marine Ecology* 4, 45–63.

——, Schroeder, H. G. and Le Campion-Alsumard, T. 1983: Algal micro-reefs – coated grains from freshwater environments. In Peryt, T. M. (ed.), *Coated Grains*, pp. 284–98. Berlin: Springer-Verlag.

Sollas, W. J. 1880: On the action of a lichen on a limestone. *Report, British Association for the Advancement of Science* 586.

Splettstoesser, J. F. 1985: Note on rock striations caused by penguin feet, Falkland Islands. *Arctic and Alpine Research* 17, 107–11.

Stanton, W. I. 1984: Snail holes in Mendip limestones. *Proceedings, Bristol Naturalists' Society* 44, 15–18.

Stephens, J. L. 1843: *Incidents of travel in Yucatan* (2 vols.). New York: Harper and Brothers.

Strecker, M. R., Bloom, A. L. and Lecolle, J. 1984: Time span for karst development on Quaternary coral limestones, Santo Island, Vanuatu. *Abstracts of Papers, 25th International Geographical Congress, Paris*, vol. 1, Th 1.75.

Sweeting, M. M. 1972: *Karst Landforms*. London: Macmillan.

Torunski, H. 1979: Biological erosion and its significance for the morphogenesis of limestone coasts and for nearshore sedimentation (northern Adriatic). *Senckenbergiana Maritima* 11, 193–265.

Trudgill, S. T. 1976a: The erosion of limestone under soil and the long-term stability of soil-vegetation systems on limestone. *Earth Surface Processes* 1, 31–41.

—— 1976b: The subaerial and subsoil erosion of limestones on Aldabra Atoll, Indian Ocean. *Zeitschrift für Geomorphologie* NF 26, 201–10.

—— 1985: *Limestone Geomorphology*. London: Longman.

Turner, B. L. 1986: Mystery of the Maya revealed: The agricultural base of a tropical civilisation. *Focus* 36 (2), 2–7.

Viles, H. A. 1984: Biokarst: review and prospect. *Progress in Physical Geography* 8, 523–42.

—— 1987a: Blue-green algae and terrestrial limestone weathering on Aldabra Atoll: An SEM and light microscope study. *Earth Surface Processes and Landforms* 12, 319–30.

—— 1987b: A quantitative scanning electron microscope study of evidence for lichen weathering of limestone, Mendip Hills, Somerset. *Earth Surface Processes and Landforms* 12, 467–74.

—— and Spencer, T. 1986: 'Phytokarst', blue-green algae and limestone weathering. In Paterson, K. and Sweeting, M. M. (eds), *New Directions in Karst*. Norwich: Geobooks.

Wall, J. D. R. and Wilford, G. E. 1966: Two small-scale solution features on limestone outcrops in Sarawak, Malaysia. *Zeitschrift für Geomorphologie* 10, 90–4.

Waltham, A. C. and Brook, D. B. 1980: Cave development in the Melinau limestone of the Gunung Mulu national park. *Journal of Geography* 146, 258–66.

Went, F. W. 1969: Fungi associated with stalactite growth. *Science* 16, 385–6.

Whitton, B. A. 1971: Terrestrial and freshwater blue-green algae of Aldabra. *Philosophical Transactions, Royal Society* 260B, 249–55.

Woo, M.-K. and Marsh, P. E. 1977. Effect of vegetation on limestone solution in a small High Arctic basin. *Canadian Journal of Earth Science* 14, 571–81.

Perspectives

H. A. Viles

A wealth of illustrations of biological influences on geomorphological processes has been presented in the preceding chapters. It is useful to try to categorize these biological effects before any attempt is made to consider their geomorphological importance. A simple classification can be made into static and dynamic effects, i.e. those due to organism communities in some sort of equilibrium with their surroundings, and those due to a change in such organism communities. The role of vegetation in conditioning soil erosion can, for example, be viewed in these terms. Vegetation communities have a static effect on soil erosion through their impact upon hydrology and soil structure; whereas the removal or alteration of vegetation has a dynamic effect, in that the equilibrium is upset and soil erosion may change dramatically. Static influences upon geomorphological processes may be subdivided further into active and passive effects. Active effects are those where an organic process is involved, i.e. where there is a flow of energy and/or nutrients; whereas passive effects are due simply to the presence of an organism. An example of a passive biological effect upon a geomorphological process might be the presence of coastal grasses encouraging dune sedimentation and growth. Active effects include such phenomena as bioturbation and indirect effects such as animal grazing on vegetation which encourages soil erosion. Some more examples of biological effects in these different categories are presented in figure 1. In all three categories, organisms and organic material may have an encouraging or subduing effect on earth surface processes.

Different types of organism, or organism community, have different kinds of geomorphological influence. Active processes may involve organisms at many trophic levels, including primary producers, consumers, and decomposers. Passive processes mainly involve primary producers and dead organic material (debris). Microorganisms and lower plants play a particularly wide range of roles in geomorphological processes, although their influences are often hard to elucidate. All the different static biological effects contribute to the geomorphological process association acting upon landforms at any one time. Change in organism communities produces dynamic effects upon geomorphological processes. Such changes to organism communities are brought about by a range of events, including succession, tectonic activity, sea level change, fire, and human disturbance. Some of these events may also affect soil and geomorphological characteristics

Static (i.e. involving an established community)		
Active (involving an organic process)	*Passive*	*Dynamic* (related to change to the community)
biochemical weathering	vegetation influence on hydrology, microclimate	Removal or change to vegetation, leading to destablization, erosion (causes of change include fire, tectonics, human activity)
biophysical weathering	stabilization of soils or sediments by vegetation	
bioturbation		
bioerosion	accumulation of organic debris, forming debris dams, peat bogs	
formation of biochemical sediments		

Figure 1. A categorization of biological influences upon geomorphological processes

directly, and so the effects due to organic change may be hard to elucidate exactly. Changes in vegetation and animal communities lead to a disruption of equilibrium within the biogeomorphological system, and a new association of processes may become dominant. For example, burning of forest vegetation may lead (with a time-lag) to accelerated soil erosion, which in turn may be followed by a further development of vegetation. In many cases vegetational change and geomorphological change are intimately related and causality becomes blurred.

There are several problems involved with any attempt to categorize biological influences upon geomorphology, and also with attempts to classify geomorphological processes themselves. It is probably more realistic to think of biological influences as occupying a spectrum between active and passive, static and dynamic effects, as the boundaries between categories are blurred in reality. Further complications to the categorization presented in figure 1 occur due to the fact that organisms are bound up in communities. One can picture this in terms of ecosystems, where individual organisms are enmeshed in the structure and functioning of the ecosystem. Biological influences upon geomorphology may come from individual organisms and the whole ecosystem (through nutrient cycling, for example). Although the categorization presented in figure 1 has limitations, it is a useful starting point from which to examine the nature of organic influences on different geomorphological processes, and in different environments.

The earth surface processes (which may be grouped into weathering, erosion, transport, and deposition processes) are affected in different ways by organisms. Organisms can be seen to play some sort of role in most earth surface processes, excluding perhaps glacial and tectonic processes. Rock weathering, for example, is influenced by all categories of biological effects recognized in figure 1. As yet, it is difficult to suggest which type of effect might be the most important, and undoubtedly this will vary with environmental conditions. Processes of erosion (including soil erosion, rock and sediment erosion) are encouraged by a whole suite of bioerosion processes. They are also subdued by the stabilizing effect of vegetation, and may be encouraged on removal of such vegetation. Transport processes are generally retarded by biological influences, as debris dams and vegetation act to impede movement of water, wind and sediment. Processes of deposition and sedimentation, conversely, are generally encouraged by biological influences (both active and passive) which produce, trap and bind sediments. These are obviously fairly gross generalizations, and many important biological effects may actually act to link these geomorphological processes. For example, biological processes often provide links between weathering and erosion of bare rock surfaces. In some cases, biological influences upon earth surface processes lead to the production of land-forms. Examples include termite mounds, animal burrows, nebkhas, bogs and the various biokarst features. However, it may be difficult in some cases to establish the exact contribution of biological processes to the creation of such forms, and these features may not be the most important manifestation of biological effects.

There are great variations in the nature of both geomorphological process associations and also biological communities over the face of the earth, leading to different biogeomorphological situations. Biological communities, for example, vary in terms of growth characteristics (seasonality etc.), biomass, and nutrient cycling. In some environments with a large and diverse biomass (such as tropical rain forests, coral reefs) the biogeomorphological effects will be very different to those in environments with a limited and stressed biomass. However, chapters 6 and 7 of this volume have indicated that even in such stressed environments as desert and tundra biomes, floral and faunal effects upon geomorphology may be important. It may be postulated that such biogeomorphological effects are different in nature to those in environments with larger biomass. Certainly hypotheses need to be tested on the variation of biological influences upon geomorphology with various ecosystem charac-teristics such as biomass and diversity.

How can the importance of biological influences upon geomorphology be assessed for any particular situation? One approach is to compare the rates of biogeomorphological processes with 'inorganic' earth surface processes. So, for example, one can compare rates of bioturbation of soils by various species with other rates of soil mixing. Similarly, one can compare rates of bioerosion of hard rock coastlines with erosion rates due to other processes such as abrasion, and solution by sea water. Also, through detailed studies of nutrient cycling within forest ecosystems, information can be obtained about the rate of organic removal of weathered material, compared with that lost to rivers. Such studies can quantify the importance of certain biological processes within overall geomorphological process associations. Information is limited, but some points can be made. Studies of coastal bioerosion, for example, indicate that on limestone coasts throughout the world erosion rates are very similar and may represent an ecological balance between biological boring and abrasion processes (see chapter 8). Comparisons of data on soil bioturbation by worms in temperate and tropical areas indicate that the geomorphological work done by worms is greater in tropical than temperate areas (as shown in chapter 5, table 5.10).

There are many biological influences upon geomorphology for which rate information is difficult to obtain and thus may not provide an accurate representation of their importance. Weathering of rock surfaces by micro-organisms and lower plants, for example, may be difficult to quantify in terms of surface lowering or volume of rock lost. Such micro-weathering processes may result in a weakened rock surface alone, or may act to encourage normal chemical weathering. In general, biological effects that are passive or indirectly active cannot easily be translated into rates of geomorphological processes. Also, it is often difficult to produce 'natural experiments' to compare the rate of geomorphological processes with and without such a biological component. The obvious and far-reaching dynamic effects upon geomorphological processes of the removal of vegetation in many areas, however, indicate the importance of such vegetation to the normal working of those systems.

There are important longer term perspectives to be considered in any assessment of the effects of organisms upon geomorphology. There have been large scale and often catastrophic changes in the world's flora and fauna over geological time. Of particular geomorphological importance are the Pleistocene extinctions of many large mammals. The effects of large mammals (both direct and indirect) upon sediment movements, pan formation, etc. have been discussed in chapter 6. One can envisage a scenario before the Pleistocene extinctions when the browsing, trampling and sediment movement caused by huge populations of large mammals would have been far greater than any effects seen today. Today, features produced by such large mammals often seem bizarre and of localized importance, but they would have been more widespread and important in those times. Environmental changes over the course of the Holocene have also led to important modifications to flora and fauna which are of geomorphological importance. In Britain, for example, there has been a marked decline in forest cover since *c.* 5,000 BP, which is presumed to have concurrent effects upon erosion and sedimentation. Within such long-term environmental change, however, where both organic communities and geomorphology are responding to external changes, causality can become blurred.

During the Holocene the 'human impact' upon the environment has become of great importance over much of the earth's surface. Human activity has triggered major changes in flora and fauna, leading to associated geomorphological changes. As Thomas indicates in chapter 6, for example, populations of large game animals in Africa have been reduced since colonial expansion. The game that is left, however, has been artificially constrained into small 'game park' areas, leading to an acceleration of animal-induced erosion in such areas. Introductions and extinctions on islands will have had similar geomorphological consequences. As discussed in chapter 8, coral islands where natural vegetation has been replaced by coconut plantations are more susceptible to hurricane damage than undisturbed islands. Agriculture and land management have obviously had a wide impact upon erosion and sedimentation rates, as described in chapter 3. Such wide-ranging changes mean that many geomorphological field investigations are carried out in such human-influenced terrain. However, it should be pointed out that human impact upon flora and fauna has been uneven in both extent and nature over the globe.

Human activity also affects flora and fauna indirectly, with ramifications for geomorphology. Atmospheric pollution leading to acid rain, for example, has influenced geomorphological systems through changes to vegetation and solute concentrations. Human activity has also caused direct geomorphological changes, for example, where mining activity has lead to the production of 'artificial' landforms such as spoil heaps. Vegetation has often been used as a means of 'reclaiming' such land, in implicit acceptance of the role played by vegetation in geomorphology. In a sense, the most extreme example of biogeomorphology is in urban areas, where human activity (involving direct and indirect active and dynamic influences upon geomorphological processes) has largely replaced natural ecosystems and geomorphological systems.

What is the future for biogeomorphological studies? It appears from the work presented in this volume that there is a vast range of individual species–earth surface process interactions which need further study and elucidation. Similarly, there are many areas of community influences upon geomorphology which deserve further study. Far more quantitative information must be produced, in order to test specific hypotheses over the importance of certain biological influences. It is vital that geomorphologists become more aware of the ecology of landforms and landscapes, in order to attempt to model the complex interactions which result.

List of Contributors

Andrew S. Goudie is Head of Department, School of Geography, University of Oxford, UK.

Kenneth J. Gregory is Professor, Department of Geography, University of Southampton, UK.

Angela M. Gurnell is Senior Lecturer, Department of Geography, University of Southampton, UK.

Carl F. Jordan is Senior Ecologist, Institute of Ecology, University of Georgia, Athens, Georgia, USA.

Peter B. Mitchell is Lecturer, School of Earth Sciences, Macquarie University, Sydney, Australia.

Tom Spencer is Lecturer, Department of Geography, University of Manchester, UK.

David S. G. Thomas is Lecturer, Department of Geography, University of Sheffield, UK.

Stanley W. Trimble is Associate Professor, Department of Geography, University of California, Los Angeles, USA.

Heather A. Viles is Lecturer in Geography, Jesus and St Catherine's Colleges, Oxford, UK.

Rendel B. G. Williams is Lecturer, Geography Laboratory, University of Sussex, Brighton, UK.

General Index

afforestation, and soil erosion, 108
Africa, 167, 176, 181, 185
agricultural land, eroded, 126–7
agriculture
  and infiltration, 110–11
  conservation tillage, 121–2
  conventional tillage, 111–12
Alaska, 227, 230–1, 241, 245–6
Aldabra Atoll, 275, 323, 326–8,
  335–6, 339–40, 343
algal mats, 277
ant bears, and termite mounds, 181
ants
  and bioturbation, 45, 47–8
  ground nests, 47
aussen Stalactiten, 343
Australia, 54–72, 171, 173, 181,
  198–200, 213, 215

bacteria, 211, 212, 344
badgers, mounds of, 102
badlands, 125–6
bank erosion, and vegetation, 27
beachrock, 278–9
beavers, dam building by, 246–8
biocrusts, 202–3
bioerosion, 2, 264–7, 269–70, 271,
  272, 354
biogeochemistry, 5
biogeography, 2–4
biokarst, 319, 322–4
biolithogenesis, of calcrete, 212
bioturbation
  of marine sediments, 292–7,
  298–300
  of soils, 46–7, 54–73, 97, 102
bird droppings, and weathering, 340–1

birds, and soil turnover, 50–1
bivalve molluscs, bioerosion by,
  265–7
Blackheath, New South Wales,
  58–62
blue-green algae, 212, 277–9, 323,
  331–6
  colonization rates, 326–8
bogs, 227, 228, 237–8, 242–5
brousse tigre, 196
Burren, the, 325–6, 329
butt hillocks, 52, 67

calcrete, formation processes,
  212–15, 343
California, Redwood Creek, 21–2
callianassid shrimps, and sediments,
  297–301
canopy cover, and soil erosion, 90–1
carbon dioxide, biogenic production
  of, 321
case-hardening, 279, 342
caves, 323, 329–31, 344
channel bars, 19–21
channel capacity, and lichen limits,
  17
channel changes
  and deforestation, 28
  and vegetation, 31–3
channel morphology
  and grass sod, 19
  and tree roots, 18
  and vegetation, 17–24
channel roughness, and vegetation,
  23–5
channelization, 29–30
chasmoliths, 325

Species Index